HOLOCENE HUNTER-GATHERERS OF THE LOWER OHIO RIVER VALLEY

HOLOCENE HUNTER-GATHERERS OF THE LOWER OHIO RIVER VALLEY

RICHARD W. JEFFERIES

THE UNIVERSITY OF ALABAMA PRESS
Tuscaloosa

The University of Alabama Press
Tuscaloosa, Alabama 35487-0380

Manufactured in the United States of America

Typeface: AGarmond

∞

The paper on which this book is printed meets the minimum requirements of American National Standard for Information Sciences-Permanence of Paper for Printed Library Materials, ANSI Z39.48-1984.

Library of Congress Cataloging-in-Publication Data

Jefferies, Richard W.
Holocene hunter-gatherers of the lower Ohio River Valley /
Richard W. Jefferies.
p. cm.
Includes bibliographical references and index.
ISBN 978-0-8173-1658-7 (cloth : alk. paper) — ISBN 978-0-8173-5541-8 (paper : alk. paper) — ISBN 978-0-8173-8241-4 (electronic) 1. Paleo-Indians—Ohio River Valley. 2. Hunting and gathering societies—Ohio River Valley. 3. Ohio River Valley—Antiquities. 4. Excavations (Archaeology)—Ohio River Valley. 5. Land settlement—Ohio River Valley—History. 6. Landscape—Social aspects—Ohio River Valley—History. 7. Paleo-Indians—Agriculture—Ohio River Valley. 8. Paleoecology—Ohio River Valley. 9. Human ecology—Ohio River Valley—History. 10. Social archaeology—Ohio River Valley.
I. Title.
E78.O4J44 2009
977.1′01—dc22

2008050481

To Parker, Lucy, Harper,
and those who come later . . . follow your dreams!

Contents

Illustrations

Tables

Acknowledgments

Holocene Hunter-Gatherers of the Lower Ohio River Valley reflects the efforts of hundreds of people who have studied the archaeological remains of the Lower Ohio Valley's earliest inhabitants. This work has spanned almost 200 years, starting with nineteenth-century settlers who found interesting "spear points" and "axes" as they cleared their fields and built their roads, canals, and towns. It continues today through the scientific investigations of dozens of highly trained professional and interested avocational archaeologists. I discuss the deeds and accomplishments of many of these people in the following chapters; however, the contributions of many others are not documented and need to be publicly acknowledged. This book is dedicated to all those who have contributed to this effort. Many colleagues, students, and friends have directly contributed to this book. Foremost among these are Kenneth E. Sassaman (University of Florida) and Marvin T. Smith (Valdosta State University) who provided valuable comments and suggestions in their reviews of my manuscript for the University of Alabama Press. Brian M. Butler (Southern Illinois University Carbondale), C. Russell Stafford (Indiana State University), William H. Marquardt (University of Florida), and Christopher R. Moore (University of Kentucky) provided additional comments and recommendations at various times during the editing process. I greatly appreciate the time and effort they spent wading through the lengthy manuscript.

Dick Gilbreath, Jeff Levy, Eric Truesdell, and Jacob Wasilkowski (University of Kentucky Department of Geography) prepared many of the maps. I greatly appreciate their extreme patience and professionalism. Sheldon R. Burdin (University of Kentucky) and Christopher R. Moore provided additional assistance with the graphics. Scott Hutson (University of Kentucky) greatly assisted in the production of the photographs. George M. Crothers (Kentucky Office of State Archaeology), Michael Wiant (Illinois State Museum), and Rick Jones (Indiana Department of Natural Resources) provided the site location data used in many of the site distribution maps. Philip Mink (Kentucky Archaeological Survey) compiled the site location data and created the databases used to make many of the maps. George M. Crothers, Donna Butler (Southern Illinois University Carbondale), and David Pollack (Kentucky Heritage Council) assisted in getting permission to use many of the figures. Frank R. Ettensohn, Department of Earth and Environmental Sciences, University of Kentucky, provided valuable assistance in the writing of the section dealing with the geology of the Lower Ohio Valley. I deeply appreciate Mr. Rex Robinson giving me permission to use his artwork for the book cover and for figure 7.12. Special thanks go to the Dean's Office, College of Arts and Sciences, University of Kentucky for providing the funds to pay for producing the maps.

Some of the contents of this book incorporate and expand on earlier discussions of the Kentucky Archaic and Lower Ohio Valley hunter-gatherers published by the Kentucky Heritage Council (Jefferies 1990a), the University Press of Kentucky (Jefferies 1996a), and the State University of New York Press (Jefferies 2008).

Finally, without the support and encouragement of my family, I never would have completed this book. I am particularly indebted to my wife, Zanne, who kept me going when things got discouraging, and my grandchildren, Parker, Lucy, and Harper, who were constant sources of inspiration. My strong interests in the past stem from lengthy discussions about history and family genealogy with my parents, William and Virginia Jefferies, and in-laws, Woodrow and Mary Parker.

HOLOCENE HUNTER-GATHERERS OF THE LOWER OHIO RIVER VALLEY

1 / Introduction

The Ohio River, one of the longest rivers in North America, extends nearly 1,600 km from the confluence of the Monongahela and Allegheny rivers at Pittsburgh, Pennsylvania to the Mississippi River at Cairo, Illinois. During the eighteenth and nineteenth centuries, the Ohio River served as an important transportation and communication route for pioneers and settlers seeking access to the interior part of eastern North America. Once they cleared the land and built their towns, Ohio Valley farmers and merchants shipped their crops and goods down the Ohio to markets in St. Louis, Memphis, New Orleans, and beyond. In the mid-nineteenth century, the Ohio River divided the warring Confederate states to the south and the Union states to the north. Even today, the Ohio River serves as a cultural boundary between distinctively different Southeastern and Midwestern ways of life.

Clearly, the Ohio River has played important roles in the historical development of the United States, serving as both a corridor through which people, goods, and information moved and as a border or boundary dividing distinct regional cultures. Given the social, economic, and political significance of the Ohio River in the history of the United States, it should not be surprising that it maintained just as prominent a role in the lives of the valley's prehistoric Native Americans.

Based on more than 100 years of archaeological research, we now know that the ancestors of modern Native Americans lived in the Ohio River Valley for at least 12,000 years prior to the arrival of Euroamerican settlers in the mid-eighteenth century. This book focuses on the earliest Native American inhabitants of the valley—those groups that practiced what anthropologists call a *hunter-gatherer* way of life. Very simply, the valley's earliest people subsisted by exploiting the naturally available plants and animals that once lived in and on the rivers, floodplains, and uplands of what is now the Ohio Valley.

In eastern North America, including the Ohio Valley, ancient Native Americans pursued this hunter-gatherer way of life for at least 9,500 years, starting more than 12,500 years ago and continuing to about 3,000 years ago, or before present (B.P.). After that time, domesticated plants, first native (goosefoot, sunflower, etc.), then tropical species (maize, beans) introduced from Mexico, became increasingly important parts of the subsistence base of Woodland (3000 to 1000 B.P.) and Mississippian (1000 to 400 B.P.) groups. Despite the increasing dietary significance of cultivated plants, however, later people continued to use and modify many aspects of hunter-gatherer technology and organization first developed by the valley's early inhabitants. In many ways, the hunting and gathering way of life never disappeared from the region—it was simply a matter of adding other subsistence strategies to the hunting and gathering core that already existed.

The Region

Along its ca. 1,600 km course to the Mississippi River, the Ohio River flows through numerous environmental and physiographic zones that once supported highly diverse plant and animal communities. Just as diverse were the hunting and gathering societies that populated this extensive and highly variable landscape, making it difficult to discuss the full range of their cultural variability. Therefore, this book focuses on one part of that vast river system—the section known as the *Lower Ohio Valley.*

The Lower Ohio Valley region consists of the western one-third of the Ohio River watershed, extending from the Falls of the Ohio River near Louisville to the Ohio-Mississippi River confluence (Figure 1.1). Previous archaeological research indicates that even within this portion of the valley, considerable cultural variability existed during much of the prehistoric past (Muller 1986:ix). The region's rich and diverse archaeological record provides an excellent opportunity to explore patterns of diachronic change and synchronic variability associated with the development of culturally complex hunting and gathering societies that lived in parts of the valley by about 5,000 years ago.

I selected the Falls of the Ohio River, known locally as the Falls, as the upriver or eastern boundary of my study area because, prior to the construction

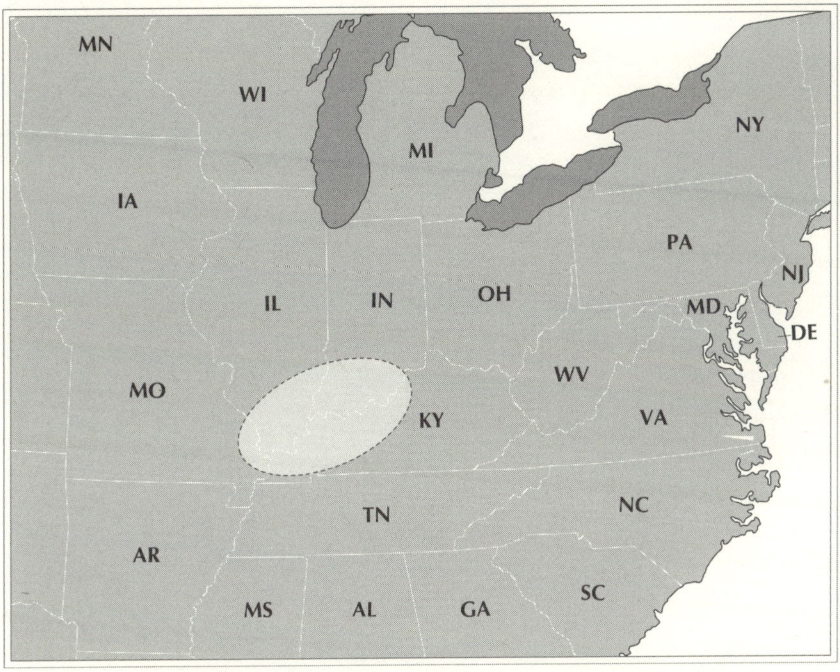

Figure 1.1. The Lower Ohio Valley region.

of nineteenth-century dams and canals, it represented the only natural impediment to peoples' movement up or down the river. In addition, the Falls appears to have marked some kind of social boundary throughout much of prehistory (Burdin 2004; Griffin 1978:551; Jefferies 1997:483), suggesting some degree of cultural difference among groups that lived up- or downriver of this prominent landscape feature. The Ohio-Mississippi River confluence, some 650 km downriver from the Falls, marks the western end of the Ohio.

The study area's northern and southern boundaries are more broadly defined, being the limits of the Ohio River watershed. However, since the watershed covers such a large part of the North American Midcontinent, I emphasize that portion within 80 to 160 km of the Ohio's main channel. Figure 1.2 shows the portion of the Lower Ohio Valley that forms the "core" of my research area. Most archaeological examples I use come from this region, but occasionally, I draw on research conducted farther afield.

The Lower Ohio Valley's physical landscape is highly diverse, ranging from the relatively rugged Knobs of west central Kentucky and the Shawnee Hills of southern Illinois to the low, marshy Coastal Plains near the Mississippi River

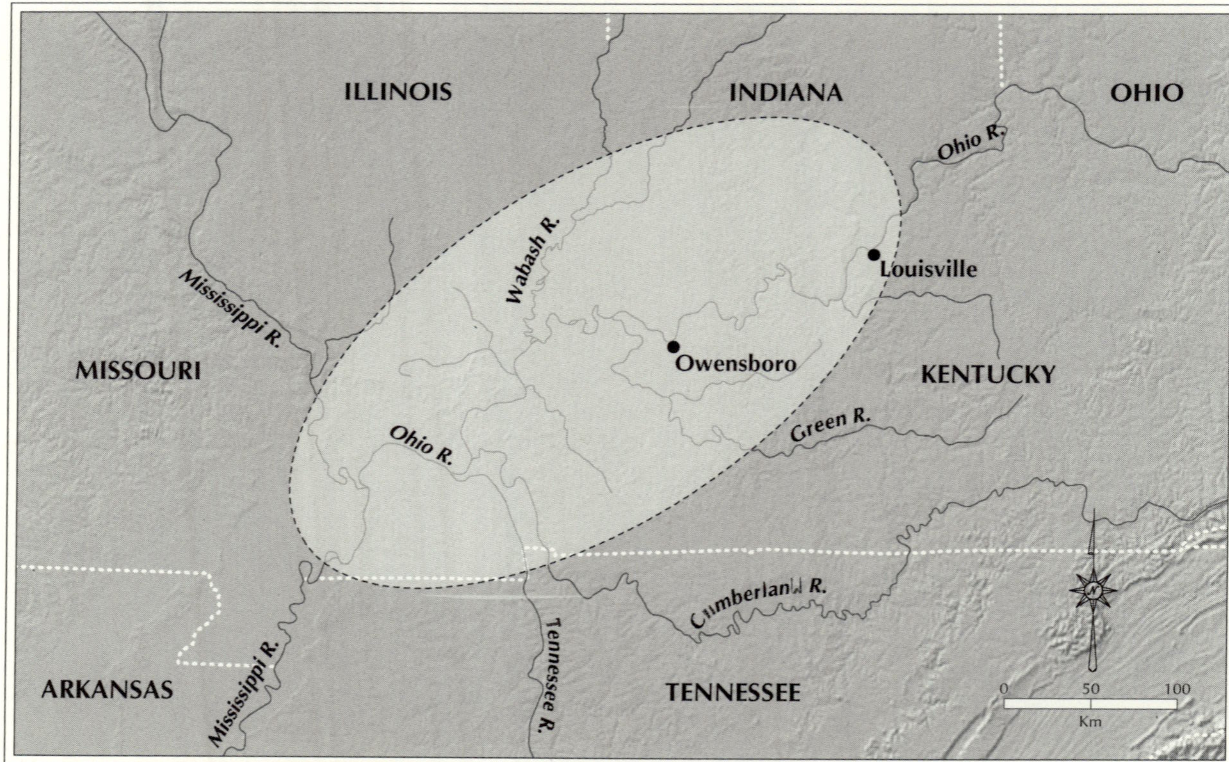

Figure 1.2. Core of the Lower Ohio Valley study area.

(Pollack 1990:7). Not surprisingly, the plants and animals that inhabited the valley's uplands and wetlands included species native to both the southern and northern parts of eastern North America. Today, for example, cypress and pecan trees, commonly found in more southerly climates, thrive in the western part of the Ohio Valley, while species more typical of the Northeast can be found near the Falls of the Ohio.

The study area includes the lower portions of many of the Ohio River's major tributaries like the Salt, Green, Wabash, Saline, Tradewater, Cumberland, Tennessee, and Cache rivers. The network formed by these rivers linked scattered hunter-gatherer groups together, facilitating transportation, communication, and exchange at many different levels. River confluences were particularly important places within this far-reaching river system (Muller 1986:25). The rivers and the diverse habitats flanking them provided hunter-gatherers with critical plant and animal resources that they used for food and to make clothing, shelters, and tools.

The book's temporal focus extends from approximately 12,500 to 3,000 years ago, representing nearly 75 percent of the Lower Ohio Valley's documented prehistory. Regional archaeologists refer to the early part of this era as the Paleoindian period (prior to 10,000 B.P.), while the remaining millennia constitute what is called the Archaic period (10,000 to 3000 B.P.). During much of this time, the Lower Ohio Valley was home to scattered groups of hunter-gatherers that survived by hunting, fishing, and collecting the region's abundant natural resources. While these groups shared a common reliance on the region's plants and animals, over time they developed increasingly distinctive ways of coping with the changing physical and cultural landscapes on which they lived.

Many of the plants and animals the earliest hunter-gatherers encountered were substantially different from those of later times. The earliest inhabitants dealt with a physical landscape that still contained remnants of colder Pleistocene conditions. Megafauna, like mammoths, mastodons, and large bison, roamed parts of the Lower Ohio Valley when the earliest humans arrived. By that time, however, these large animals, along with many other Pleistocene mammals, were well along the road to extinction. The role humans had in their disappearance is still, however, open to question (Dillehay 2000; Grayson 1991; Grayson and Meltzer 2002). The physical landscape continued to evolve over the next 10,000 years. Gradually, as the climate warmed, Holocene plant and animal communities similar to today's replaced those of the Late Pleistocene.

For purposes of this book, I have divided the Holocene into temporal units (Early, Middle, and Late) conforming to the chronological framework used by Smith (1986) in his archaeological overview of the Southeastern United States. The beginning of the Early Holocene (12,500–8000 B.P.) is marked by the climatic moderation that followed the retreat of the Laurentide glacier; the beginning of the Middle Holocene (8000–5000 B.P.) coincides with the start of the warmer, drier Hypsithermal interval; and the start of the Late Holocene (5000 B.P. to the present) coincides with the end of the Hypsithermal (Smith 1986:6).

Along with environmental change, the increasing intensity of human activity also helped to shape the landscape's appearance and character. Like the physical environment, the cultures and traditions of the region's hunter-gatherer societies also changed over time, producing a rich and complex cultural landscape. Subsequent generations continually altered these landscapes, each modifying what earlier groups established, while at the same time adding their own touches.

The People

Current archaeological evidence suggests that prior to about 12,500 years ago, few, if any, people lived in the Lower Ohio Valley. If people were present before that time, their impacts to the landscape were so slight that, at least so far, they remain archaeologically invisible.

The first hunter-gatherers to enter the Lower Ohio Valley in significant numbers were Paleoindian groups that began to filter in at the end of the Pleistocene (ca. 12,500 B.P.). The style and distribution of their fluted projectile points suggest that these people were organized into small, highly mobile, and widely scattered groups that shared a generally similar cultural tradition based, at least in part, on the pursuit of late Pleistocene megafauna (Tankersley 1990a:91, 1996:38).

The arrival of Paleoindian groups coincided with major changes to the physical landscape associated with the end of the Pleistocene. These groups adapted to the rapidly changing landscape by altering how and where they hunted and gathered. Early Paleoindian people focused their subsistence efforts on areas of low relief and open vegetation that attracted game. They also frequented places where they could monitor game movement and collect high-quality lithic material needed to make hunting and butchering tools (Tankersley 1996:37–38). Later Paleoindian groups adjusted their subsistence strategies to the rapid extinction of megafauna by focusing on a wider variety of game and plants (Tankersley 1990a:91). In parts of the Lower Ohio Valley, this shift in landscape use is marked by the first regular occupation of upland rockshelters (Tankersley 1996:35; Walthall 1998b).

During the Early Holocene, sometime after 10,000 years ago, the Lower Ohio Valley experienced a significant population increase as small hunter-gatherer groups filled the riverine and upland landscapes (Milner 2004:28), continuing the trend started by Late Paleoindian groups. During the succeeding 7,000 years, Lower Ohio Valley hunter-gatherers followed variable cultural trajectories, resulting in a diverse cultural landscape filled with groups having similar, but regionally distinct, cultural traditions. This regional variability is reflected by the differential distribution of stylistically distinct projectile points across the North American Midcontinent (Justice 1987).

During the Middle Holocene (ca. 6000 B.P.), the archaeological records of some parts of the Lower Ohio Valley suggest that hunter-gatherers were becoming more sedentary, repeatedly occupying strategic spots on the landscape, increasing in population, and participating in more interregional interaction, collectively pointing to an overall increase in cultural complexity (Brown 1985; Marquardt 1985). In contrast, other regional hunter-gatherer societies appear to have remained relatively mobile, maintaining a less complex way of life comparable to that of the past (Sassaman 1995:181). As with modern societies, these hunter-gatherers did not live in isolation, but maintained social connections of variable intensity with other nearby and distant groups. Sometimes these relationships were cooperative; other times they were confrontational.

By the Late Holocene (ca. 3000 B.P.), these variable cultural trajectories, reflecting thousands of years of interaction between hunter-gatherers and their evolving physical and social worlds, produced a diverse cultural landscape occupied by groups of variable size and complexity. Some of these groups were even experimenting with plant cultivation and domestication, a pursuit that would eventually contribute to even more significant cultural changes (Smith 1989, 1995; Watson 1989).

The Research

The material remains of the Lower Ohio Valley's hunter-gatherer societies are well represented in the archaeological record, but the vast majority occur as surface artifact scatters having no stratigraphic integrity. Sites with intact cultural deposits are rare, mostly deeply buried beneath more recent alluvial deposits or under thick colluvial sediments at the bases of bluffs and slopes. Fortunately, important exceptions do exist (Bader and Granger 1989; Collins 1979b; Jefferies 1982b; Nance 1987; Stafford 2004).

Our present understanding of hunter-gatherer settlement in the Lower Ohio Valley suggests that people were differentially distributed across a dynamic, evolving landscape. Butler's (2009) study of Archaic site distributions in southern Illinois demonstrates some of these trends. For example, hunter-gatherer settlement in the southern Till Plain along the northern margin of the study area is marked by numerous small sites, but few large ones. Here, major rivers and streams developed in poorly drained glacial lake basins resulting in repeated flooding and very wet conditions. However, hunter-gatherer activity was more common on higher areas adjacent to the wetlands and a few rockshelters were available for short-term habitation.

Sites are much more common to the south in the Hill Country where hunter-gatherers frequently occupied ridgetops, slopes, and rockshelters; however, thick alluvial and colluvial deposits cover many valley bottom sites. Farther to the south, hunter-gatherers frequently occupied ridges in the Cache River lowlands, but cultural deposits at most sites are unstratified. Evidence of hunter-gatherer activity seldom occurs on or near the ground surface in the Mississippi River floodplain because the constantly meandering Mississippi River has destroyed or buried many of the sites that once existed (Butler 2009).

In contrast, sites are relatively common on high terraces and ridgetops near the Ohio-Mississippi River confluence. Unlike in Kentucky (Nance 1987), archaeologists have not documented any deeply stratified sites along the Illinois side of the Ohio River, probably because of the lack of extensive floodplain settings. However, hunter-gatherers occasionally occupied floodplain ridges as evidenced by the Fitzgibbons site (Robison 1986) situated on a prominent alluvial ridge (Butler 2009).

Hunter-gatherer site distributions also varied throughout other parts of the Lower Ohio Valley. The specifics of these are discussed in subsequent chapters. The point to be made here is that many different cultural and natural processes, both ancient and modern, affect our knowledge of hunter-gatherer site distributions and, therefore, hunter-gatherer societies. Understanding these processes is critical to explaining the synchronic and diachronic variability observed in these distributions.

The rich archaeological record left by nearly 10,000 years of hunter-gatherer activity has attracted archaeologists to the Lower Ohio Valley for nearly 100 years (Schwartz 1967). However, the intellectual motivation behind this interest has varied considerably. Untrained antiquarians who were primarily interested in the artifacts as antiquities and curiosities did some of the earliest work. Gradually, scholars became more interested in what those artifacts could tell them about human activities and how those activities varied temporally and spatially. The development of new field techniques made data recovery more controlled and systematic, while more rigorous laboratory procedures helped ensure that the resulting data were of the highest possible quality.

Throughout the early and mid-twentieth century, archaeological investigations, particularly those associated with the federal work relief programs, like the Works Progress Administration (WPA), yielded abundant data on the Lower Ohio Valley's hunter-gatherers (Jefferies 1988a, 1988b; Milner and Smith 1986). Collectively, this research provided the basic chronological, technological, subsistence, settlement, and social information for a fundamental understanding of the region's early culture history. In addition, many of these studies also contributed to the development and refinement of the proposed Archaic concept in eastern North American prehistory (Ritchie 1932).

More recent investigations, many conducted as part of state or federally mandated cultural resource management studies, have added to this growing database, helping to identify and clarify patterns of culture change and variability. Some projects focused on deeply stratified floodplain and/or midden-bearing sites that yielded detailed chronologies and fine-grained data on culture change (Collins 1979a; Jefferies 1982b; Marquardt and Watson 2005; Milner and Jefferies 1998; Nance 1987). Others surveyed areas or investigated sites in the uplands adjacent to the Ohio River, along tributary streams and rivers, or in interfluvial settings between the rivers. Although upland sites seldom had deeply stratified deposits like some floodplain sites, they yielded complementary information on activities that took place away from the main course of the Ohio River (Canouts et al. 1984; Driskell et al. 1984; Granger 1985; Hargrave and Butler 1994; Munson 1980a; O'Malley et al. 1980; Smith 1997). Large-scale archaeological surveys provided important insights on regional settlement strategies and how they changed over time (Jefferies et al. 2005; Stafford 1994).

The combined efforts of archaeologists working in southern Illinois, southern Indiana, and western Kentucky during the past 100 years have produced literally thousands of books, technical reports, monographs, book chapters, journal articles, and meeting papers on different aspects of Lower Ohio River valley hunters and gatherers. Some of these publications are widely accessible to researchers, scholars, and the public. Others appear in professional journals found only in university libraries or are part of the extensive archaeological "gray" literature. Because of funding and/or temporal constraints, the results of many projects remain unintegrated into our view of the Lower Ohio Valley's prehistoric past. The goal of *Holocene Hunter-Gatherers of the Lower Ohio River Valley* is to synthesize the more than 100 years of archaeological research on regional hunting and gathering societies and to provide a better understanding of their temporal, spatial, and cultural variability.

Hunter-Gatherer Landscapes

As highly mobile hunter-gatherers initially filtered into the Lower Ohio Valley at the end of the Pleistocene, they encountered a highly diverse physical environment created by millions of years of climatological, geological, hydrological, and glacial activity. The number, diversity, and distribution of plants and animals on which they depended were closely linked to the region's environmental character. As hunter-gatherers spread throughout the region, they superimposed the technological, organizational, and cosmological aspects of their cultures on the physical landscape. Over time, they gradually transformed physical spaces into meaningful places, beginning the construction of an ever-changing landscape that reflected processes of behavioral change across space and over time (Anschuetz et al. 2001:161).

In recent years, a growing number of archaeologists have adopted a "landscape" approach to investigate how humans interact with the environment (Ashmore and Knapp 1999; Tilley 1997). Landscapes are seen as products of interaction between culture and nature, not simply a superficial imposition of culture on nature. Each inhabiting group superimposes its own patterns of material and nonmaterial occupation, adding layers to the material remains of earlier cultural groups (Anschuetz et al. 2001:185).

According to Anschuetz et al. (2001:188), the archaeological landscape consists of a palimpsest of cultural material attributable to both cultural and natural processes that operate at different temporal and spatial scales. Therefore, the archaeological landscape represents a patterned, if somewhat convoluted, distribution of archaeological traces across space. The natural environment consists of the sum of the geological and biological elements in a specific space. The elements' spatial and temporal distributions determine the structure of the physical environment, which influences people's decision-making and behaviors that ensure their survival.

The temporal and spatial distributions of archaeological materials left by a region's inhabitants can be used to study and evaluate variation in the patterning of cultural traditions as realignments of human behavior. Such a research approach helps to identify and explain the interactive economic, social, and ideational strategies that people used to ensure their continued existence in lieu of constantly changing environmental conditions (Anschuetz et al. 2001: 187–191).

Three areas of research appear to be particularly relevant to the study of landscape archaeology—settlement ecology, ritual landscapes, and ethnic landscapes (Anschuetz et al. 2001:176). These three approaches share the view that people are more than passive recipients of change dictated from outside the cultural system. Collectively, they contribute useful, relevant insights into a society's past behavioral dynamics and patterns of change (2001:181).

When dealing with mobile hunter-gatherers, some of these concepts are more clearly reflected in the archaeological record than others. We can garner information related to the study of settlement ecology from the distribution and variation of hunter-gatherer activity loci and the material correlates of their associated behaviors. In contrast, insights into hunter-gatherer ritual and ethnic landscapes may be more difficult to discern than for more complex late prehistoric societies. For example, architectural data from Mississippian mound centers and the stylistic attributes of Mississippian pottery have the potential to provide important insights into late prehistoric ritual and ethnic landscapes (Lewis and Stout 1998). Similar indicators of hunter-gatherer ritual practices and ethnicity are much more difficult to detect in the archaeological record. Nevertheless, careful examination of various hunter-gatherer cultural characteristics, such as mortuary practices and the spatial distribution of artifact styles (e.g., bone pins, atlatl weights, shell items, etc.) may provide clues about the ritual and ethnic aspects of hunter-gatherer societies. I will return to the concept of hunter-gatherer landscapes in the final chapter of this book.

2 / Physical Landscape

Introduction

The first hunter-gatherers to enter the Lower Ohio Valley encountered a world unmodified by human activity. The physical landscape consisted of a variety of environmental features that included climate, landforms, plants, animals, mineral and lithic resources, and water. At the time the first hunter-gatherers arrived, their physical world was strongly influenced by the retreating Pleistocene glacier. Over time, the climate gradually ameliorated with Holocene environmental conditions becoming more prevalent. Changes in the kind, number, and distribution of the region's natural resources created new opportunities for people to make choices about how to best respond to those changes.

Today, most people would agree that the environment did not determine the nature of these societies; however, the distribution and frequency of the region's natural resources did influence where people lived, what they did there, and how long they stayed. As hunter-gatherers spread across the Lower Ohio Valley landscape, they gradually modified their physical world, creating a cultural landscape that reflected their particular cultural preferences and beliefs. The Lower Ohio Valley's cultural landscape continued to evolve throughout the period covered by this book (and long after) as its Native American inhabitants slowly changed the physical landscape into a place that reflected their technological, social, and ideological worlds.

If we are to understand how hunter-gatherers created their cultural landscape, we need to know about the physical world in which they lived and how it varied temporally and spatially. This chapter discusses the character of the late Pleistocene and Holocene environments, drawing on information from geology, geography, paleoecology, and the historical record.

Landforms and Topography

The Ohio River flows south and southwest for nearly 1,600 km before joining the Mississippi River at Cairo, Illinois, making it the one of the longest rivers in eastern North America. Although shorter than the Mississippi River, at their confluence, the Ohio typically carries a much greater volume of water than does the Mississippi. The Ohio River system (including the Tennessee and Cumberland rivers) drains nearly 520,000 km^2 of eastern North America, including parts of New York, Pennsylvania, Maryland, North Carolina, Ohio, West Virginia, Kentucky, Indiana, Illinois, Tennessee, Virginia, Georgia, Alabama, and Mississippi (U.S. Fish and Wildlife Service 1999).

The Ohio River basin is often subdivided into three units based on its physiographic variability. The eastern portion lies within the Appalachian Plateau. The region's rugged topography is largely attributable to the erosion of thick beds of sandstone and shale. The Central Lowlands comprise the basin's middle one-third. Its flat to slightly rolling topography was created by repeated Pleistocene glaciation. Finally, the Lower Ohio River valley, comprising the western one-third of the river, is dominated by sandstone and limestone formations of the Interior Low Plateau province. A small area of the Gulf Coastal Plain province is located near the confluence of the Ohio and Mississippi rivers (U.S. Fish and Wildlife Service 1999).

For much of its course, the Lower Ohio River flows along the northern edge of the Interior Low Plateau. Bedrock geological formations, consisting of thick beds of Mississippian and Pennsylvanian limestone, sandstone, and shale, underlie the river, dipping gently to the northwest from the Appalachian highlands in the east to the coal basin of western Kentucky, Indiana, and Illinois in the west (Fenneman 1938:411–413). Much of the bedrock valley floor lies below deep deposits of alluvial and glacial clays, sands, and gravels of Holocene or Pleistocene age. Bedrock is exposed only where the river flows against the valley wall, and at the Falls of the Ohio River (Powell 1999:9).

Geological processes that spanned millions of years created the landscape of the Lower Ohio Valley. Between the Falls of the Ohio and the Mississippi River, the Ohio flows through at least four major physiographic regions containing geological formations dating from the Ordovician (ca. 490 million years ago) to the Quaternary (ca. less than one million years ago) geological periods (McGrain 1983:Figures 9 and 10) (Figure 2.1). Pleistocene glaciers and their melt waters shaped or modified many of the valley's river and stream channels. In addition, thick layers of windblown glacial silt, known as loess, cap ridgetops and hilltops throughout the western part of the region.

The Falls of the Ohio River, located at Louisville, Kentucky, was perhaps the most impressive natural feature of the Lower Ohio River (Figure 2.1). Now inundated by waters impounded by twentieth-century dam construction (Figure 2.2), the rapids that formed the Falls once extended for 3.2 km over which river elevation fell 8.0 m (Figure 2.3). Like much of the Ohio Valley topography, the Falls' origin is attributable to glacial activity. Floodwaters from the melting glaciers filled an earlier river channel with thick sand, gravel, and clay deposits, forcing the river to find a new route across the underlying limestone formations, thereby creating the rapids. The previous channel is now buried under modern-day Louisville (Indiana Geological Survey 1998; McGrain 1983: 17–21).

As the Lower Ohio River flows toward its confluence with the Mississippi River, it passes through a constantly changing landscape (Figure 2.1). The eastern portion of the Lower Ohio flows through a gently rolling region characterized by extensive outcrops of Ordovician and Silurian limestones and shales. In Kentucky, this region is known as the Outer Bluegrass and has a generally low to moderate relief. Erosion has cut numerous steep valleys, resulting in little flat land (McGrain 1983:66). Much of the western Outer Bluegrass is drained by the Salt River, which flows directly into the Ohio River.

The geological formations forming the Outer Bluegrass extend north of the Ohio River into southeastern Indiana where they are known as the Muscatatuck Regional Slope. To the west, an area of relatively little topographic relief, known as the Scottsburg Lowlands, developed on soft shales. Much of this part of southern Indiana is covered by as much as 45 m of glacial outwash deposits. The East Fork of the White River drains much of this part of southern Indiana (Hall 1989:19–29).

The western edge of the Outer Bluegrass region is marked by a 16 to 24 km wide band of uplands known as the Knobs. Hundreds of isolated, steep conical hills that are detached erosional outliers of the Mississippian Plateau to the west dot the landscape (McDowell 1986:66–67; McGrain 1983:46–48). In southern Indiana, this geologic feature is known as the Norman Uplands (Braun 1950: 136) and is demarcated by the Knobstone Escarpment. Just to the west of the Scottsburg Lowlands, the knobs rise to up to 190 m in elevation. The escarp-

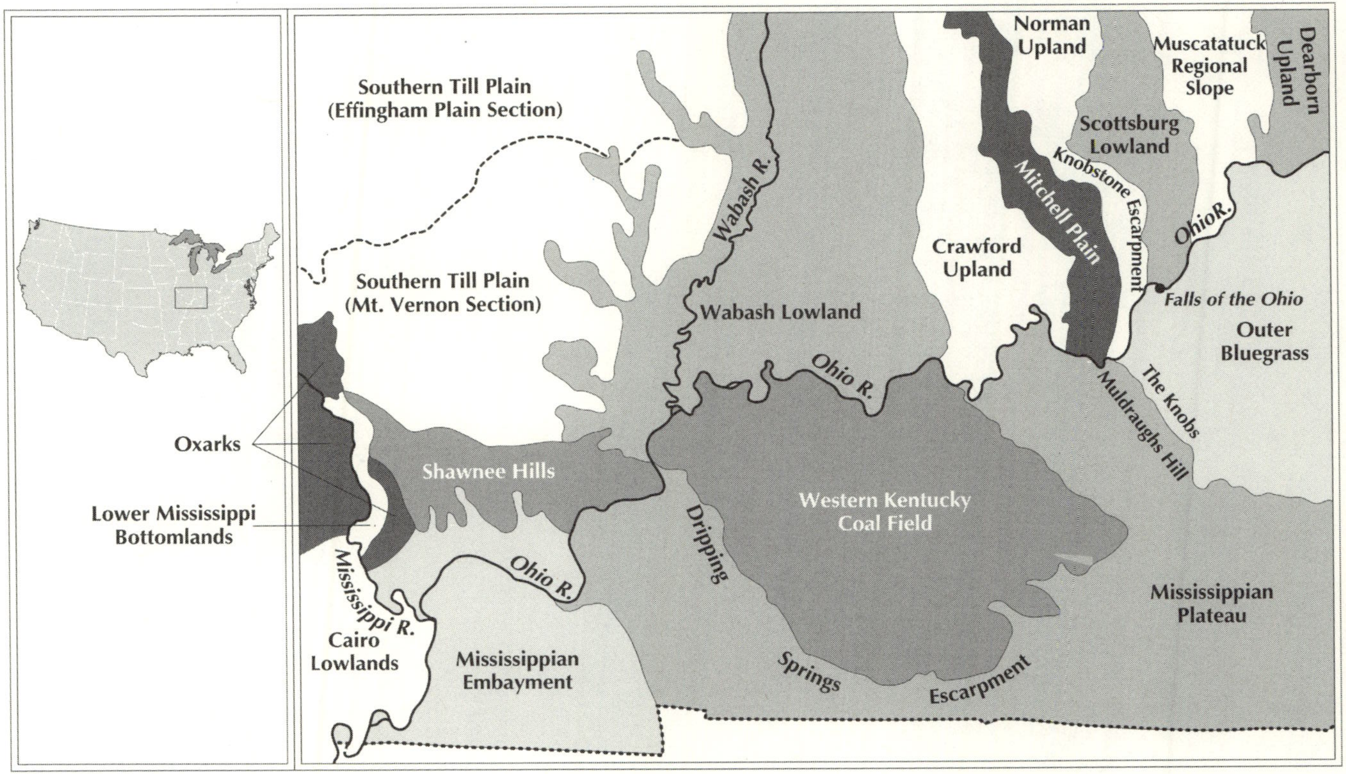

Figure 2.1. Physiography of the Lower Ohio Valley (adapted from Lopinot et al. 1982:Figure 4; White 2001:Figure 2).

Figure 2.2. Photograph of the Falls of the Ohio taken in 1922 from the fossil beds below Louisville on the Indiana shore looking to the southeast (Caufield & Shook Collection, Special Collections, University of Louisville).

ment gradually decreases in height to the north, eventually disappearing below thick glacial deposits (Hall 1989:27).

The topography of the Mississippian Plateau region dominates much of south central Indiana and west central Kentucky. The region's Mississippian-age limestone plain is covered by thousands of sinkholes, sinking streams, springs, and caves, creating karst topography with a poorly developed surface drainage system. In Kentucky, the Muldraugh Hills separate the Mississippian Plateau region from the Knobs region (Kentucky Geological Survey 2002). Before the introduction of modern agricultural practices, portions of the Mississippian Plateau were covered by *barrens,* a pioneer-era term for grassland prairies. Much of the Mississippian Plateau is drained by the Green River and its tributaries (McDowell 1986:67). Considerable debate exists over whether barrens represent static intermediate habitats or relic prairies in the process of being reforested (Braun 1950:483).

In south central Indiana, the Mitchell Plain, also characterized by sinkholes

Figure 2.3. Map of the Falls of the Ohio by Henry McMurtrie (1819) (courtesy of the University of Kentucky Geology Library and Map Collection).

and karst valleys, marks the northward extension of the Mississippian Plateau across the Ohio River. The Mitchell Plain is flanked by the Norman Uplands on the east (Hall 1989:27–28). To the west, the Chester Escarpment separates the Mitchell Plain from the Crawford Uplands, a belt of relatively rugged limestone and sandstone hills (Gray et al. 1983:175). Much of south central Indiana is drained by the East Fork of the White River. Many rivers and streams in the southernmost areas, like the Blue River, flow directly into the Ohio River (Hall 1989:27–28).

Kentucky's Western Coal Field region (Pennsylvanian age) marks the southern extent of the Illinois or Eastern Interior Basin, which also includes the southern Indiana and southern Illinois coalfield. In Kentucky, this region is denoted by hilly uplands dissected by streams flowing through wide, swampy valleys. A narrow band of sandstone ridges, known as the Dripping Springs Escarpment, marks the Coal Field's boundary with the Mississippian Plateau. The Western Coal Field is drained by the Tradewater River and part of the Green River drainage system (McDowell 1986:68).

Pennsylvania geological formations extend northward across the Ohio River into southwestern Indiana and southeastern Illinois. In southwestern Indiana, the Crawford Upland gradually merges with the Wabash Lowland, the largest physiographic unit in southern Indiana. The lowlands contain extensive glacial lake deposits, while uplands are capped with windblown loess. The Wabash and White rivers form the major drainages in southwestern Indiana (Hall 1989:39–41). The floodplains of the Wabash River and its tributaries, comprising the Wabash Border physiographic section, were once covered by oxbow lakes, marshes, and sloughs (Schwegman 1975:35).

In southeastern Illinois, Pennsylvanian deposits underlie the flat to rolling uplands of the Southern Till Plains. Prior to modern cultivation, much of the Southern Till Plains was covered by a mix of forests and tallgrass prairies, with prairie becoming more prevalent as one moves to the north. Glacial till deposits are common throughout this region where they are often buried under thick layers of loess. In some parts of southern Illinois, loess deposits are more than six meters thick. The southern margin of the Till Plains marks the approximate southern limit of maximum Pleistocene glaciation (Illinoisan ice sheet). The valleys and low cliffs of the more rugged Mt. Vernon Hill Country lie along the Till Plain's southern margin (Harris et al. 1977:7–8). The region is dissected by numerous rivers and streams with broad, poorly drained bottomland (Schwegman 1975).

The westernmost portion of the Ohio River Valley is located near the junction of several different geographic provinces (Till Plains, Wabash Border, Shawnee Hills, and Mississippi Embayment), resulting in the presence of markedly diverse natural habitats in very close proximity. This part of the valley has been fairly stable throughout the Holocene. The active channel often directly abuts high terraces and uplands, particularly on the Illinois side of the river (Butler 2009). Downriver from the Pennsylvanian deposits of the Western Kentucky Coal Field, the river cuts through the western section of the Mississippian Plateau. In Kentucky, this area is known as the Pennyrile region, marked by a series of sandstone-capped hills or "knobs" of the Dripping Springs Escarpment.

To the west in southern Illinois, the Shawnee Hills are formed by a series of long, asymmetrical ridges or cuestas stretching from the Ohio to the Mississippi River. The highest ridges have elevations exceeding 300 m. Extensive areas of bedrock exposed in the Shawnee Hills provided hunter-gatherers with numerous rockshelters for short-term occupation and many different chert types that were quite suitable for making flaked stone tools. As in much of the Lower Ohio River, the backwater sediments from glacial outwash formed the broad floodplains (Harris et al. 1977:3–4).

South of the Shawnee Hills, the Cache River–Bay Creek drainage, a former channel of the Ohio River, extends east to west across southern Illinois, connecting the Ohio with the Mississippi River valley. Prior to modern drainage efforts, this broad alluvial corridor was forested swampy lowland containing many shallow ponds and lakes. The Cache River–Bay Creek drainage is separated from the Ohio River by a band of low rolling uplands that are bordered on the south by a series of Pleistocene terraces along the Ohio River (Butler 2009; Harris et al. 1977:3–4).

The confluence area of the Ohio and Mississippi rivers is comprised of very complex alluvial terrain that is made even more complicated by the proximity of the western end of the Cache River valley (Butler 2009). This part of the Lower Ohio Valley, consisting of extreme western Kentucky and the southern tip of Illinois, lies in the northeastern extension of the upper Mississippian Embayment, part of the Gulf Coastal Plain. Similar topography occurs in southeastern Missouri, just west of the Ohio-Mississippi River confluence. Geological deposits consist of unconsolidated Cretaceous, Tertiary, and Quaternary gravel, sand, and clay. In southern Illinois, the Cache River marks the northern limit of the Mississippian Embayment (Frankie et al. 1997:55–58). The New Madrid Fault, the most active earthquake zone in the central United States, underlies this part of the Lower Ohio Valley (Kentucky Geological Survey 2002).

The landscape of much of the Mississippian Embayment is generally flat, with many lakes, ponds, sloughs, and swamps. Uplands seldom rise more than 30 m above the surrounding landscape. Numerous meander loops and oxbow lakes occur along the Ohio River floodplain, and extensive flooding is common. This part of western Kentucky is drained by the Tennessee and Ohio rivers (Kentucky Geological Survey 2002). The major drainage in southern Illinois is the Cache–Bay Creek system. Marking the western margin of the embayment, the Loess Hills rise some 40 to 80 m above the Mississippi River floodplain. This area of ravines and slopes sharply contrasts with the low rolling hills to the east (Braun 1950:151).

Clearly, the Lower Ohio River flows through a geologically and topographically diverse landscape. In turn, this environmental diversity created many different kinds of habitats that, prior to recent times, supported a great diversity of plant and animal communities.

Paleoenvironment

In the time since the first hunter-gatherers entered the Lower Ohio Valley more than 12,000 years ago, many changes have taken place with respect to the climate, the physical landscape, and the resident plant and animal communities. This section discusses the Lower Ohio River's environmental past as a way of creating a natural context for examining the development of the region's hunting and gathering societies. Discussion focuses on (1) the geomorphological processes that literally shaped the river valley and its surrounding landscape, (2) regional and local climatic change since the end of the last glaciation, (3) the diversity and distribution of lithic resources that were so economically important to the region's hunter-gatherer populations, and (4) diachronic trends in the composition and distribution of plant and animal communities exploited by these hunter-gatherers.

Pleistocene Processes

As a way of setting the stage for a discussion of the Lower Ohio Valley's hunter-gatherers, it is first necessary to discuss the region's landscape prior to the appearance of the earliest people. The Wisconsin glaciation reached the Ohio River in only a few places; its associated glaciofluvial processes were responsible for shaping much of the landscape the first hunter-gatherers encountered when they entered the region. Perhaps the greatest glacial impact was on the rivers and streams that served as sluiceways for vast quantities of glacial meltwater and its load of clay, sand, and gravel. The landscapes and natural habitats created by these processes played critical roles in the development of later hunter-gatherer societies.

Prior to the start of Pleistocene glaciation about two million years ago, the drainage pattern of what is now the Ohio River basin was much different from today. Initially, many rivers flowed to the north, eventually emptying into what is now the St. Lawrence Valley. Later, glacial activity created a major waterway, known as the Teays River, that flowed from West Virginia, through central Ohio, northern Indiana, and central Illinois, where it joined the ancient Mississippi River. During this time, only water downstream from Madison, Indiana flowed down the Ohio River (Indiana Geological Survey 1998; Stein 1980:44).

During the early Pleistocene, advancing glaciers buried the Teays River under deep deposits of till. Afterward, water from the Kentucky River and other rivers in eastern Kentucky, southern Ohio, and West Virginia backed up until the ponded water eventually cut through a divide near Madison, Indiana. Once the water breached the divide, it quickly established the Ohio River's modern channel (Indiana Geological Survey 1998).

As the glaciers melted and retreated northward, the runoff transported a huge sediment load of clay, sand, and gravel downstream, filling much of the Ohio Valley with deep alluvial deposits. At the same time, many of the Lower Ohio River's unglaciated tributaries were flooded by backwater from the glacial runoff, creating large slackwater lakes and filling the valleys with deep deposits of fine-grained silt. These deeply alluviated valleys are now common features in western Kentucky, southern Indiana, and southern Illinois (Gray 1971; Stein 1980:43–51).

Particularly good examples of these lakes once covered parts of the Saline River and Green River valleys well into the nineteenth century (Frye et al. 1972; Heinrich 1982; Lopinot et al. 1982; Shaw 1915; Stein 1980). For example, in southern Illinois, Glacial Lake Saline was created by the blockage of the Saline River's normal flow by the glacially flooded Ohio and Wabash rivers (Butler 1994:13; Heinrich 1982). During its maximum extent around 20,000 years ago, Lake Saline covered parts of five counties. Although the lake ceased to be an open water lake around 13,000 B.P., the former lakebed supported extensive swamp and shallow wetlands well into the historic period. During the Holocene, hunter-gatherers made frequent use of the plants and animals that inhabited the wetlands (Breitburg 1982; Lopinot 1982a).

Pleistocene glaciation also shaped the Lower Ohio Valley landscape in other ways. Prior to the end of the Wisconsin glaciation, water from the Ohio, Tennessee, and Cumberland rivers flowed through what is now the Cache River valley. Just prior to the start of the Holocene, a large lake on the Tennessee River cut through a low divide, diverting the Ohio River to the former course of the Tennessee-Cumberland rivers. Once the Ohio River stopped flowing through the Cache Valley, sediments from tributary drainages started to fill the valley bottom. Floodplain ridges that remained above the quickly deepening sediments were favorite places for hunter-gatherer settlement (Butler 2009; Frankie et al. 1997:55–58).

Paleoenvironmental Change

Reconstruction of past climatic conditions and their associated vegetation are largely based on palynological and other kinds of paleoecological investigations. With a few exceptions, these kinds of data are lacking for much of the Lower Ohio Valley. Some environmental data for the recent past can be gleaned from General Land Office (GLO) survey records in Indiana and Illinois, but the GLO did not survey Kentucky.

Much of the direct evidence for what environmental conditions were like in the past comes from botanical and faunal specimens from archaeological sites and pollen profiles collected from ancient swamps and ponds (Smith and Mocas 1995:17–18). Some of the best information on Lower Ohio Valley paleoenvironmental conditions comes from Jackson Pond in Larue County, Kentucky. Jackson Pond, a seven-acre spring-fed sinkhole, is situated within 1 km of the Dripping Springs Escarpment that separates the Mississippian Plateau from the Western Coal Field physiographic province (Wilkins et al. 1991:225).

Perhaps the most significant factor affecting Late Pleistocene and Early Holocene vegetation was proximity to the glacial front. While never reaching the Lower Ohio River during the Wisconsin glaciation, the approach of the front brought colder conditions that affected regional vegetation patterns. Glacial retreat brought milder conditions and the return of plants that preferred warmer, often drier conditions.

Today, many researchers contend that Pleistocene plant and animal communities had no modern analogs. Plant communities did not migrate as discrete bands or zones, but instead taxa responded independently, creating variable mosaic associations. Lower Ohio Valley plant communities have experienced a number of broad changes since the last major glacial advance some 18,000 years ago, transitioning from boreal taxa in the Late Pleistocene, to a mesic deciduous forest during the Early Holocene, to drier deciduous forest in the

Middle Holocene, and eventually to modern conditions during the Late Holocene (Smith and Mocas 1995:18–19).

For most of eastern North America, vegetation patterns at 18,000 B.P., prior to known human presence, bore little resemblance to modern conditions. Pine forests were restricted to the Southeast United States while in the Midwest spruce trees dominated a woodland landscape of scattered trees and tundra understory. To the east, trees were absent south of the ice sheet, creating an open tundralike landscape. Oaks grew only in a small area of the Southeast and hickories were restricted to the immediate Gulf coast (Jacobson et al. 1987:280).

More specifically for the Lower Ohio Valley, the Jackson Pond pollen samples indicate that during the full glacial interval from approximately 20,000 to 17,000 B.P., the area surrounding the pond consisted of closed boreal forests dominated by spruce, with lesser quantities of jack pine, fir, and tamarack. In many ways, these forests were like those that grow in parts of Ontario, Canada today (Wilkins et al. 1991:236). Other botanical data suggest that precipitation rates were relatively high.

By 14,000 B.P., the ice sheet started to thin, opening up new areas for vegetation colonization. For the most part, vegetation changes between 14,000 and 12,000 years ago were more a difference in detail than the rearrangement of major vegetation patterns on the landscape. In the Southeast, oaks, along with hickory and beech, formed an increasingly rich deciduous forest along the Gulf coast (Jacobson et al. 1987:281). A spruce–jack pine forest expanded into the Lower Ohio Valley from the east, with white spruce forests in the lowlands and spruce–jack pine forests in the uplands. Increased flooding caused by the glacial meltdown scoured the Ohio floodplains, making much of the valley inhospitable for human settlement (Delcourt and Delcourt 1981:147; Muller 1986:48–50).

The greatest vegetation change in eastern North America during the past 18,000 years occurred at the end of the Pleistocene. As the ice sheet disappeared from eastern North America, conifer forests retreated to the north. Throughout much of the Midwest, forests of oak, hickory, ash, elm, and ironwood trees were established where conifers once grew (Jacobson et al. 1987:282). This time also marked the disappearance of many of the large Pleistocene mammals from the Lower Ohio Valley.

At Jackson Pond, spruce continued to dominate during the late glacial interval (16,800 to 11,300 B.P.). However, declining pollen values for spruce and pine, combined with increasing percentages for upland herbs, suggest that the closed boreal forest was evolving into a more open, taigalike boreal woodland (Wilkins et al. 1991:234).

Vegetation changes during the Early Holocene (12,500 to 8000 B.P.) were almost as dramatic as those during the Late Pleistocene (Jacobson et al. 1987: 282). The cool, moist climate promoted the widespread expansion of species-rich mixed hardwood forests in the area from 34 to 37 degrees north latitude (Delcourt and Delcourt 1981:147) or basically the zone between Tupelo, Mississippi and the Ohio-Mississippi River confluence. Oak-hickory forests continued to dominate most of the Southeast; however, in the Midwest, populations of elm, ash, and ironwood were far less abundant and extensive than during the Late Pleistocene (Jacobson et al. 1987:282). As for the Lower Ohio Valley, oak-hickory forests were established in southern Illinois, mixed conifer–northern hardwood forests spread across the Midwest, and mixed hardwood forests grew south and east of the Ohio Valley (Muller 1986:50).

Around 11,000 B.P., oaks, black ash, and other temperate deciduous trees gradually replaced spruce and jack pine near Jackson Pond, but regional temperature oscillations continued for several thousand more years. A woodland with mesic, open-canopy deciduous trees gradually developed during the Early Holocene (ca. 10,000 and 8000 B.P.), dominated by cool-temperature trees like oaks and hornbeam (Wilkins et al 1991:235).

For parts of the Midwest and Southeast, Middle Holocene (8000 to 5000 B.P.) changes in regional temperature and rainfall distributions resulted in significant vegetation changes. By approximately 6,000 years ago, the last glacial remnants disappeared from eastern North America and vegetation approached modern conditions. Oak and hickory grew south of the Great Lakes, increasing in prevalence to the south and southeast. Oak savannahs, consisting of grasslands with stands of oak, grew to the north of the Ohio Valley; mixed hardwoods were restricted to sheltered valley and slope habitats to the south in the Cumberland Plateau (Muller 1986:51).

Increased warmth and dryness in the Great Plains, associated with the mid-Holocene Hypsithermal interval (Deevy and Flint 1957), resulted in an eastward shift of prairie grasslands. These dryer and warmer conditions contributed to decreased upland vegetation, increased surface erosion, and aggraded floodplains. Newly created riverine shoal habitats promoted the expansion of shellfish populations, while the establishment of backwater slough environments increased floodplain productivity (Anderson 2001:158; Smith 1987:22).

Middle Holocene precipitation and temperature changes may have contributed to the increased strategic importance of site locations near rivers, swamps, and other permanent wetland habitats (Brown 1985; Brown and Vierra 1983). The prairie-forest boundary reached its easternmost limit approximately 7,000 years ago, gradually retreating some 400 km to the west as conditions became moister during the Late Holocene (ca. 5000 B.P.) (Delcourt and Delcourt 1987:98; Jacobson et al. 1987:282). At Jackson Pond, this change is reflected by the disappearance of mesic trees from upland forests and their replacement by oak-hickory forests. The effects of the Hypsithermal interval are indicated by seasonally fluctuating water levels, warmer water temperatures, and reduced water depth in Jackson Pond (Wilkins et al. 1991:236).

Essentially, modern vegetation patterns were established throughout the Lower Ohio Valley by 5,000 years ago and, despite minor temperature and precipitation fluctuations, have remained roughly the same throughout the Late Holocene. As Muller (1986:51) cautions, however, this does not mean that the overall nature of the environment has gone unchanged, only that the general resources available to regional hunter-gatherers did not significantly change.

At Jackson Pond, the Late Holocene (post-5000 B.P.) environment was characterized by increasing percentages of grass pollen and the appearance of prairie indicators like prairie clover. This suggests that central Kentucky's vegetation mosaic of mixed prairie and deciduous forest encountered by west central Kentucky's first Euroamerican settlers was established early in the Late Holocene (Wilkins et al. 1991:236).

Historic Vegetation

Clearly, the vegetation that once covered the Lower Ohio Valley landscape has undergone considerable change during the past 12,000 years, going from the predominance of peri-glacial species to those that thrive in today's warmer, drier climate. Because of this diachronic variability, it is inappropriate to make direct comparisons between the composition and coverage of modern plant communities and those of the distant past. However, since essentially modern forests were present in the Lower Ohio Valley by at least the early Late Holocene, we can use historic vegetation data to get a *general* idea of the kinds of plant resources that were available to the region's hunter-gatherers.

The vegetation that covered the Lower Ohio Valley prior to extensive nineteenth- and twentieth-century modification was as diverse as the physical landscape. It is often difficult to reconstruct the composition of ancient plant communities because of extensive disturbance by modern agriculture, land leveling, logging, or urban development; historical documents and analyses of relic plant communities provide some clues (Braun 1950).

In her classic work, Lucy Braun (1950:122) designated the Lower Ohio Valley as part of the Western Mesophytic Forest Region, a transition zone containing a complex mosaic of vegetation types. She attributed that variability to past and recent natural and cultural influences on the vegetation, as well as underlying geology.

Braun distinguished six sections within the Western Mesophytic Forest Region, five of which include parts of the Lower Ohio Valley. The western edge of the Bluegrass section, which includes the Outer Bluegrass, extends into the Falls of the Ohio area. Outer Bluegrass forests were dominated by white and red oak, sugar maple, tulip tree, beech, black walnut, white ash, shagbark hickory, and red elm. Early travelers also reported an abundance of cane, along with wild rye and clover, suggesting that some forests were not very dense and that sizeable areas of open grassland existed (Braun 1950:125–131).

The area of Illinoisan glaciation east of the flat, undissected, and poorly drained Norman Uplands in southeastern Indiana was vegetationally distinct from the Bluegrass to the south (Braun 1950:124). Topographic conditions created a favorable habitat for wet-tolerant trees like beech, pin oak, red maple, white elm, white oak, and shellbark hickory. Dissected, drier parts of this area, situated adjacent to the larger streams, support an oak-ash-maple forest, along with beech, buckeye, and white basswood (Braun 1950:132–136).

Based on the similarity of their forests, Braun (1950:136) included the Knobs of Kentucky, all of unglaciated southern Indiana (including parts of the Norman Upland, Mitchell Plain, Crawford Upland, and Wabash Lowland), the Ozark Hills of southern Illinois, and Kentucky's Western Coal Fields and Dripping Springs Escarpment, in the Hills section of the Western Mesophytic Forest Region. The Kentucky Knobs once hosted oak, oak-hickory, and oak-chestnut communities on its drier slopes and uplands. Oak–tulip tree forests grew in areas of low relief, while drier areas hosted pines, along with patches of red cedar and prairie vegetation. In southern Indiana, mixed mesophytic forests, largely consisting of beech, tulip tree, sugar maple, and walnut, grew on northerly slopes, and oak or oak-hickory forest on drier slopes and ridges. The wetter parts of the Wabash Lowlands and the Green River valley hosted southern swamp forests (cypress, poplar, cottonwood, black willow, pecan, pin oak, swamp white oak, willow oak, swamp chestnut oak, and hackberry) (Braun 1950:150–151); drier areas were typically oak-hickory. Beech grew in the most favorable areas.

In the Shawnee Hills of southern Illinois, upland areas hosted black and white oaks, while beech, sugar maple, tulip tree, wild cherry, mulberry, and species of hickory grew on the sheltered slopes (Braun 1950:146–151). In contrast, communities of widely spaced red cedar and prairie species grew where extensive limestone outcrops occurred, forming "cedar barrens" (Braun 1950: 150–151).

The Mississippian Plateau vegetation was a mosaic of diverse plant communities. In the dissected and hilly areas to the east, beech-dominated mixed mesophytic forests grew on many slopes, while the drier slopes and ridges hosted oak, oak-hickory, and oak-chestnut forests. To the west, oak forests inhabited much of the undulating plateau surface; oak-hickory occupied the isolated hills. Areas of prairie vegetation (barrens) also existed. Cedar barrens grew in the drier slopes, while swamp forests developed around depressions and wet flatlands (Braun 1950:151–152).

At the western end of the valley, the Mississippi Embayment section hosted vegetation types commonly found on the Gulf Coastal Plain. Oak-hickory forests were common in the uplands. Prairie communities resembling those of the Mississippian Plateau also grew there. The Embayment's broad alluvial valleys once were covered by dense, luxuriant forests consisting of willow, swamp chestnut, pin, overcup, and water oaks; downy poplar; cottonwood; winged and white elm; hackberry; river birch; sycamore; and pecan trees. Many of these species comprised the swamp forests found in wet areas. Bald cypress occurred in deep swamps (Braun 1950:157–159).

The vegetation of the Loess Hills contrasts sharply with that of the lowlands. Beech was the most dominant species, often comprising one-half of the forests. Tulip trees, red oak, hickories, cucumber trees, sugar maples, and basswood were also common (Braun 1950: 160).

The diverse plant communities that grew across the Lower Ohio Valley were associated with a wide range of habitats that included ridgetops, slopes, floodplains, rivers, streams, and swamps. In addition to supporting highly varied plant communities, these habitats also provided living space for a high diversity of large and small mammals, birds, reptiles, amphibians, and fishes. Like the plant communities, the composition of these animal communities also experienced considerable change during the late Pleistocene and Holocene.

Paleofauna

The rapidly changing vegetation conditions found in the Lower Ohio Valley at the close of the Pleistocene were accompanied by equally dynamic changes in regional animal populations. Eastern North America's Pleistocene fauna consisted of a combination of now extinct megafauna, such as mammoth, mastodon, ground sloth, horse, ancient bison, camel, and dire wolf, along with modern species like the white-tailed deer, rabbit, raccoon, and squirrel (Semken 1983:192). By the end of the Pleistocene, more than 30 genera of megafauna had become extinct, including all examples of North American horse, camel, and elephant (Anderson 2001:153). The mass extinction of megafauna has led some researchers to characterize Holocene fauna as "depauperate" or "impoverished" with respect to the high species density of the Late Pleistocene (Semken 1983:192).

The earliest humans to enter the Lower Ohio Valley during the Late Pleistocene encountered the remnants of the once numerous species of Pleistocene megafauna. The extent to which early hunter-gatherers pursued megafauna remains unclear (Haynes 2005), but there is solid archaeological evidence that they were occasionally targeted as sources of food and raw materials to make tools (Dunbar and Webb 1996; Graham et al. 1981; Webb et al. 1984). Likewise, the role of early hunter-gatherers, if any, in the extinction of the megafauna is still being investigated (Grayson 1991; Martin 1967; Martin and Klein 1984; Mosimann and Martin 1975). Migrating megafauna may have taken advantage of the Falls of the Ohio's shallow water to cross the Ohio River while following trails that led to northern Kentucky's salt licks, much like bison did during the early historic period (Boisvert 1984; Smith and Mocas 1995: 21–22).

One of the best sources of information on the kinds of Pleistocene fauna that once inhabited the Ohio River Valley comes from Big Bone Lick, an ancient marsh and spring complex in Boone County, Kentucky (Boisvert 1984; Sanders et al. 1974; Tankersley 1985:27–49). Located approximately 130 km northeast of the Falls of the Ohio and approximately 3 km east of the Ohio River, more than 200 years of research (Carr and Shaler 1876; Jillson 1936; Schultz et al. 1963, 1967) have yielded bones from an assortment of both grazing and browsing mammals including mammoth, mastodon, bison, horse, caribou, bear, tapir, moose, musk ox, and ground sloth (Kurten and Anderson 1980:66; Smith and Mocas 1995:22). Associated radiocarbon dates range from 17,200 to 10,600 B.P. (Tankersley 1985:Table 1). Researchers have found early forms of projectile points (Clovis, Quad, Beaver Lake) at Big Bone Lick, but so far none of these artifacts have been directly associated with Pleistocene megafauna (Tankersley 1996:28).

The beginning of the Holocene witnessed a reduction in the diversity of the Lower Ohio Valley mammal population (Semken 1983:192). By approximately 10,000 years ago, animal species that inhabited the Lower Ohio Valley landscape were essentially the same as those at the time of European settlement. Variation in regional climate and vegetation patterns, however, is reflected by diachronic changes in the distribution and frequency of some species. For example, the bones of northern and boreal animal species found in a vertical fissure along the Mississippi River bluff in southwestern Illinois suggest a cool moist period during the Early Holocene (Kurten and Anderson 1980:63). Likewise, skeletal elements of *Geomys bursarius,* the plains pocket gopher, at the Black Earth site in Saline County, Illinois, suggest a nearby prairielike habitat from approximately 5000 to 1400 B.P. By the mid-twentieth century, the plains pocket gopher's range extended no farther south than north central Illinois, some 500 km to the north of the Black Earth site (Breitburg 1982:905–907).

Analyses of faunal collections associated with Early to early Late Holocene components indicate that hunter-gatherers exploited a wide range of habitats. They exploited these places using a variety of hunting, fishing, and collecting strategies to capture many kinds of large and small mammals (white-tailed deer, elk, woodchuck, squirrel, rabbit), terrestrial and aquatic birds (turkey, quail, duck, goose), fishes (various), shellfishes, reptiles (box turtle and snake), and amphibians. Most of these species served as important sources of dietary protein and fat; however, many also were important for their skins and furs used to fabricate clothing, shelter covers, and containers; and bone, shell, and antler to make tools, containers, and items of personal adornment (Breitburg 1982:861–957; Webb 1946:281–314; White 2005:339–350).

Lithic Resources

Due to preservation issues, much of what we know about Lower Ohio Valley hunter-gatherers is based on the study of their lithic industries. It is abundantly clear that these groups employed lithic technology to make many different kinds of weapons, general utility tools, fabricating and processing implements, ritual/ceremonial objects, and items of personal adornment (Justice 1987; Winters 1969). During the Early Holocene, groups maintained small inventories of portable equipment, much of which they manufactured using flaking technology. By the Late Holocene, artifact assemblages included a diverse assortment of flaked stone items, along with a substantial and varied inventory of ground stone objects.

Analyses of hunter-gatherer lithic assemblages suggest that numerous technological changes occurred during the Holocene. Primary among these was the gradual transition from a largely curated technology emphasizing biface production and maintenance to one focusing more on expediently produced flake tools (Amick and Carr 1996:44).

Obtaining the raw materials to manufacture stone tools required knowledge of the kinds of stone best suited to perform specific tasks and where to get those materials. Analyses of Lower Ohio Valley lithic assemblages indicate that hunter-gatherers collected most of their lithic material from the local area; they also used stone that came from sources that are more distant. Identification of the kinds of lithic materials used, their geological sources, and the spatial distribution of artifacts made from those materials has provided important information on the organization of hunter-gatherer lithic technologies, mobility strategies, economic organization, and interaction patterns (Andrefsky 1994; Morrow and Jefferies 1989; Munson and Munson 1984; Wiessner 1982, 1983).

Lower Ohio Valley hunter-gatherers used chert, a lithic material associated with some regional limestone formations, to make most flaked stone artifacts. Although variable in quality, the crystalline structure, hardness, and fracture characteristics of many Lower Ohio Valley cherts made them well suited for making projectile points, scrapers, knives, drills, and other kinds of implements requiring sharp edges.

Chert consists largely of microcrystalline quartz and occurs as nodules, lenses, or tabular strata throughout much of the region. It comes in many colors (black, gray, white, red, green), reflecting the material's specific geological origin and mineral content (Bassett 1980:77). In the Lower Ohio Valley, much of the chert is associated with various limestone formations dating to the Mississippian geological period; cherts also occur in older Ordovician, Silurian, or Devonian formations, or as secondary deposits in glacial till or river/streambeds.

Hunter-gatherers collected chert from natural exposures or quarries at their geological sources, from secondary deposits (gravel bars, glacial tills), or occasionally, through trade with other groups (Gibson 1996:292; Johnson and Brookes 1989). Nonlocal chert transported by water and deposited in gravel bars is represented by nodules having a highly patinated, water-rolled cortex (Granger et al. 1992:218).

Investigations of prehistoric chert utilization have demonstrated that some types were preferred over others, primarily because of different technological characteristics and/or accessibility. Some researchers have interpreted diachronic changes in chert selection and use as reflecting shifts in hunter-gatherer mobility strategies (Amick and Carr 1996; Conaty 1985, 1987; Moore 2008; Morrow et al. 1992; Munson and Munson 1984).

The following discussion identifies the varieties and sources of many of the cherts used by Lower Ohio Valley hunter-gatherers. Because of their ubiquity and diversity, only the major types are discussed.

In the upper portion of the Lower Ohio Valley, analysis of chert artifacts from deeply stratified sites near the Falls of the Ohio revealed that, except for some early horizons, most flaked stone implements were manufactured from Wyandotte (Ste. Genevieve), Muldraugh, or various Ohio River gravel cherts. The percentage represented by each type was consistent throughout the Early and Middle Holocene (Collins and Driskell 1979:1040). Both Muldraugh and Wyandotte cherts occur locally—Muldraugh coming from the Knobs regions just to the south and east of the Falls, and Wyandotte from across the Ohio River in southern Indiana. Many other chert types identified in the assemblages (St. Louis, Paoli, Boyle, and Brassfield) were attributable to cobbles collected from gravel bars in the Ohio River or from nearby tributaries; other cherts came from more distant locations (Boisvert et al. 1979:60).

At the Railway Museum site (15Jf630), situated just upriver from the Falls of the Ohio, flaked stone artifacts dating to ca. 4700 B.P. were manufactured from four chert types—Ste. Genevieve, Haney, Boyle, and Laurel. Based on debitage attributes, site inhabitants collected Haney, Boyle, and Laurel cherts from sources located very near the site, while they obtained Ste. Genevieve chert from more distant sources and brought it to the site in a reduced form. Inhabitants probably collected Boyle chert from river gravels near the site (Anslinger et al. 1994:79–82).

In southern Indiana, hunter-gatherers collected chert from exposed Mississippian-age strata in the Norman Uplands, Mitchell Plain, and Crawford Upland physiographic provinces (Munson et al. 1983:189). As in the Falls area, chert knappers collected some materials from their primary sources and others from nearby gravel bars (Bassett 1980:72). Southern Indiana's prehistoric hunter-gatherers commonly used Allen's Creek, Muldraugh, Harrodsburg, Upper St. Louis, Wyandotte (Ste. Genevieve), Bryantsville, and Haney cherts (Munson et al. 1983). Middle to Late Holocene hunter-gatherers preferred Allen's Creek and Muldraugh chert. Wyandotte, a high-quality blue-gray chert found in western Harrison and eastern Crawford counties, was used throughout the Holocene. Implements made from Wyandotte chert have been found on sites as far as 300 km from the Harrison-Crawford county area, having been transported there by wide-ranging hunter-gatherer groups or passed along through regional exchange networks (Munson et al. 1983:194).

Thomas Gatus's survey of chert resources in the Big Bend of the Green River indicated that Vienna, St. Louis, Paoli, Girkin, Haney, and Menard cherts are suitable for tool manufacturing, but that only Vienna and St. Louis were of sufficient quality and quantity to be considered dependable lithic resources (Gatus 2005:448).

In the Lower Tennessee–Cumberland region (Land between the Lakes) of far western Kentucky, Middle to Late Holocene hunter-gatherers most commonly used Upper St. Louis and Lower St. Louis/Salem cherts, along with lesser quantities of Fort Payne, Warsaw, Ste. Genevieve/Fredonia, Tuscaloosa, Degonia, and Kinkaid (Conaty 1987:143–144). Upper St. Louis deposits are more commonly exposed in the Cumberland River drainage, while Lower St. Louis outcrops are more prevalent in the Tennessee River basin (Conaty 1987:146; Gatus 1979; Nance 1980, 1984). Most of the other chert types can be found in the local area, but Tamms chert probably originated from across the Ohio River in southern Illinois (Conaty 1987:146).

Southern Illinois hunter-gatherers obtained much of their chert from the Shawnee Hills' chert-rich Mississippian and Devonian limestone formations that run east to west between the Ohio and Mississippi rivers. In contrast, chert sources are scarce north of the Hills, so chert found there had to be imported from the south. Commonly exploited southern Illinois cherts include St. Louis, Mounds Gravel, Kinkaid, Cave-in-Rock, Kaolin, Mill Creek, Cobden/Dongola, Fort Payne, Elco, and miscellaneous cherts found in glacial till. In addition, Burlington chert from west central Illinois and Dover chert from western Tennessee were occasionally acquired (May 1984; Morrow 1982).

Hunter-gatherer groups that inhabited portions of the Gulf Coastal Plain physiographic province in far western Kentucky heavily exploited chert gravels found on river and stream terraces and in gravel bars. On some sites, Mounds Gravel accounts for more than 95 percent of all Archaic flaked stone artifacts. Hunter-gatherers also used minor amounts of St. Louis, Mill Creek, Dover, Kaolin, Kincaid, and Fort Payne chert (Butler et al. 1981:37–38).

In addition to chert, hunter-gatherers also used other types of lithic resources for various technological, and, occasionally, social or ideological purposes (Butler 1994:23; Smith and Mocas 1995:16). They used sandstone for many heat-related activities, primarily hot-rock cooking (Butler 1994:23; Hensley 1994:82–84; Smith and Mocas 1995:16). Sandstone was obtained from several different geological settings—exposed Pennsylvanian geological strata, colluvial deposits found along the base of uplands, and in stream gravel. The abundance of Pennsylvanian sandstone in the Lower Ohio Valley makes it difficult to determine where the inhabitants of a specific site collected sandstone, but it probably came from the source that required the least effort in terms of collection and transport to the site (Hensley 1994:84).

North of the Ohio River, glacial cobbles collected from till deposits served as important lithic resources (Butler 1994:23; Smith and Mocas 1995:16). In addition to being used to construct various kinds of features and facilities, glacial cobbles were also selected to make many kinds of ground stone implements such as hammerstones, grinding stones, axes, atlatl weights, mauls, and pestles (Butler 1994:23). Ornamental and ceremonial items, like beads and pipes, were also occasionally made from glacial cobbles (Boisvert et al. 1979: 390–392).

Although the geological source of many flaked and ground stone artifacts can be identified using a variety of analytical techniques, their ultimate distribution across the landscape is determined by a variety of social, economic, and political factors. In many cases, differences in chert use may simply reflect the selective actions of toolmakers with respect to available lithic resources and the social environment in which they operated.

In the following chapters, I examine synchronic patterns and diachronic trends in hunter-gatherer resource acquisition and use and the interrelationships between the Lower Ohio Valley's evolving physical landscape and the emerging cultural landscape that was constructed by the Native Americans who lived there. The story begins more than 12,000 years ago.

3 / Hunter-Gatherer Archaeological Research in the Lower Ohio Valley

The Archaic Concept

Recognition and study of Lower Ohio Valley hunter-gatherer societies has been closely linked to the development of the Archaic Concept (Ritchie 1932). Although earlier Paleoindian groups were present, until very recently little was known about their way of life. Therefore, this chapter focuses on that portion of the Lower Ohio Valley's archaeological record associated with Archaic period hunter-gatherers.

As early as 1917, archaeologists working in the Lower Ohio Valley recognized that the region had once been inhabited by groups of people that did not make pottery or practice agriculture, and that lived there prior to the better-known agriculturally based groups that were commonly referred to as Mound Builders (Nelson 1917:68–69). Nevertheless, there was no widely accepted name for them or their way of life.

This all began to change in 1932 with the publication of William Ritchie's (1932) report on the results of his archaeological investigations at the Lamoka Lake site in New York State. For the first time, Ritchie used the term *Archaic* when referring to the way of life he documented at Lamoka Lake—one based on hunting and gathering and lacking agriculture and pottery.

A short time later, Ritchie (1936) proposed the Archaic as the third *pattern* in the recently devised Midwestern Taxonomic System. Initially, James B. Griffin rejected Ritchie's proposal, but eventually he relented because many archaeologists were already using the term when referring to the aceramic cultures in eastern North America (Griffin 1952a; Stoltman 1978:708). In the late 1930s, William Webb and David DeJarnette (1942) applied the term *Archaic* to sites and artifacts associated with preceramic hunter-gatherers in Kentucky and northern Alabama, expanding its use to most of the Eastern Woodlands (Stoltman 1993:4).

By the early 1940s, James A. Ford and Gordon R. Willey (1941:332) were referring to the earliest known cultures in eastern North America as Archaic. They regarded these early societies as lacking pottery and agriculture, and as producing an artifact inventory that was inferior to those of later cultures. Archaeological sites assigned to their Archaic category included one of the now famous Green River shell middens known as Chiggerville (15Oh1) (Figure 3.1). Diagnostic Archaic artifacts included full-grooved axes, bannerstones or atlatl weights, expanded base chert drills, leaf-shaped and stemless flaked stone knives, and expanded-head engraved bone pins. Ford and Willey (1941:332–333) maintained that while the eastern United States was thinly populated by small groups of Archaic hunter-gatherers, sites like Chiggerville represented places where conditions permitted them to aggregate.

The temporal and cultural parameters of the *Archaic* continued to be topics of discussion for much of the twentieth century (Caldwell 1958; Griffin 1946, 1952a, 1967; Stoltman 1978, 1993; Willey 1966; Willey and Phillips 1958). Archaeologists had initially used *Archaic* to label eastern North American hunter-gatherer societies (Webb and DeJarnette 1942:319), but as regional archaeologists gradually refined their culture chronologies, they redefined the Archaic as a *period.* Perhaps the "coming of age" for the Archaic was its being included as one of five pan-eastern periods in Griffin's (1952a) "Green Bible." At the same time, archaeologists continued to use it to designate a cultural *stage* (Stoltman 1978:708).

In 1966, Willey (1966:247) included the Archaic as one of his four cultural traditions of the Eastern Woodlands. According to Willey, the Archaic tradition, which followed the Big-Game Hunting tradition, evolved locally in the early post-Pleistocene. It was characterized by a hunting-fishing-plant-collecting economy, large and broad-bladed dart points, and ground- and polished-stone tools and ornaments. Despite all the confusion about how they used the term, eastern North American archaeologists quickly adopted it and continue to use it today (see Emerson et al. 2009).

The advent of radiocarbon dating in the late 1940s, combined with the excavation of deeply stratified sites (Broyles 1971; Coe 1964; Fowler 1959), helped archaeologists to refine their estimates for the length of the Archaic pe-

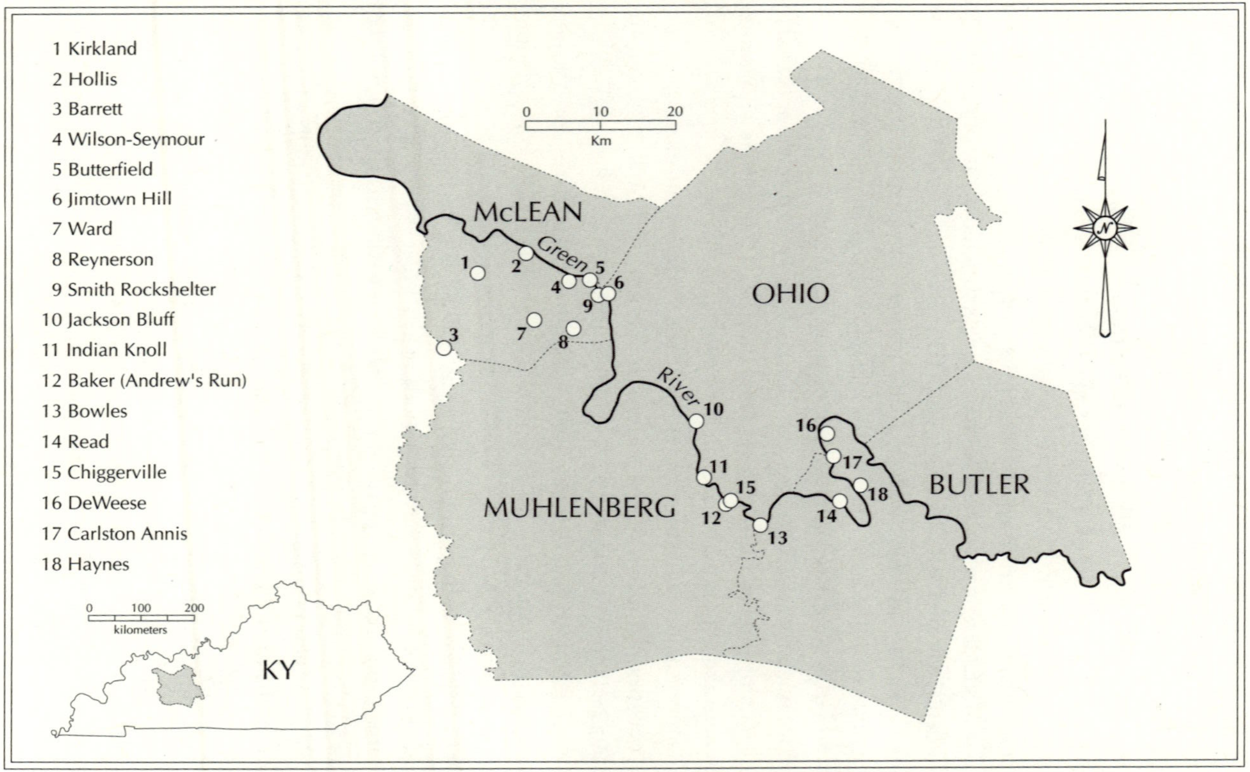

Figure 3.1. Important Green River sites.

riod. By the late 1960s, the consensus among most archaeologists working in the Lower Ohio Valley and nearby parts of the Southeast and Midwest was that the Archaic period extended from approximately 10,000 to 3,000 years ago.

Primarily based on diachronic changes in projectile point styles, researchers further divided the Archaic into early, middle, and late subdivisions that had both temporal and cultural significance. Unfortunately, considerable debate ensued over the temporal boundaries of those subdivisions (compare Griffin 1967:178–179; Muller 1986; Stoltman 1978:708; Willey 1966:249). We now recognize that the regionally variable timing of the first appearance of traits used to distinguish the subdivisions was responsible for much of the disagreement (Stoltman 1993).

Today, Lower Ohio Valley archaeologists generally define the Early Archaic subdivision as extending from 10,000 to 8000 B.P., coinciding with the Early Holocene. The Middle Archaic subdivision covers the time from 8000 to 5000 B.P., overlapping with the Middle Holocene, while the Late Archaic subdivision runs from 5000 to 3000 B.P., conforming to the early part of the Late Holocene (Anslinger 2001; Jefferies 1996b:39–40). In some areas, archaeologists distinguished the Terminal Archaic (ca. 4500 to 3000 B.P.) from the Late Archaic (Granger 1988).

Another potentially confusing situation with the Archaic chronology stems from the difference between the dates archaeologists have traditionally assigned (based on radiocarbon years before present) and calendar dates based on the most recent radiocarbon calibration techniques (Milner 2004:34). The actual calendar dates for the "start" and "end" of the Early, Middle, and Late Archaic subdivisions, after calibration, are from 1,300 years (start of the Early Archaic) to 200 years (end of Late Archaic) older than the traditional dates that appear in many archaeological publications (Table 3.1). In order to avoid undue confusion, dates in this book are usually presented as radiocarbon years before present (B.P.), corresponding to dates that have long appeared in the archaeological literature (see Milner 2004:8–9). In reality, the starting and ending dates for the Archaic subdivisions are, at best, only estimates. For the most part, these somewhat arbitrary divisions reflect the needs of archaeologists to categorize data rather than times of major changes in prehistoric human behavior.

Hunter-Gatherer Research in the Lower Ohio Valley

The recognition of early hunter-gatherers in the Lower Ohio Valley has been a long and tedious process, paralleling events in other parts of eastern North

Table 3.1. Comparison of Archaic Period Chronologies for Eastern North America (dates are uncalibrated).

Temporal Unit	Eastern North America (Griffin 1967)	Eastern North America (Willey 1966)	Lower Ohio Valley (Muller 1986)[1]	American Bottom (Higgins 1990)
Archaic Period	ca. 8000–1000 B.C.	8000–1000 B.C.	8000 to 1000–600 B.C.	7900–600 B.C.
Early Archaic	ca. 8000–6000 B.C.	8000–5000 B.C	8000 to 6000 B.C.	7900–6000 B.C
Middle Archaic	6000–4000 B.C.	5000–2000 B.C.	6000 to 4000–3000 B.C.	6000–3000 B.C.
Late Archaic	4000–1000 B.C.	2000–1000 B.C.	4000 to 3000–1000 to 600 B.C.	3000–600 B.C.

[1] Muller (1986:57) recognized that the criteria archaeologists used to define the span of the different Archaic subperiods varied depending on location within the Lower Ohio Valley. All dates are uncalibrated.

America. Although hunter-gatherer groups were present in the valley during the earlier Paleoindian period, evidence of their presence is usually quite ephemeral and often mixed with those of later groups. Therefore, what we knew about hunter-gatherers, until very recently, largely came from the study of the Archaic period archaeological record.

The sites and artifacts of the Ohio Valley's ancient Native Americans have piqued the interest of Euroamericans for more than 200 years (Atwater 1820; Janzen 1972; McCulloh 1829). As early as the 1780s, George Rogers Clark, one of the founders of Louisville, Kentucky, reported several Indian burial grounds near the Falls of the Ohio River. In the early 1800s, human bones, stone axes, and projectile points were reported from several mounds in Louisville that were destroyed to get fill dirt (Janzen 1972:306–307). In his 1829 publication, Dr. J. H. McCulloh, Jr. (1829:503) noted that considerable differences existed in the placement and construction of mounds and earthworks built in the Ohio Valley, specifically Kentucky and Ohio, compared with mounds in Georgia and Florida. There is even a report of a mastodon tooth found at a depth of 8 m below ground surface among human bones and "implements of the Stone Age" (Durrett 1893:10–11, cited in Janzen 1972:307).

In 1848, Ephraim G. Squier and Edwin H. Davis published the results of their archaeological investigations in *Ancient Monuments of the Mississippi Val-*

ley. Like most early studies, their work focused on the mounds and earthworks of the region's later Woodland and late prehistoric societies, not on sites attributable to Holocene hunter-gatherers. Squier and Davis concluded that the mounds and earthworks were built by the mythical lost race of Moundbuilders, not the ancestors of modern Native Americans.

By the 1870s, geologists from Kentucky, Illinois, and Indiana were busy conducting geological surveys of the Lower Ohio Valley. Some of their reports contained information on archaeological sites and artifacts, as well as descriptions of chert sources. Construction projects near the Falls of the Ohio River exposed deeply buried cultural deposits, but little or no subsequent investigations were conducted. Geological reports for Floyd and Clark counties, Indiana, reported what might have been an extensive shell midden near Clarksville (Anslinger 2001; Janzen 1972). This was probably the same site, now known as the Old Clarksville site, that George Rogers Clark described almost 100 years earlier (Janzen 1972:313). The first description of Falls area chert sources also dates to the 1870s. These quarries were the source of the blue-gray Wyandotte or Harrison County chert that was so widely used by Falls area hunter-gatherers (Collett 1879, cited in Janzen 1972:315).

The first widely distributed description of a Lower Ohio Valley Archaic site was Clarence B. Moore's 1916 report on his Green River expedition (Moore 1916). Moore, an independently wealthy member of the Antiquarian Society of Philadelphia, was an avid artifact collector. In November 1915, Moore steamed off on his boat, the *Gopher,* to investigate sites along Kentucky's Green River (Polhemus 2002:1–27; Schwartz 1967:21–23).

During the winter of 1915–1916, Moore excavated portions of at least ten Green River sites in Edmonson, Butler, Henderson, McLean, and Ohio counties (Figure 3.1). He spent nearly a month at "the Indian Knoll" opposite the town of Paradise, Kentucky. While there, he excavated nearly 300 human burials and their associated grave goods. Moore noted the frequent co-occurrence of artifacts that he interpreted as "netting needles and sizers" used to make nets. The perforated stone or shell weights and socketed antler hooks, sometimes accompanied by socketed antler handles, are now recognized to be parts of atlatls used to throw spears or javelins. Moore's contemporary, Charles C. Willoughby, thought that these objects were parts of atlatls, but Moore rejected the idea because the *Gopher*'s captain had made a net using the bone and antler artifacts (Polhemus 2002:25–26).

C. B. Moore's Green River excavations brought to light the existence of the preceramic shell and earth middens of the Lower Ohio Valley. While his report was largely descriptive, its many photographs and drawings probably stimulated the interest of William S. Webb and other archaeologists who have studied the Green River shell middens since Moore's forays (Polhemus 2002:26).

About the same time that Moore was excavating Indian Knoll, Nels C. Nelson of the American Museum of Natural History conducted archaeological investigations at Mammoth Cave in Edmonson County, Kentucky. Years earlier, mummies found in the cave had attracted national attention. The mummies, plus the discovery of ancient artifacts in European caves, led Nelson to think that Mammoth Cave would yield important information on the antiquity of humans in North America (Schwartz 1967:27–31).

Nelson's (1917) Mammoth Cave excavations yielded artifacts that he attributed to relatively recent "Stone Grave and Mound Builder tribes"; however, he also found evidence of what he described as "simple hunting and gathering people." Nelson (1917:68–69) suggested that these "primitive" people ate naturally available foods and that they were in some way ancestral to later agriculturally based Mound Builder groups. Unfortunately, the existence of these more ancient hunter-gatherers went largely unnoticed and had little impact on the development of Ohio Valley culture history (Schwartz 1967:29–31).

During the 1920s, the pace of research on what archaeologists would later call *hunting and gathering societies* quickened. University of Kentucky professors William S. Webb and William D. Funkhouser conducted much of that work. In the summer of 1924, the two budding archaeologists traveled to west central Kentucky to visit several of the Green River shell middens, including Indian Knoll (15Oh2), the "De Weese Mound" (15Bt6), and a site they simply called "the mound at Chiggerville" (15Oh1) (Funkhouser and Webb 1928:156). They described Chiggerville as being "100 yards long, 75 yards wide, and when we visited it about ten feet high" (1928:156). Funkhouser and Webb reported that large numbers of chert artifacts, stone pestles, hammerstones, bannerstones, and other kinds of artifacts covered the ground. Like Moore (1916), Funkhouser and Webb (1928:186, 265) wrestled with determining the functions of the bannerstones (atlatl weights) and netting needles (atlatl hooks).

Funkhouser and Webb excavated several trenches through the Chiggerville midden, revealing an "almost solid bed of mussel shells for a depth of six feet" (1928:157). Among the shells were human, elk, wildcat, and wild turkey bones, along with distinctive stone, bone, antler, and shell artifacts (1928:157).

Funkhouser and Webb noted that since the sites containing artifacts like those found at Chiggerville and Indian Knoll were located along the rivers and had artifacts that suggested their inhabitants were fishermen and boatmen, they decided to call these ancient groups the "River People" (1928:153). Based on his analysis of Indian Knoll human skulls, Ales Hrdlicka, a physical anthropologist with the Smithsonian Institution, described the "River People" as having crania that were "typical of undeformed Algonquin skulls, evidently not Shawnee, although coming from the region ascribed in general to that tribe" (Funkhouser and Webb 1928:155).

In 1928, Funkhouser and Webb published their summary of Kentucky prehistory in *Ancient Life in Kentucky.* In it, they divided Kentucky's archaeological record into six "cultures" that they named after different archaeological (Stone Grave), linguistic (Algonquin), or ethnic (Pre-Cherokee) entities. The prehistoric record had little time depth to it, since most people thought the world had been created only a little more that six thousand years earlier (Schwartz 1967: 31–34). Although none of Funkhouser and Webb's "cultures" fully equated with what archaeologists later would describe as hunter-gatherer societies, some of the sites attributable to the Algonquin culture included the "great shell mounds on the Green River" (Funkhouser and Webb 1928:67).

In 1932, Funkhouser and Webb (1932) reported the findings of their statewide archaeological survey, the final chapter of which included an updated version of Kentucky's prehistoric culture areas. At that time, they included the part of the Green River inhabited by their River People in the "Shell Mound Area." They suggested that the shell mounds' inhabitants subsisted entirely by fishing and hunting since they had found no evidence of agriculture. Webb and Funkhouser also speculated that "these mounds may be among the oldest evidence of mound occupancy in this state" (1932:425).

In retrospect, the early investigations of Moore, Nelson, Webb, and Funkhouser provided the first archaeological evidence from the Lower Ohio Valley of the preagricultural societies that would eventually be called *Archaic.* As Douglas Schwartz (1967:80) noted, although Moore published the results of his work at Indian Knoll and other Green River shell middens, and Nelson proposed the existence of pre–Mound Builder groups, it was not until Ritchie's (1932) work at the Lamoka Lake site in New York that the two were linked.

Hunter-gatherer research in Kentucky greatly benefited from the passage of the Emergency Relief Appropriation Act in 1935 and the subsequent formation of the Works Progress Administration (WPA). Under the direction of William S. Webb, WPA workers made the first large-scale excavations of the Green River shell midden sites using standardized field and laboratory techniques (Milner and Smith 1986).

Webb's use of standardized procedures allowed the artifacts, features, and stratigraphy from these sites to be easily compared and similarities recognized (Winters 1974:vii). Lower Ohio Valley Archaic sites that were investigated using New Deal–era labor included five each in Ohio and Butler counties, seven in McLean County, two in Hopkins County, and one in Muhlenberg County. Fieldwork for these sites was directed by site supervisors like Ralph D. Brown, Albert G. Spaulding, C. T. R. Bohannan, Marion H. Baugh, John B. Elliot, David Stout, and James C. Greenacre (Milner and Smith 1986:Appendix 1). Their excavations at Indian Knoll, Chiggerville, Butterfield, Read, Ward, and Carlston Annis (Figure 3.1) yielded tens of thousands of artifacts and hundreds of features and human burials that Webb used to define the culture of their hunter-gatherer inhabitants (Jefferies 1988a:14–25; Milner and Smith 1986). Webb's (1950a) Carlston Annis report was among the first to provide stratigraphic evidence demonstrating that midden accumulation in eastern North America was attributable to the activities of many groups of people over a long period of time (Winters 1974:vii).

The first Lower Ohio Valley Archaic site to be investigated by New Deal–era archaeologists was the Read shell midden (15Bt10), where fieldwork extended from December 1937 to March 1939 (Webb 1950b). In spring and summer 1938, Webb initiated excavations at the Andrews Run or Baker site (15Mu12), followed by work at Ward (15McL11), Bowles (15Oh13), Chiggerville (15Oh1), and Jackson Bluff (15Oh12). By the fall of 1938, Webb's crews were excavating at Jimtown Hill (15Oh19), Kirkland (15McL12), and Barrett (15McL4), followed by Butterfield (15McL7) in March 1939. Carlston Annis was excavated from mid-1940 to late 1941 (Jefferies 1988a:Table 1; Milner and Smith 1986).

In their Chiggerville report, William Webb and William Haag (1939:60–61) distinguished the prepottery occupation responsible for most of the midden from a later occupation containing shell-tempered ceramics (Figures 3.2 and 3.3). They compiled a 70-item trait list for the newly defined "shell mound" complex that included artifacts (side- and corner-notched projectile points; straight-sided drills; bell-shaped pestles; bone awls, pins, and projectile points; antler tip projectile points; shell beads and gorgets; atlatl weights, hooks, and handles; and full-grooved axes), burial characteristics (flexed and extended burials, associations of utilitarian items, etc.), and features (burned areas, prepared clay floors, tool caches) found in the shell middens.

Webb and Haag (1939:59) speculated that, in the terms of the Midwestern Taxonomic System, archaeologists would eventually define a Green River *focus* based on the occurrence of similar artifact inventories at other Green River

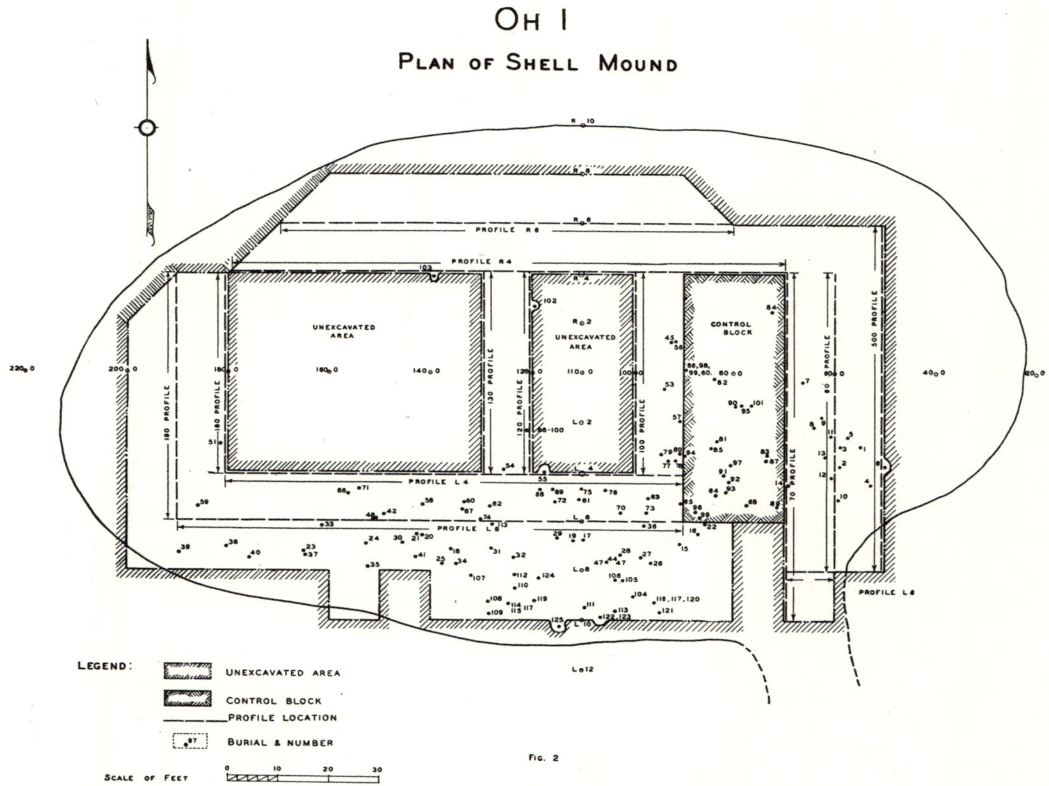

Figure 3.2. Plan view of Chiggerville (Oh1) excavations (from Webb and Haag 1939:Figure 2. William S. Webb Museum of Anthropology, University of Kentucky).

Figure 3.3. Excavations at Chiggerville (Photograph no. 980H1, William S. Webb Museum of Anthropology, University of Kentucky).

sites. The association of artifacts made from nonlocal materials, like copper and marine shell, suggested to them that Late Archaic hunter-gatherers participated in long distance exchange (Goad 1980; Winters 1968).

Perhaps Webb's best-known work, at least in the Green River region, was his excavation of the already famous Indian Knoll site (Webb 1946). Although extensively dug by Moore in 1915, Webb hoped that additional excavation would supplement Moore's findings (Webb 1946:123). Webb moved Marion Baugh and his crew from Jimtown Hill to Indian Knoll in May 1939, where they spent the following 19 months excavating the midden. To Webb's surprise, Moore's digging had only affected a small part of the site.

By the end of 1940, field crews had excavated more than 65,000 square feet of Indian Knoll (Figure 3.4). Excavation was accomplished by digging parallel trenches that were 10 feet wide, removing the midden in six-inch-thick levels. As one trench was finished, field workers filled it using the backdirt from the next trench to be excavated (Herrmann 2002:37; Webb 1946). Some of the trenches were at least eight feet deep, reflecting the immense size of the midden. These efforts exposed nearly 800 burials, along with numerous pits, hearths, and artifact caches. Administrative problems and WPA labor shortages eventu-

Figure 3.4. Excavations at Indian Knoll (15Oh2) (Photograph 3050H2, William S. Webb Museum of Anthropology, University of Kentucky).

ally resulted in closing down the Indian Knoll project before the entire site was excavated (Webb 1946:125; Jefferies 1988a:18–19).

The Indian Knoll skeletal collection consists of approximately 1,100 individuals, including the 298 that Moore removed, making it one of the largest North American hunter-gatherer skeletal collections excavated from a single site. Although occupation of Indian Knoll extended from the Archaic to the Mississippian period (ca. 10,000 to 400 B.P.), the skeletal sample dates from the late Middle to the Late Archaic (ca. 6000 to 4000 B.P.) (Herrmann 2002:37).

Dozens of researchers have analyzed the Indian Knoll skeletal collection to investigate many different research questions. Moore (1916) reported basic burial data including a partial description of associated artifacts and a summary of Hrdlicka's typological assessment of the crania. Webb's (1946) *Indian Knoll* report contained basic demographic data (age and sex), information on burial treatment (positioning), and a listing of associated artifacts. Charles Snow (1948) provided basic age-at-death information, described pathologies, and tabulated metric and discrete observations. Over the past 40 years, the collection has served as the basis for many other osteological studies (Haskins and Herrmann 1996; Powell 1996).

Prior to the development of radiocarbon dating techniques in the late 1940s, archaeologists greatly underestimated the antiquity of hunter-gatherers in the Lower Ohio Valley. In their report on the Cypress Creek villages (Ward—15McL11 and Kirkland—15McL12) of west central Kentucky, Webb and Haag (1940:109) suggested that people inhabited these sites from 1,000 to 2,000 years ago. They characterized them as hunters and fishers who lived in small family groups; hunted with the atlatl; ate deer, shellfish, roots, and berries; lacked agriculture and cooking vessels; and built simple structures. The advent of radiocarbon dating pushed back the temporal boundary of the Archaic nearly 5,000 years, radically changing the amount of time over which these hunter-gatherer societies evolved (Schwartz 1967:85).

In 1942, Haag (1942:221–222) published a paper summarizing Archaic period research in the Southeast. He suggested that the "shell heap" horizon postdated the Folsom-Yuma horizon, that groups of related hunter-gatherers lived at the midden sites, and that similar shell middens occurred throughout the Southeast.

The start of World War II abruptly ended the need for public works programs like the WPA and its extensive archaeological projects. By early 1942, virtually all of the projects had closed down and most personnel had joined the war effort. In less than a decade, however, New Deal archaeologists had excavated more than 20 major sites containing Archaic components. Their efforts produced thousands of feature and burial forms, maps, profile drawings, and photographs, not to mention the huge inventory of artifacts and human skeletal materials that Webb used to define the Shell Mound Archaic or Indian Knoll culture along the Green River (Jefferies 1988a:25; Milner and Smith 1986).

It is hard to overstate Webb's impact on the development of the Archaic concept and our understanding of eastern North America's hunter-gatherers (Schwartz 1967; Winters 1974). In the years immediately following World War II, he published a number of reports on his prewar Green River excavations including those at Indian Knoll (1946), Carlston Annis (1950a), Read (1950b), and several sites in McLean County (Webb and Haag 1947). Before Webb and his colleagues published their reports, we knew very little about the Lower Ohio Valley's hunter-gatherers. These promptly published volumes provided critical data that helped to fill this void. Before long, the efforts of Webb and his colleagues led to the recognition of several regionally distinct Archaic "cultures" across eastern North America (Griffin 1946; Lewis and Kneberg

1947; Sears 1948), reflecting the cultural/temporal variability of these hunting and gathering societies.

Webb's excellent organizational and administrative skills allowed him to keep several large archaeological projects running at the same time. In addition, he developed new techniques for excavating mounds, cemeteries, middens, and villages, and for cataloging and analyzing the resulting artifacts. His efforts in these areas contributed to the consistent reporting of archaeological findings over an area that stretched from the Tennessee Valley in northern Alabama to the tributaries of the Ohio River (Winters 1974:vii). As succinctly stated by Winters (1974:v), Webb was not particularly concerned about archaeological theory; however, his concern for producing solid, if sometimes brief, excavation reports provided future generations of archaeologists with good data for addressing new research issues. For the time in which Webb and his colleagues wrote them, their reports represented the state of the art.

Webb's reports emphasized time-space systematics and the construction of trait lists for comparative studies (Winters 1974:vi). The culmination of his comparative efforts appeared in the Indian Knoll report (1946), in which he used a lengthy trait list to characterize the Green River Archaic assemblage. Diagnostic artifacts included corner- and side-notched projectile points; bell-shaped pestles; bone awls, pins, and projectile points; shell beads and gorgets; atlatl weights; and full-grooved axes. Unfortunately, most scholars felt that Webb's trait lists and his criteria for inclusion in the Archaic pattern were too general to make meaningful comparisons to similar materials from other parts of eastern North America (Sears 1948:123).

Over time, Webb gradually shifted his research focus from presence-absence trait lists and comparative studies to data synthesis and interpretation of hunter-gatherer lifeways (Winters 1974:vi). He was an early advocate of functional analysis, conducting detailed studies of the manufacture and use of the atlatl (Webb 1957) and bone fishhook production (Webb 1950a; Winters 1974:viii).

In retrospect, even though William Ritchie used the term Archaic in 1932 in reference to a prepottery site in New York, it was not until Webb published the results of his Green River research that much was really known about Archaic hunter-gatherers. Perhaps his greatest contribution to hunter-gatherer/Archaic period research was the standardized and systematic way in which he excavated the sites, analyzed the artifacts, and reported his findings. Webb's attention to these details has ensured the continued study of the Green River data using new theoretical approaches and analytical techniques (Winters 1974).

Admittedly, archaeologists focused a lot of their time and resources on the Green River sites during the first one-half of the twentieth century; however, researchers were also investigating hunter-gatherers in other parts of the Lower Ohio Valley. During the 1930s, E. Y. Guernsey, under the direction of Glenn A. Black, conducted an extensive survey of the Falls area (Janzen 1972:317) (Figure 3.5). Although his final report, *An Archaeological Survey of Clark, Floyd, and Harrison Counties,* was never published, Guernsey did write several papers relating to Falls area prehistory (Guernsey 1937, 1939, 1942, cited in Janzen 1972:317). Guernsey constructed a three-stage cultural sequence for the Falls area based on his excavations at the Prather, Koons, Willey, Aydelotte, Clark's Point, and Elrod sites, The oldest cultural stage, represented by shell deposits in the lower levels of the Clark's Point and Elrod sites, shared many attributes with Webb's Indian Knoll focus (Janzen 1972:317). Reports from "amateur" archaeologists that appeared in state archaeological journals supplemented Guernsey's work (Burnett 1963; Matthews 1958, cited in Janzen 1972:319).

Rex Miller's investigations at the McCain site in Dubois County provided important data on Middle to Late Holocene shell middens in southwestern Indiana (Miller 1941). Slightly more than a decade later, James Kellar (1956) reported on his research at the Crib Mound in nearby Spencer County. Despite the contributions of these early- to mid-twentieth-century researchers, archaeologists continued to extrapolate notions about hunter-gatherers in this part of the Lower Ohio River valley from Webb's work in the Green River region (Janzen 1977:126–127).

By the mid-twentieth century, a growing number of archaeologists questioned the research strategies used by Webb and his colleagues during the preceding decades. In particular, they were critical of his heavy reliance on presence/absence trait lists as a way of comparing the similarities and differences among sites. In 1948, Walter Taylor acknowledged that Webb had produced some of the best archaeological reports to date, but criticized him for his limited interpretations and for not consistently providing detailed artifact descriptions, quantitative data, or good contextual information. Taylor placed much of the blame on Webb's heavy reliance on the Midwest Taxonomic System and its focus on comparative studies (Taylor 1948:77).

Unlike in west central Kentucky where the many Green River shell middens attracted the attention of early excavators, hunter-gatherer research in southern Illinois started much later. The University of Chicago conducted the first systematic archaeological investigation of a southern Illinois Archaic component at the Faulkner site (Mx^{v}43), located near the Ohio River in an area known as

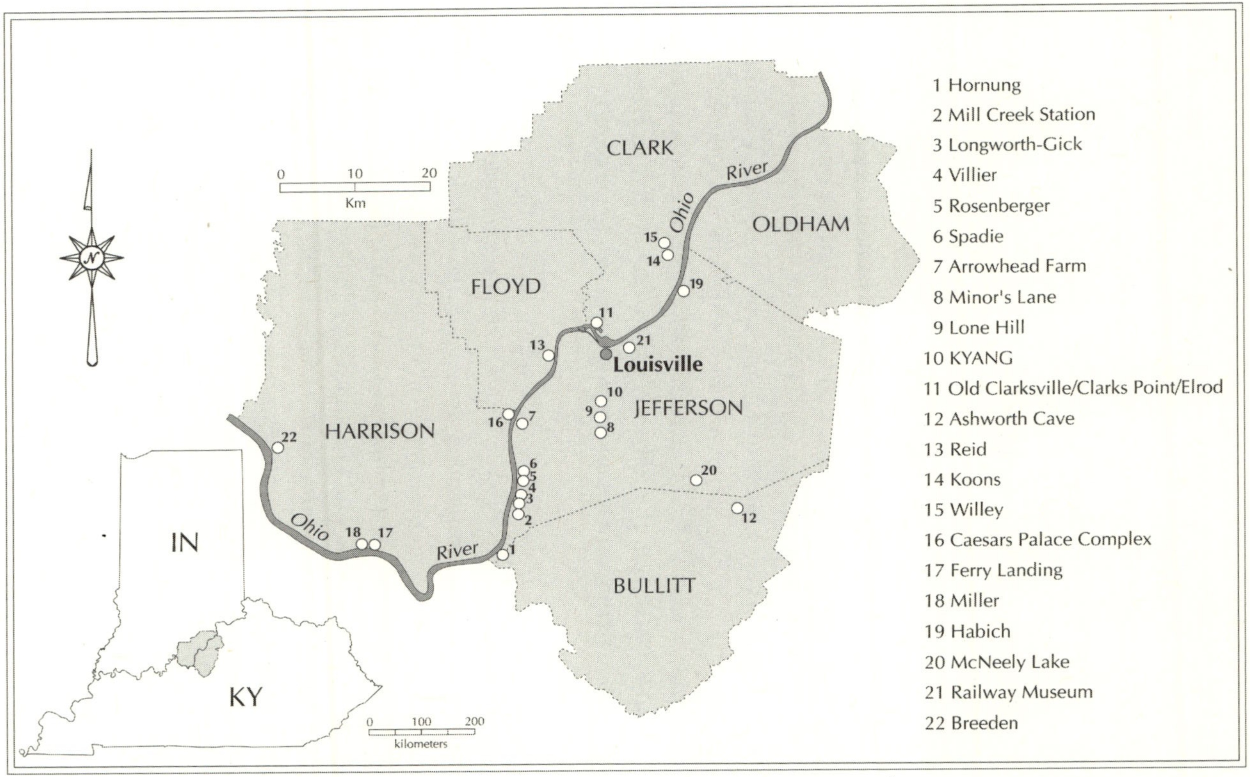

Figure 3.5. Important Falls area sites.

the Black Bottom (Cole et al. 1951; MacNeish 1948). The Faulkner work was part of a larger University of Chicago project conducted from the late 1930s to the mid-1940s that focused on the now famous Kincaid Mississippian mound center (Cole et al. 1951; Muller 1986:59–60). Although cultural deposits were poorly stratified and mixed with those of later occupations, MacNeish's efforts represented a "first step" in learning more about the hunter-gatherer societies that lived near the Ohio River floodplain (Butler 2009).

MacNeish's excavations disclosed artifacts, features, and burials that he assigned to the Faulkner focus. The absence of ceramics led Richard MacNeish (1948:239–240) to conclude that the Faulkner "people" predated the region's pottery-making cultures. At the time, archaeologists had not identified other prepottery cultural assemblages in Illinois.

In keeping with practices of the time, MacNeish compiled a trait list characterizing the Faulkner culture, and then used it to examine the Faulkner peoples' relationships to other prepottery groups (1948:240). He suggested that "the Faulkner site represents the remains of a people who were an early hunting and food-gathering group, living along the banks of streams in southern Illinois" (1948:243). MacNeish (1948:243) concluded that the Faulkner material most closely resembled that from the Indian Knoll focus, but that they also were similar to other eastern North American prepottery groups assigned to the Archaic pattern.

In 1952, James B. Griffin (1952b:228) lamented that there was very little evidence for hunter-gatherers in the Central Mississippi Valley. Part of the reason for this dearth of information was because archaeologists, as usual, had concentrated their efforts on the region's larger, more spectacular mound sites. Those sites were easier to find and had many spectacular burials and elaborate artifacts that attracted the attention of museums and the public. Griffin also recognized that the shifting course of the Mississippi River and its tributaries probably destroyed many of the valley's older sites. However, he was confident that the discovery of such sites "awaits intelligently directed effort toward finding early sites which will take into account the physiographic factors which have helped to obscure the earliest period" (1952b:228).

Moreau Maxwell's (1952) overview of Lower Ohio Valley archaeology published in 1952 provided a mid-twentieth-century perspective on the region's hunter-gatherer research. He maintained that southern Illinois seemed to have been "peripheral in character" throughout the majority of prehistory, becoming a "minor cultural center" only in the late prehistoric period (1952:176). Early cultural influences spread into southern Illinois from adjacent parts of the

Southeast and Midwest along the major rivers, moving west along the Ohio River, west and north along the Green and Cumberland rivers, north along the Tennessee River, and south along the Wabash, Illinois, and Mississippi rivers.

Maxwell believed that Paleoindian hunter-gatherers similar to those discovered in western North America (Clovis and Folsom) had entered the Lower Ohio Valley relatively early in the region's culture history. The numerous fluted projectile points found in surface contexts across southern Illinois provided the evidence for that presence (1952:176). Maxwell noted that the fine pressure flaking observed on southern Illinois fluted points was very similar to that seen on Paleoindian points from the Plains. Like the western Folsom culture, these early southern Illinois groups were primarily hunters that lived in small nomadic bands, seldom camping in one place long enough to leave a refuse midden (1952:177).

Based on a detailed comparison of the cultural assemblages from preceramic shell middens (Indian Knoll, Lauderdale, and Eva foci components), Maxwell (1952:178) concluded that there were no centers of the shell mound cultures (Archaic) in Illinois, but that some cultural traits associated with these foci had diffused to the north. Evidence of this diffusion into southern Illinois consisted of artifacts, like biwinged atlatl weights, full grooved axes, and long narrow side-notched projectile points, which resembled those of the preceramic Indian Knoll focus to the south. At the time, the best example of this trait diffusion came from the Faulkner site where MacNeish had defined the Faulkner focus and pointed out its similarity to the Indian Knoll focus (MacNeish 1948).

Maxwell (1952:179) concluded that "shell mound traits" diffused northward into southern Illinois from centers along the Lower Ohio River to marginal sites like Faulkner and other campsites in the Carbondale area. However, he could not decide whether the presence of these traits indicated the movement of people from the shell mound core area or just the diffusion of those traits to indigenous people.

In 1954, Melvin Fowler took leave from his important research program at Modoc Rock Shelter to investigate an unusual site in southeastern Illinois. Known as the Ferry site (Fowler 1957; Muller 1986:68–70), it consisted of an eroded hilltop lithic scatter along the Saline River near its confluence with the Ohio (Butler 2009).

Excavation of the largely plow-disturbed site yielded numerous projectile points, hammerstones, pitted cobbles, and atlatl weights. Fowler classified most of the points (68 percent) as Late Archaic straight and expanding stemmed types, closely conforming to Winters's Saratoga point cluster (Muller 1986:69;

Winters 1967:Figure 4). The predominance of stemmed points and other evidence led Fowler to conclude that the Ferry site was "occupied late in the Archaic occupation of southern Illinois" and that it was closely related to the Shell Mound Archaic sites of Kentucky, such as Carlston Annis, Indian Knoll, Read, and Parrish Village (1957:30). However, Butler (2009) has recently suggested that many of the projectile points (large, corner-notched-expanding stemmed points and a possible White Springs specimen) date to the early Middle Archaic, seriously challenging Fowler's temporal placement.

By the mid-1950s, the increased availability of radiocarbon dating brought about the need for a total revamping of regional chronologies and cultural sequences. Just fifteen years earlier, Webb and Haag suggested that the Green River Archaic occupations were no more than 2,000 years old (1940:109). Now, archaeologists had an additional three or four thousand years to play with as they tried to untangle the temporal, spatial, and cultural variability of the region's hunter-gatherer societies.

In addition to the impact of radiocarbon dating, excavations at several deeply stratified sites in regions surrounding the Lower Ohio Valley revealed sealed Archaic components that had to be interpreted in light of the new temporal framework (Broyles 1971; Coe 1964; Fowler 1959; Lewis and Lewis 1961). Because of these and other factors, archaeologists working in and near the Lower Ohio River valley spent much of the late 1950s and early 1960s revising their culture histories and writing new regional syntheses. Regional settlement and subsistence models were developed and evaluated, with the Green River data playing a critical role in the process (Herrmann 2002:24).

In 1958, Joseph Caldwell (1958:vii–viii) introduced the concept of "primary forest efficiency" to the archaeological literature of eastern North America. According to Caldwell, this adaptive strategy, based on greater hunting efficiency and a growing knowledge of locally available food sources, was achieved in eastern North America during the Archaic stage (1958:18). He (1958:14) considered the hunter-gatherers who lived at shell middens, like Indian Knoll, to be excellent examples of primary forest efficiency adaptive strategy. They had achieved an increasingly sedentary and materially richer life, establishing some level of security, at least in the search for food (1958:18).

Fowler's (1959) study of hunter-gatherers who lived at Modoc Rock Shelter in southwestern Illinois also incorporated information from the Green River shell middens. Fowler proposed three Archaic adaptive stages (Initial Archaic—10,000 to 8000 B.P., Localized Adaptation—8000 to 6000 B.P., and Specialized Adaptation—6000 to 4000 B.P.) based on changes in food exploitation strategies. Adaptive stages generally increased in complexity and took kit specialization through time. Fowler compared the Modoc artifact assemblages with those from other regional Archaic sites, including several of the Green River middens (Parrish, Read, Carlston Annis, Ward, Butterfield, Kirkland, and Barrett). Fowler concluded that assemblage variability reflected seasonally specific occupations at these sites and that Archaic hunter-gatherers had developed a seasonal cycle of food resource exploitation by 5500–4000 B.P. (1959:57).

Lewis and Kneberg (1959) also considered Lower Ohio Valley hunter-gatherers when writing their regional synthesis, *The Archaic Cultures of the Middle South.* Through statistical analysis of 83 traits from 22 Archaic sites in Kentucky, Tennessee, Alabama, and Georgia, they proposed the existence of two contemporaneous Archaic traditions. Late Archaic groups at Read, Carlston Annis, Indian Knoll, and early occupations at Chiggerville, coeval with the Eva and Three Mile phases in Tennessee, represented a Midcontinent tradition. In contrast, the late Chiggerville component, a contemporary of the Kays phase, possibly indicated contact with people of an Eastern tradition (Lewis and Kneberg 1959:178).

In 1966, Martha Rolingson and Douglas Schwartz published the results of their study of Paleoindian and Archaic artifacts from four western Kentucky sites excavated between 1939 and 1958—Henderson (15Ly27), Roach (15Tr10), Morris (15Hk49), and Parrish (15Hk45). Morris and Parrish were on a tributary of the Tradewater River in the Western Coalfield; Henderson and Roach were in the Tennessee-Cumberland River valleys (1966:145). The goal of their research was to synthesize Paleoindian and Archaic data from these sites, examine the temporal and cultural relationships between the two cultural manifestations, and explore their relationships with contemporary cultures (1966:2). They also investigated the possibility of contemporaneous Paleoindian and Archaic cultures in the Midcontinent.

Using artifacts from Henderson and the lower levels of Morris and Roach, Rolingson and Schwartz defined the hunting-based Henderson phase. The assemblages included both bifacial and unifacial Dalton tools, which they dated from 9900 to 7500 B.P. (1966:162) and resembled assemblages from the lower levels of Modoc Rock Shelter that Fowler associated with his Initial Archaic adaptive stage (1966:154). They proposed that by 5500 B.P., Henderson phase groups had developed into seasonally specialized hunter-gatherers identified archaeologically as the Late Archaic Green River and Ledbetter phases in Kentucky and Tennessee, respectively (1966:162).

Rolingson's (1967) analysis of the Green River shell midden data represented its first systematic examination in nearly three decades. The primary goal of her research was to determine if shell middens accumulated gradually or if they formed quickly because of short-term, intensive occupations. She also attempted to define discrete cultural units and to establish a regional chronological sequence (1967:393).

After examining the distribution of projectile point types and clusters, Rolingson concluded that the middens contained artifacts spanning much of the prehistoric era, but that most of the deposits and artifacts dated prior to the introduction of pottery. She also proposed that the middens formed due to repeated occupations by hunter-gatherer groups (Rolingson 1967:418–419).

As a consequence of her exhaustive efforts, Rolingson (1967:409–410) defined the Indian Knoll phase, proposing that it dated from 4500 to 3500 B.P., and that its distribution was limited to the middle Green River region, including the Read, Bowles, Chiggerville, Indian Knoll, Ward, Kirkland, Carlston Annis, and Barrett sites. Diagnostic Indian Knoll phase traits included stemmed projectile points, a diversified economy, seasonal site occupations, and elaborate grave goods associated with a small percentage of burials.

During the middle and late 1960s, Howard D. Winters, perhaps best known for his research along the lower Wabash River (1967, 1969), was an influential figure in the study of Lower Ohio Valley hunter-gatherers. Winters, affiliated with the Southern Illinois University Museum from 1955 to 1960, conducted extensive surveys in the Cache River valley, recording many Archaic sites in the process (Butler 2009).

Perhaps Winters's greatest contributions to Lower Ohio Valley archaeology was establishing a regional projectile point typology based on the many points found during his Cache River survey. After nearly 40 years of additional research, most of Winters's point types (Thebes, Cache, Saratoga, Karnak, Cypress) have stood the test of time. Unfortunately, while Winters described the point types and collected abundant metric data, he never published detailed descriptions of them, resulting in considerable latitude in their definitions by later researchers (Butler 2009).

Winters's ongoing interest in hunter-gatherer settlement strategies resulted in his investigation of Archaic settlement in the lower Wabash (1967, 1969) and Green River (1974) valleys. For these studies, Winters assigned artifacts to different tool groupings (general utility, weapons, fabricating and processing, etc.) based on their "functional" characteristics. In turn, he assigned sites to different types (settlements, base camps, transient camps, etc.) by the proportion of different tool groups represented. For example, in his study of the lower Wabash valley's Riverton culture, Winters classified the Robeson Hills site as a "settlement" based on the presence of a deep midden, clay floors, postholes, storage pits, many burials, few weapons, an emphasis on hide working, and many fabricating and processing tools. Environmental indicators suggested to him that Robeson Hills was occupied during the winter. In turn, Winters considered the Riverton site to be a "base camp" because it lacked houses, had few storage pits or burials, and had more weapons than Robeson Hills. He proposed a May to September occupation for this site. "Transient camps," exemplified by the Swan Island site, were occupied in the spring and fall (1969:108–137). These sites were incorporated into a seasonal cycle of settlement during which hunter-gatherers occupied each site for a specific part of the year.

Unfortunately, Winters's assessment of the Riverton settlement system was severely impaired by his overemphasis of tool function based on an artifact's morphological attributes. Analytical techniques now used to determine tool use(s), like microscopic use-wear analysis, were still in their developmental stages (Semenov 1964). His research also suffered from problems of small artifact sample size and issues of site contemporaneity. Nevertheless, Winters's efforts helped bring the concepts of seasonality and settlement patterns/systems to the forefront of archaeological research.

Winters (1974) also addressed the issue of Archaic settlement organization in his Introduction to the New Edition of Webb's Indian Knoll report. Using a similar approach to that used in his Riverton project, he characterized the Indian Knoll Culture as a harvesting economy. Winters classified Read, Chiggerville, and Ward as "settlements"; Kirkland, Barrett, and Carlston Annis as "base camps"; Butterfield as a "transient camp"; and Kirkland as a "hunting camp" (1974:xii–xvi).

Winters's (1968) analysis of the distribution of different artifact types among burials at several of the Green River shell middens was one of the first large-scale, systematic attempts to look at hunter-gatherer mortuary programs, social status, and long distance exchange. He (1968:209) noted that artifacts made from nonlocal materials, like marine shell and copper, were associated with only very few of the burials. Winters interpreted this limited distribution of exotic items as reflecting the differential distribution of wealth among living members of these societies. The association of exotic items with infants and children, as well as adults, suggested to Winters (1968:203) that these objects were indicative of ascribed rather than achieved status. In addition, he interpreted the restricted distribution of shell objects and shell-working kits found in some burials as evidence for the beginning of craft specialization in Indian Knoll culture (1968:209).

Starting in 1971, William Marquardt and Patty Jo Watson initiated the Shell Mound Archaeological Project (SMAP) at the Carlston Annis site (15Bt5) in the Green River's Big Bend. The initial focus of the SMAP research program was to compare Carlston Annis's subsistence strategies with those of the prehistoric mineral miners of Mammoth (Watson 1974) and Salts (Watson 1969) caves (Marquardt and Watson 2005a:630). Although radiocarbon dates associated with the cave materials, including native domesticated plants, were more recent (Early Woodland) than most dates from the Green River shell middens, hunter-gatherers did enter the caves during the Late Archaic (Muller 1986:80). Few open-air habitation sites of the proper age (Late Archaic) were known for the immediate vicinity of Mammoth Cave, so Marquardt and Watson looked downriver some 50 to 70 km to the Big Bend area's Archaic middens for evidence of the development of horticulture (Herrmann 2002:33–34; Muller 1986:79–80).

To do their work, SMAP personnel developed new equipment, strategies, and techniques to investigate the shell middens and their contents. Much of SMAP's early work focused on the shell midden's complex stratigraphy and the need for more intensive study of the process of midden formation (Marquardt and Watson 2005b:xix).

SMAP researchers had two primary goals for their Carlston Annis investigations—to collect charcoal for radiocarbon dating and to collect charred plant material to study prehistoric plant use (Marquardt and Watson 1983:324). Of particular interest was the question of whether or not Green River Late Archaic hunter-gatherers were experimenting with native plant cultivation. Watson's (1974; Watson 1969) and Yarnell's (1969, 1974) cave investigations had identified squash (*Cucurbita pepo*) and gourd (*Lagenaria siceraria*) associated with cultigens like sunflower (*Helianthus annuus*), sumpweed (*Iva annua*), goosefoot (*Chenopodium* sp.), and maygrass (*Phalaris caroliniana*). Unfortunately, finding the suite of cultigens represented in the caves proved to be difficult at the shell middens. More than 90 percent of the botanical remains from the initial work at Carlston Annis were hickory nut and acorn shell; however, analysts did identify several small cucurbit rind fragments. At the time, these specimens were among the oldest cucurbits from eastern North America (Crawford 2005:Figure 10.2).

Marquardt and Watson have continued to expand the scope of the Shell Mound Archaeological Project since its 1971 inauguration. SMAP's ongoing investigations at Carlston Annis, along with new work at the Bowles (15Oh13) site, revealed the very complex depositional histories of these large middens. To learn more about site formation processes, Marquardt and Watson developed a multifaceted research design to address the issue and assembled a multidisciplinary research team to operationalize it. Julie Stein's (1980, 2005) geomorphological research at Carlston Annis focused on site formation processes, providing an interpretative base for analyzing artifacts, radiocarbon dates, and botanical samples. Other SMAP researchers, some of whom were Watson's graduate students, investigated site stratigraphy (Gorski 2005; May 2005; Wagner 2005), lithic resources and artifacts (Gatus 2005; Hensley 1996; Marquardt 2005), faunal materials (Crothers 2005; White 1990, 2005), botanical remains (Crawford 1982, 2005), human skeletal biology (Belovich 2005; Haskins and Herrmann 1996; Herrmann 1990, 2002; Mensforth 2005; Ward 2005), and settlement/subsistence organization (Crothers 1999, 2005; Hensley 1994; Wagner 1996, 2005) (Carstens and Watson 1996; Herrmann 2002: 34–35).

The mid-twentieth century brought with it a continuation of hunter-gatherer research in the Falls of the Ohio region. In the early 1950s, construction of the Ford assembly plant in Louisville destroyed the Lone Hill site (15Jf10) and its extensive Archaic deposits. Unlike most large Falls area Archaic sites, Lone Hill was situated away from the Ohio River. The site's destruction, along with a very similar one in south Louisville, meant the loss of important information on hunter-gatherer adaptive strategies away from the immediate river environs (Janzen 1972:319–320). During the 1960s, James Bellis and John Dorwin excavated part of the Breeden site (12Hr11), a Late Archaic shell midden in Harrison County, Indiana (Bellis 1968, 1982).

Donald E. Janzen (1977) initiated the Falls of the Ohio River Archaeological Project in the late 1960s to study Holocene hunter-gatherer adaptation within a ca. 50 km radius of the Falls. This region exhibited considerable environmental diversity with five distinct physiographic provinces merging within the project area (1977:123).

Prior to Janzen's work, archaeological research near the Falls area had been, at best, sporadic and very few sites had been professionally documented. In fact, Funkhouser and Webb (1932) listed just 16 sites for the five counties surrounding the Falls. By the mid-twentieth century, this lack of information led some archaeologists to say that there was not much in the way of archaeology around the Falls (Janzen 1972:316). Janzen's research clearly demonstrated that this was not the case.

Archaic sites investigated by Janzen's Falls area project included Old Clarksville (12C11), Reid (12F11), Ferry Landing (12Hr3), and Miller (12Hr5) in Indiana, and Hornung (15Jf60) and Lone Hill in Kentucky. These sites consisted of extensive, deep middens containing abundant features, burials, and artifacts that indicated intensive site use during the late Middle and Late Archaic (Janzen 1977).

One of Janzen's (1977) research goals was to obtain radiocarbon dates from a sample of Falls area Archaic sites. By the end of 1974, he had 19 dates from nine sites, most falling between 6000 and 5000 B.P. Based on excavations at seven sites, Janzen proposed that this period of hunter-gatherer activity was characterized by intensive occupation of certain parts of the landscape. Thick, homogeneous, and stratigraphically uninterrupted middens suggesting considerable cultural continuity marked these places. Janzen maintained that the high diversity of food resources concentrated within the Falls area minimized the need for seasonal movement and that groups living near the Falls at that time practiced a "hub-and-spoke" or semisedentary wandering settlement model. He attributed a sharp drop in activity after 5000 B.P. to a decrease in regional population density (1977:138–141).

In 1969, the University of Louisville established the Archaeological Survey, under the direction of Joseph Granger, to study the archaeology of the Falls area. Since that time, Granger and his associates have investigated a number of important Archaic sites, including Mill Creek Station (15Jf206), Arrowhead Farm (15Jf237), Minor's Lane (15Jf36), Habich (15Jf550), Lone Hill (15Jf10), KYANG (15Jf267), and McNeeley Lake (15Jf200) (Bader and Granger 1989; Granger 1985, 1988; Granger et al. 1992; Mocas 1976). Excavations at these sites were specifically designed to collect data to develop and test hypotheses concerning the ecology, dietary patterns, lithic technology, settlement patterns, chronology, and climate (Granger 1988:157).

Granger's (1988) exhaustive research on Falls area Archaic settlement-subsistence systems eventually led to the definition of two phases dating to the late Middle Archaic (Old Clarksville) and Late Archaic (Lone Hill) subdivisions. Distinctions between the two were based on differences in flaked stone tool morphology and mortuary practices (1988:179).

The Old Clarksville phase (6000–4400 B.P.), originally defined by Granger as a Late Archaic cultural manifestation, is now considered to be late Middle Archaic (Granger et al. 1992). Major Old Clarksville components are known for many Falls area sites including Old Clarksville in southern Indiana, and Hornung, KYANG, McNeeley Lake, Mill Creek Station, and Arrowhead Farm in Kentucky. Deep midden deposits having high densities of features, burials, and artifacts characterize many of the Old Clarksville components (Granger 1988:154). Traits diagnostic of the Old Clarksville phase include side-notched projectile points resembling the Godar, Big Sandy II, or Matanzas types (Justice 1987) and small, bowl-shaped pits containing flexed burials and associated engraved bone pins; deer, bear, and wolf tooth necklaces; atlatl hooks; and red ocher (Anslinger 2001; Granger 1988:175).

Archaeologists have documented Lone Hill phase (4400–3200 B.P.) components at many Falls area sites including Lone Hill, Arrowhead Farm, McNeeley Lake, Durrett, KYANG, Minors Lane, Spadie, Villier, and Rosenberger, all located in Jefferson County, Kentucky. Stemmed points, often made from high-quality Wyandotte chert, are diagnostic of the Lone Hill phase. The technology used to manufacture Lone Hill points is distinctive from that used to produce the Old Clarksville side-notched points. Lone Hill burials often contain items made from exotic material (Anslinger 2001; Granger 1988:153–156).

Based on years of survey and excavation, Granger (1988:199) defined six clusters of Old Clarksville and Lone Hill phase sites in the Kentucky portion of the Falls area. Future investigation of the temporal, functional, and ecological variation of these sites will further clarify the differences in the settlement-subsistence systems of these two hunter-gatherer cultural manifestations.

In the late 1970s, archaeological investigations conducted prior to the construction of the Southwest Jefferson County Flood Protection Project yielded important data on Early to Late Holocene hunter-gatherers who lived immediately adjacent to the Ohio River (Collins 1979a). Excavations at the Longworth-Gick site (15Jf243) proved to be particularly informative because of the numerous stratigraphically discrete Early Holocene cultural deposits encountered. The association of radiocarbon dates with distinct projectile point styles revealed a similar sequence to those seen at St. Albans (Broyles 1971) in West Virginia and Icehouse Bottom (Chapman 1973, 1976) in Tennessee, but dates for the Falls area specimens differed to some extent (Collins and Driskell 1979:1024–1038). Archaeological investigations at the nearby Rosenberger (15Jf18), Spadie (15Jf14), and Villier (15Jf110) sites revealed significant Middle to Late Holocene cultural deposits, complementing the information on earlier Falls area hunter-gatherers derived from the Longworth-Gick work (Boisvert 1979; Driskell 1979; Robinson and Smith 1979).

In the late 1970s, Jack Nance started the Lower Cumberland Archaeological Project (LCAP) to investigate the cultural and natural histories of the Purchase area of far western Kentucky. The study area included the lower portions of the Cumberland and Tennessee rivers, as well as the extreme lower Ohio River. The long-term, multidisciplinary research project had a regional focus and was comprehensive in scope, but researchers placed particular emphasis on the study of Holocene hunter-gatherers (Nance 1987:95).

Archaeological resources in this part of the Lower Ohio Valley had received considerable, if somewhat unsystematic, attention since the early twentieth century (Moore 1916; Webb and Funkhouser 1929, 1931, 1933). Additional investigations were carried out from the late 1950s through the early 1970s in connection with the construction of Barkley Dam on the Cumberland River (Clay and Schwartz 1963; Coe and Fischer 1959; Morse 1962) and Interstate Highway 24 to the north of the dam (Allen 1976; Mocas 1977). By the late 1970s, archaeologists had compiled a tentative regional culture history; however, because of the spotty nature of the research and an emphasis on later ceramic horizons, substantial gaps still existed (Nance 1987:95).

The LCAP's long-range objectives included developing a detailed regional culture history, with emphasis on the Archaic period; documenting and explaining prehistoric human-environmental relationships; and developing methods for studying the many small, limited activity sites found in the region. Nance and his colleagues addressed these issues through informal site surveys employing areal sampling, excavating selected sites, documenting regional lithic resources, investigating geomorphic/environmental history, and conducting bioarchaeological research (Nance 1987).

Prior to the formal start of the LCAP, Nance's (1975, 1977) archaeological survey of the Land Between the Lakes National Recreation Area, situated between the lower Tennessee and Cumberland rivers, located numerous upland Archaic sites. A primary goal of the Land Between the Lakes survey was to document upland sites largely ignored by earlier floodplain-focused projects (Clay and Schwartz 1963; Coe and Fischer 1959; Schwartz and Sloan 1958). Based on his research in the Crooked Creek drainage, Nance suggested that upland and floodplain Late Archaic sites hosted complementary activities. Tasks at upland sites were associated with hunting, while floodplain activities included hunting, plus the butchering of game, plant food processing, and general maintenance activities. Some floodplain sites served as base camps from which Late Archaic groups exploited both floodplain and upland resources (1977:13).

Although the Land Between the Lakes survey provided a general overview of Archaic hunter-gatherer settlement in the region, most of the data came from surface contexts. Other than projectile points, little chronological information was available for more fine-grained examination of hunter-gatherer adaptive strategies. The work of the LCAP quickly changed this situation.

Perhaps the most significant insights on hunter-gatherer activity in the LCAP study area came from Nance's excavations at the Morrisroe site (15Lv156). The site, situated on the edge of the Tennessee River approximately 18 km upstream from its confluence with the Ohio River, contained deeply stratified deposits dating from the Early to the Late Holocene (Nance 1986a). The deepest cultural levels (Strata 2–4), containing Early through Middle Holocene materials, were about 3 m thick and overlain by 2 m of sterile sediments. A series of nine radiocarbon dates demonstrated that the earliest occupation dated to about 8300 B.P. and continued until at least 5500 B.P.

Although Morrisroe contained evidence of more or less continuous use of the Tennessee River floodplain, the nature of hunter-gatherer activities sharply differed through time (Nance 1988:129). Excavation of the deepest cultural zone, Stratum 4, revealed a thin late Early Archaic midden containing Kirk Stemmed and Kirk Serrated projectile points. Nance suggested that these materials were indicative of ephemeral, nonintensive hunter-gatherer activity at Morrisroe (1988:129). Stratum 3, an overlying 1-m-thick Middle Holocene (ca. 7800 and 6500 B.P.) midden, yielded Cypress Creek and Eva projectile points, abundant charred plant and animal remains, and flaked and ground stone tools and debitage. The density of cultural materials in this stratum indicated the intensive and repeated use of the site at that time (Nance 1986a:23). Finally, Stratum 2, dating to ca. 5500 B.P., contained a much lighter concentration of cultural remains, suggesting a return to less intensive occupation (Nance 1987:135).

LCAP investigations at the nearby Whalen site (15Ly48), located on the lower Cumberland River below Barkley Lake, disclosed cultural deposits extending more than 3 m below surface. A lower zone contained Early Holocene material, including Kirk and Cypress Creek projectile points, dating from 8500 to 7600 B.P. Impressions of mussel shells found in these deposits represent the only evidence for Early Holocene shellfishing in the lower Tennessee-Cumberland region. The Whalen site's upper midden zone (130–250 cm below surface) contained abundant Middle to Late Holocene cultural material, including side-notched and straight-stemmed projectile points associated with radiocarbon dates that clustered around 7100 B.P. (Nance 1986b, 1988, 2001).

Starting in the mid- to late 1970s, cultural resource management projects contributed greatly to our knowledge of Lower Ohio Valley hunter-gatherers. The previously discussed Southwest Jefferson County Flood Protection Project (Collins 1979a) was one of the first of these important projects. Other examples are briefly discussed below, but more specific data resulting from these projects are incorporated in the subsequent chapters.

In the late 1970s, archaeologists from Southern Illinois University Carbondale (SIUC) initiated large-scale excavations at Carrier Mills in southeastern Illinois (Jefferies 1987; Jefferies and Butler 1982). More than nine months of intensive fieldwork focused on a cluster of three large and complex sites (11Sa86, 11Sa87, and 11Sa88) located along the South Fork of the Saline River. Excavation units placed near the center of the Black Earth site's (11Sa87) Area A revealed more than 1 m of organically rich midden, most of which was attributable to the site's intensive late Middle Archaic occupation. A contemporary, but less well-preserved midden at nearby 11Sa88 provided important supplementary data on late Middle Archaic site organization and mortuary practices (Butler 2009).

Roughly 15 years later, SIUC archaeologists initiated the West Harrisburg Project (Hargrave and Butler 1994b) situated only 10 km northeast of the Carrier Mills sites. The project focused on three sites—Wasson No. 1 (11Sa50), Funkhouser (11Sa217), and Kottmeyer (11Sa234)—on Bankston Fork of the Saline River. These sites had the same occupation sequence as those at Carrier Mills. Although the West Harrisburg sites lacked the depth and preservation of the Black Earth site, they yielded data that both clarified and contrasted with the Carrier Mills findings (Butler 2009).

Excavation of the Little Muddy Rock Shelter (11J814), located along the southern margins of the till plains region, revealed a deeply stratified sequence extending back as far as Dalton times (Moffat et al. 1992). Particularly important were the shelter's well-stratified Middle and Late Archaic deposits, which provided an important diachronic perspective not found in most sites (Butler 2009). Other recent southern Illinois projects contributing significant data on hunter-gatherer activities include work done at the Olive Branch site in Alexander County (Gramly 2002), the Hills Branch Rock Shelter in Pope County (Wagner and Butler 2000), and the Fitzgibbons site in Gallatin County (Robison 1986).

Archaeological research conducted in parts of Illinois adjacent to the study area has also contributed greatly to our understanding of Lower Ohio Valley hunter-gatherers. In particular, multifaceted research programs at Lower Illinois River valley sites like Koster (Brown and Vierra 1983; Cook 1976), Napoleon Hollow (Wiant et al. 1983), and Elizabeth Mounds (Charles et al. 1988) and sites in the American Bottom such as Modoc Rock Shelter (Ahler 1993; Fowler 1959; Styles et al. 1983), George Reeves (McElrath and Finney

1987), and Missouri-Pacific #2 (McElrath and Fortier 1983) have had a major impact.

Knowledge of hunter-gatherer activity in southern Indiana also has greatly benefited from numerous recent research projects. Prior to the 1970s, most of what archaeologists knew was based on excavations at a few shell midden sites, like McCain, in southwestern Indiana. In the late 1970s, Indiana University archaeologists excavated four Late Archaic sites prior to the construction of Patoka Lake (Munson 1980a). Based on their findings, Cheryl Munson and Thomas Cook (1980c) defined the French Lick phase and developed a settlement model (Munson and Cook 1980b) for the region's Middle to Late Holocene hunter-gatherers (Stafford and Cantin 2009).

In the 1980s, Indiana State University archaeologist Robert E. Pace recorded numerous Archaic sites in southwestern Indiana while conducting large-scale surveys for proposed surface coal mines (Stafford 1994; Stafford et al. 1988). Impending strip mining resulted in the excavation of several important sites, like Bluegrass (Stafford and Anslinger 1988; Stafford et al. 2000) and Millersburg (Levy n.d., cited in Stafford and Cantin 2009). About the same time, archaeologists investigated an important Early Holocene (Kirk) component at Swan's Landing (Smith 1995a), located along the Ohio River in Harrison County (Stafford and Cantin 2009).

In the late 1990s, Indiana State University archaeologists initiated the Caesars Archaeological Project in southern Indiana, just downriver from the Falls of the Ohio (Stafford 2004). Field investigation, continuing for more than 39 months, focused on four sites containing deeply buried Early to Late Holocene cultural deposits. The association of more than 70 radiocarbon dates with stratified components spanning the Holocene has provided a solid baseline for hunter-gatherer activity in that part of the Lower Ohio Valley (Stafford and Cantin 2009).

Although analyses of the Caesars findings are ongoing, resulting information is helping to clarify diachronic trends in hunter-gatherer technology, settlement/subsistence, and economics (raw material acquisition and use). One of the most significant has been identifying projectile point forms associated with the poorly understood early Middle Archaic components (Stafford 2005; Stafford and Cantin 2009).

New information about Kentucky hunter-gatherer societies has come from both CRM-related projects and university/museum-based research. Excavations at the Railway Museum site (15Jf630) on the Ohio River floodplain just upstream from the Falls exposed an extensive midden and numerous features, burials, and artifacts thought to be associated with a Middle to Late

Holocene base camp. Radiocarbon dates place the occupation at about 4750 B.P. (Anslinger et al. 1994).

Large-scale surveys conducted along Cypress Creek (Jefferies et al. 2005; Thompson 2001) and test excavations at the Ward site (Jefferies et al. 2008; Webb and Haag 1940) provided a diachronic perspective on hunter-gatherer land-use practices in the Green River region, particularly those areas away from the main river valley. Numerous chenopodium seeds from the lower part of the Ward site midden suggested that intensive use of starchy seeds first occurred at upland sites, not at river-edge shell middens.

Several projects in far western Kentucky have provided important new information on Late Holocene hunter-gatherers that lived along the extreme western section of the Ohio River floodplain. Archaeological investigations at the Shawnee Atmospheric Fluidbed Conduction Power Plant identified intact cultural deposits containing Late Archaic straight-stemmed Saratoga Cluster and Terminal Archaic Trimble Side Notched points, along with abundant chert debitage (Butler et al. 1981). Study of the assemblage provided important information on lithic technology relating to the acquisition and use of Mounds Gravel obtained from nearby Ohio River gravel bars.

An increased understanding of Late Holocene plant utilization came from work done at the Hedden (15McN81) and Highland Creek (15Un127) sites (Rossen 2000, 2006). Hedden, located in a marshy portion of the Tennessee River floodplain near its confluence with the Ohio River, was most intensively occupied from ca. 5100 to 3800 B.P. (Rossen 2000:Table 1). Highland Creek, also occupying a wetland setting, is located approximately 140 km upriver from Hedden near the Ohio-Wabash confluence (Maggard and Pollack 2006). Charcoal samples collected from Highland Creek's thick midden yielded dates ranging from ca. 4500 to 4100 B.P. (Maggard and Pollack 2006:Table 4.2).

Clearly, archaeologists have learned a lot about the Lower Ohio Valley's earliest inhabitants, people who followed a hunter-gatherer way of life, over the past century. As we have learned more, our understanding of the different aspects of their cultures—technology, social organization, economy, political structure, and, even to some extent, their ideology—has become clearer. Several recent publications have drawn on this continually expanding database to develop excellent overviews of the hunting and gathering societies that once inhabited the Lower Ohio Valley (Marquardt and Watson [eds.] 2005), Phillips and Brown 1983; Prufer et al. 2001).

Archaic Hunters and Gatherers in the American Midwest (Phillips and Brown 1983) represented the first truly comprehensive summary of hunter-gatherer societies of the North American Midcontinent (Illinois, Missouri, and Kentucky) since the publication of Griffin's (1952 [ed.]) "Green Bible" in 1952. Indicative of its time, the volume incorporated the findings of many large-scale, multidisciplinary cultural resource management projects (Emerson and McElrath 1983; Fortier 1983; Jefferies and Lynch 1983; O'Brien and Warren 1983; Phillips and Gladfelter 1983; Wiant et al. 1983), as well as those of more traditional archaeological research programs (Brown and Vierra 1983; Charles and Buikstra 1983; Kay 1983; Marquardt and Watson 1983; Styles et al. 1983). The data presented in this seminal volume provided a much-revised diachronic perspective on Holocene hunter-gatherer cultural complexity and diversity in the region.

Following in the footsteps of Phillips and Brown (eds.) (1983), organizers of the 2004 *Urbana Conference on the Archaic Societies of the Midcontinent* brought together scholars from throughout the eastern United States to discuss issues relating to Holocene hunter-gatherer social and political complexity (Emerson et al. 2009). Data spanning the area from the Gulf Coast to Canada, including Lower Ohio Valley (Butler 2009; Jefferies 2009; Stafford and Cantin 2009), suggest that these groups were highly diverse, ranging from those that were socially, politically, and economically complex to groups that had much simpler forms of social organization. Like the Phillips and Brown volume, publication of the conference proceedings (Emerson et al. 2009) has had a major impact on our understanding of these early hunting and gathering societies.

Of a narrower geographic focus, recent publications on specific parts of the Lower Ohio Valley have incorporated detailed new data and/or new interpretations of previously collected information on regionally specific hunter-gatherer societies. Of particular note is Marquardt and Watson's (eds.) (2005) publication, *Archaeology of the Middle Green River Region, Kentucky*, containing results of their multidisciplinary Green River Shell Mound Archaeological Project (SMAP), initiated in 1971. Also, contributors to Prufer et al.'s (2001) edited volume, *Archaic Transitions in Ohio and Kentucky Prehistory*, complement more recent investigations in the Green River region through their analysis and reinterpretation of data collected by WPA archaeological projects.

The story of the Lower Ohio Valley's hunter-gatherers that unfolds in the subsequent chapters of this book is based on the efforts of hundreds of archaeologists who, over the past century, have devoted their lives to learning about the past. Through their collective efforts, we are starting to learn about the first people of the Lower Ohio Valley and how these hunting and gathering societies changed over the next 10,000 years.

4 / Peopling of the Valley

The Late Pleistocene–Early Holocene Transition

The question of when and how the first people entered the western hemisphere has been of interest to archaeologists for many years. Until recently, many researchers believed that they were "big game" hunters who migrated to North America from northeast Asia about 11,000 years ago. Once in North America, they seem to have quickly migrated to the south and east, perhaps in pursuit of now extinct herds of Pleistocene megafauna, and slowly expanded into the vast, unoccupied space of Late Pleistocene America (Anderson 1990).

Evidence for these early hunter-gatherers, known as Paleoindians, first came to light in New Mexico during the 1920s and 1930s when archaeologists documented distinctive "fluted" Clovis projectile points associated with extinct forms of Pleistocene fauna (Cotter 1937; Howard 1935). Since then, archaeologists have found similar fluted points throughout most of North America, representing the oldest evidence for human presence in much of the continent.

In recent years, archaeologists working throughout the Americas have gradually accumulated new evidence demonstrating that hunter-gatherers were present prior to the well-documented Clovis groups. The most widely accepted evidence for a pre-Clovis occupation comes from the Monte Verde site in southern Chile where archaeologist Tom Dillehay has dated human occupation to at least as early as 12,500 years ago (Dillehay 1997, 2000), roughly 1,000 years older than most Clovis dates. Possible eastern North American sites having pre-Clovis occupations include Cactus Hill (McAvoy and McAvoy 1997) in southern Virginia, Topper (Goodyear 2005:107–111) along the Savannah

River in South Carolina, and Meadowcroft Rockshelter (Adovasio et al. 1990, 1998; Adovasio and Pedler 2005) in the highlands of western Pennsylvania.

As in other parts of North America, we do not know exactly when or how the first hunter-gatherers entered the Lower Ohio River valley. If the early groups who lived in Chile 12,500 years ago entered the Americas via the Bering Land Bridge that once connected northeast Asia and North America and then migrated to the south, it is likely that similar groups of Paleoindians came to the Lower Ohio Valley about the same time.

The search for evidence of the Lower Ohio Valley's first hunter-gatherers is hampered by many factors, primarily the meager amount of material remains left by these highly mobile groups and the dynamic changes to the physical landscape during the Late Pleistocene and Holocene that eroded or buried much of the Late Pleistocene landscape. More than eighty years of archaeological research have yielded many clues about the Lower Ohio Valley's early Paleoindian hunter-gatherers, but so far there is no compelling evidence for a pre-Clovis occupation. I suspect that it is there; we just have to find the right combinations of conditions for its preservation.

Whenever the first hunter-gatherers entered the Lower Ohio Valley, the climate, plant communities, animal populations, and physical landscape that they encountered, as shown in the previous chapter, were quite different from those of today. For example, we know that the climate was considerably cooler than today's and that the remnants of the Pleistocene ice sheet still existed just to the north of what would become the Great Lakes. Rivers that flowed from the north, like the Wabash, were flooded with sediment-laden glacial meltwater. Jon Muller maintains that, "Between the barren floodplains scoured by massive flooding and dust storms dwarfing those of the Dust Bowl, the valley would have been very unattractive to primitive hunters" (1986:49–50).

Paleoindian hunter-gatherers preceded those of the Holocene by several thousand years and are not the central focus of this book. However, the similarity of Late Paleoindian and Early Holocene material culture suggests a high degree of cultural continuity (Tankersley 1996:38). In many respects, the way of life of many Early Holocene hunter-gatherer societies more closely resembled that of Late Pleistocene groups than that of subsequent Middle and Late Holocene societies. Because of this, we need to take a quick look at Lower Ohio Valley Paleoindian culture to establish a cultural-historical context for subsequent developments during the Holocene.

Paleoindian assemblages from most of eastern North America are assigned

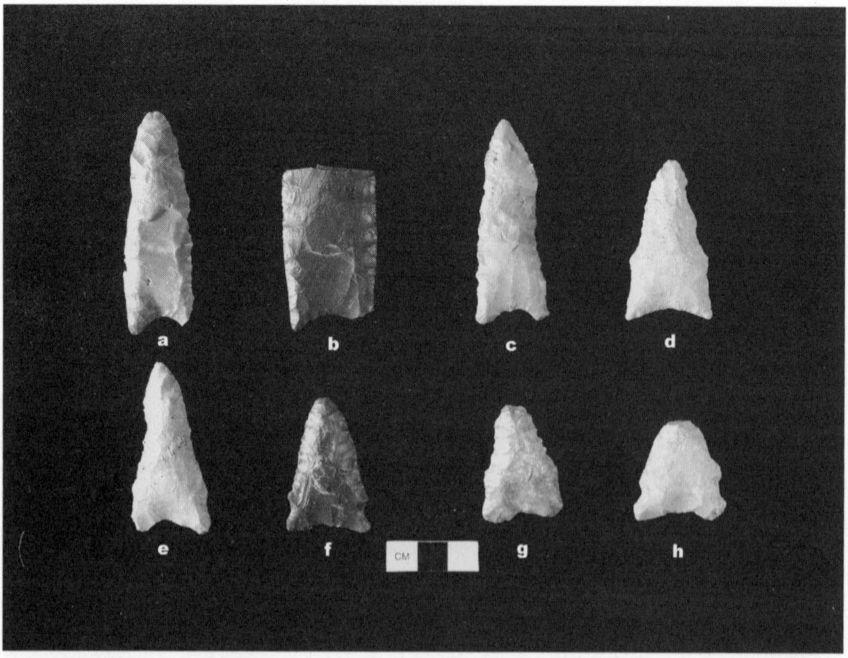

Figure 4.1. Examples of Lower Ohio Valley Paleoindian hafted bifaces—Clovis (a–b), Beaver Lake (c), Dalton (d–h) (William S. Webb Museum of Anthropology, University of Kentucky).

to one of three subperiods—Early (ca. 11,500–11,000 B.P.), Middle (11,000–10,500 B.P.), and Late (10,500–10,000 B.P.)—largely based on differences in projectile point style and lithic technology (Tankersley 1996). Early Paleoindians manufactured lanceolate fluted projectile points resembling Clovis forms, Middle Paleoindians made fluted and unfluted points with broad blades and constricted hafts like Cumberland and Quad types, and Late Paleoindians fashioned lanceolate side- and corner-notched points resembling Dalton points (Figure 4.1).

Many archaeologists consider the Early Paleoindian era to have been a time of exploration and colonization in eastern North America, followed by the establishment of regional populations and cultural variants during Middle Paleoindian times. Late Paleoindian Dalton societies are credited with developing a new way of life based on essentially modern plant and animal communities that would serve as the foundation for subsequent Holocene hunter-gatherer societies of the Archaic period (Anderson et al. 1996:7–8). So far, no solid evidence exists for a pre-Clovis population in the Lower Ohio Valley. However, investigations farther up the Ohio Valley at Meadowcroft Rockshelter in southwestern Pennsylvania suggest a pre-Clovis occupation of the shelter dating prior to 14,000 years ago (Adovasio et al. 1990; Adovasio et al. 1998).

Figure 4.2 shows the distribution of recorded Lower Ohio Valley Paleoindian sites in the study area based on site file data from Kentucky, Illinois, and Indiana. Much of the area covered by the map, particularly away from the Ohio River, is characterized by dense vegetation and rough terrain, so site frequencies are undoubtedly underrepresented. Despite this and other biases in the site data, the map provides a general picture of site frequency and distribution. Figure 4.3 shows the location of major Lower Ohio Valley Paleoindian sites.

Early Paleoindian

The earliest *known* Paleoindian hunter-gatherers to inhabit the Lower Ohio Valley arrived around 11,500 years ago. Climatic conditions were generally cooler and more moist than today's; however, rapid environmental changes were underway. During the subsequent 1,500 years, mean temperature increased and vegetation constantly changed, resulting in a varied mosaic of forests and grasslands. The megafauna native to the region during the glacial period either became extinct or followed the retreating glacial front north (Tankersley 1996:22).

The high density of Early Paleoindian sites in the Lower Ohio Valley, as well as in the lower Cumberland and Tennessee River valleys, led David Anderson (1990) to suggest that these loci served as "staging" areas during the initial migration of humans into eastern North America. These environmentally rich river valleys would have offered Early Paleoindian "colonizers" an abundant and diverse food supply, high-quality lithic materials for making tools, and easily accessible transportation/communication routes (1990:187–189).

Early Paleoindian groups probably stayed in these areas for an extended period, creating a stable social environment within which groups could grow in size and familiarize themselves with their surroundings. As group size increased over time, fissioning would have occurred to help alleviate intragroup dissension. Over time, regional Paleoindian cultural traditions emerged, reflecting strengthening social ties among affiliated regional groups. This higher level of social integration is reflected in the archaeological record by the spatial distribution of shared cultural traits like projectile point styles. Higher frequencies of Middle and Late Paleoindian projectile points suggest that the number and

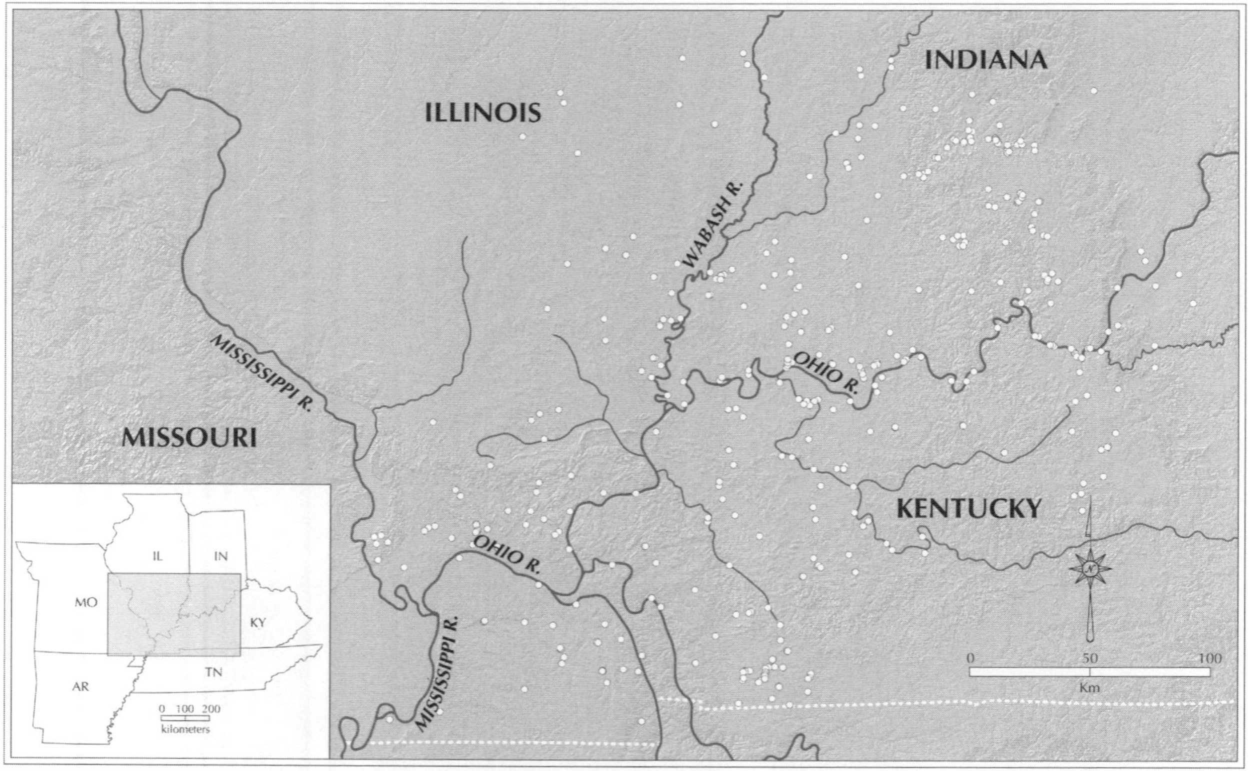

Figure 4.2. Distribution of Paleoindian sites.

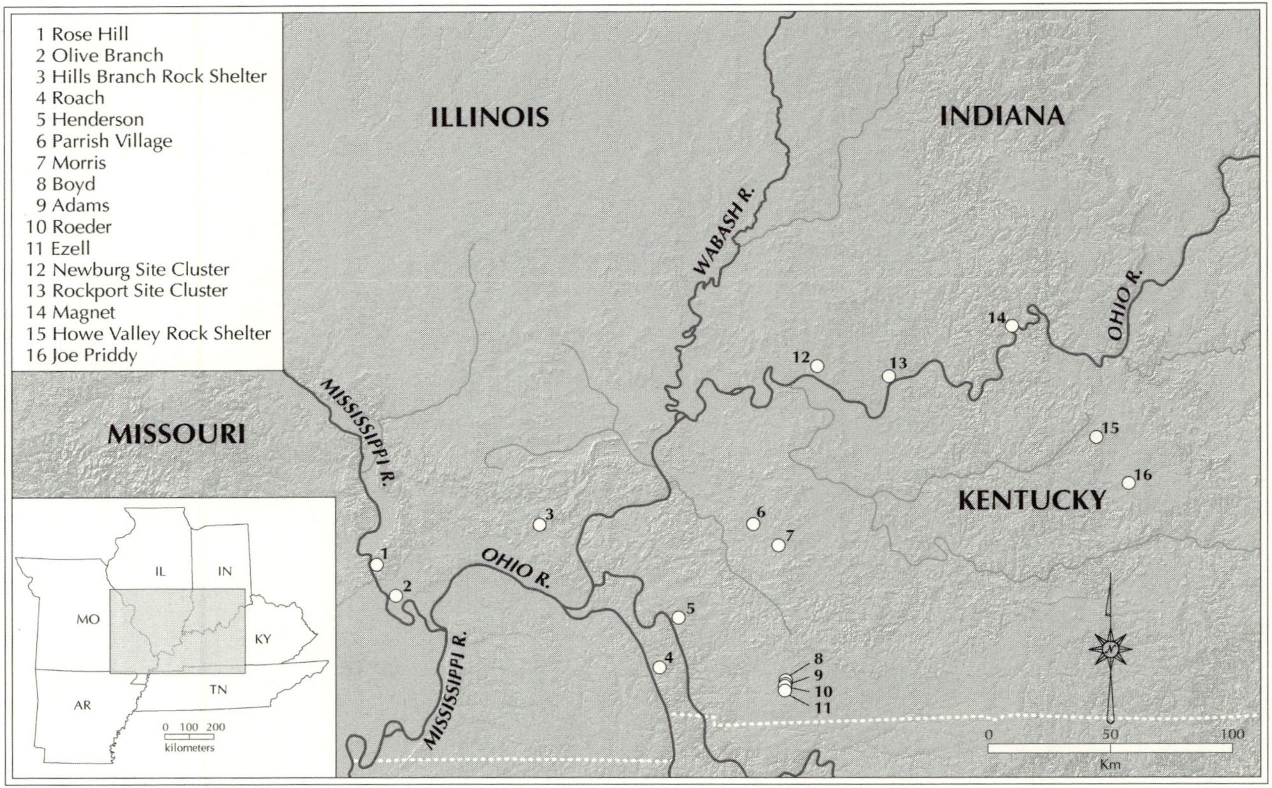

Figure 4.3. Important Lower Ohio Valley Paleoindian sites.

size of Paleoindian groups rapidly increased during the remainder of the period (Anderson 1990:204). Figure 4.4 clearly illustrates a higher frequency of Late Paleoindian sites in Kentucky and Indiana. Comparable data were unavailable for Illinois.

No radiocarbon dates currently exist for the Lower Ohio Valley's Early Paleoindian occupation, but artifacts from early sites have been cross-dated with radiocarbon-dated material from sites in adjacent parts of eastern North America. So far, the oldest artifacts found in the Lower Ohio Valley are Clovis projectile points dated to 11,500 to 11,000 years ago in other states.

Little is known about the social organization of these early groups, but they probably consisted of small, highly mobile bands of 20 to 50 related people who were organized along egalitarian lines. Most archaeological sites attributable to these groups are small, contain few artifacts, and offer little evidence for the kinds of shelters or other facilities that they constructed (Tankersley 1996:21).

The Early Paleoindian tool kits are similar to those found in some parts of Europe and Asia, the home of their distant ancestors. Knives and scrapers made from flakes and prismatic blades are the most common Early Paleoindian tool types. However, the most widely recognized implement is the fluted Clovis projectile point, many of which were used as spear or lance tips (Tankersley 1996:24). Many Clovis points were made from high-quality stone that was acquired hundreds of kilometers from where they were eventually discarded, most likely reflecting the great distances that some groups traveled.

Clovis blades are triangular or prismatic in cross section and struck from large conical-shaped cores of high-quality stone. Once made, many blades were snapped into sections to make endscrapers (Freeman et al. 1996:389). Early Paleoindian tool kits also included an assortment of implements made from ivory, bone, and antler. Other items were probably made from wood and plant fibers (Tankersley 1996:24). The similarity of Early Paleoindian tools and tool kits found throughout much of eastern North America suggests that these scattered hunter-gatherer groups shared a common origin and a similar way of life.

While no longer considered to have been exclusively "big game" hunters, the association of some Clovis-like fluted points with the remains of mammoth, mastodon, tapir, elk, caribou, peccary, and ground sloth indicates that they did occasionally hunt such animals. Many of these associations occur at "kill" sites where Paleoindian hunters killed single, or in some cases, multiple animals at one time (Graham et al. 1981). Smaller animals, along with nuts,

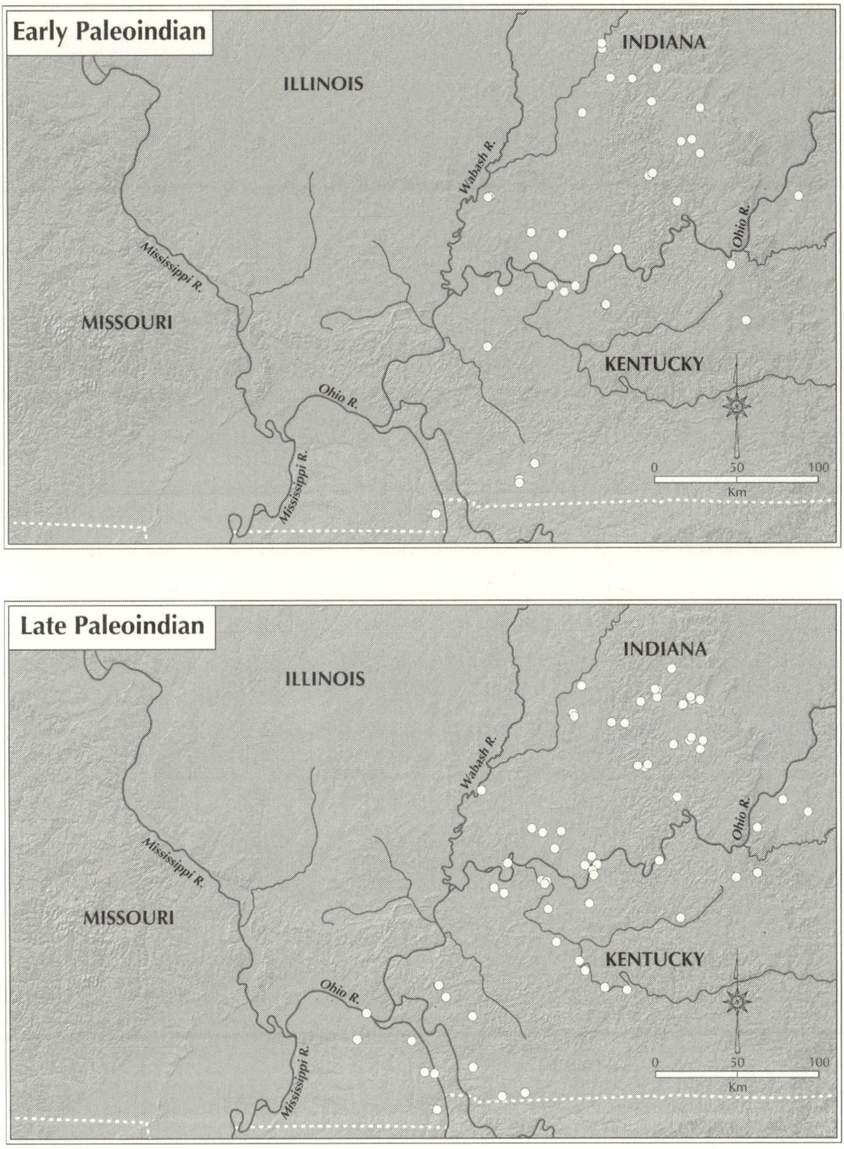

Figure 4.4. Distribution of Early and Late Paleoindian sites.

berries, fruits, and other kinds of plants were also consumed (Meltzer 1988; Meltzer and Smith 1986), but evidence of these foods in the distant past is more elusive. The dietary importance of big game probably varied throughout eastern North America, depending on local environmental conditions (Tankersley 1996:26).

So far, archaeologists have not found any solid evidence for an Early Paleoindian "kill" site in the Lower Ohio Valley. This is probably attributable to differences in sediment exposures compared to western North America, where these kinds of sites are more common. Kill sites probably exist in the region, but are deeply buried below the floodplains of the Ohio River and its tributary rivers and streams (Tankersley 1996:26).

The Big Bone Lick site, located in the Middle Ohio River valley, is a good candidate, although direct evidence for Pleistocene mammal hunting has yet to be found (Tankersley 1985). Recent archaeological investigations at the Kimmswick site in Missouri support the probable existence of nearby Lower Ohio Valley kill sites (Graham et al. 1981). At Kimmswick, investigators found Clovis projectile points, unifaces, biface fragments, and debitage in direct association with the bones of an American mastodon (*Mammut americanum*). Although Kimmswick is located approximately 175 km northwest of the Mississippi-Ohio confluence, these kinds of Quaternary deposits exist in similar microenvironments throughout the Central Mississippi River valley and its tributaries (Graham et al. 1981:1115). Kenneth Tankersley (1989:263–264) has proposed that Early Paleoindian sites can be assigned to one of two broad categories based on artifact assemblages and topographic settings—lithic-related and game-related sites or clusters. Lithic-related sites are habitation sites located near high-quality lithic sources and are characterized by abundant debitage. Hunting-related sites occur near areas attractive to game, like saline springs or river crossings, and are marked by small assemblages consisting of fluted points, unifacial tools, and debitage. Game-related sites or site clusters include the Newburg (Vanderburgh and Warrick counties) and Rockport (Spencer County) site clusters, and the Raaf site (Spencer County) in southern Indiana (Tankersley 1989:264).

Most evidence for Early Paleoindian activity consists of scattered Clovis points found in surface contexts. Unfortunately, later Archaic, Woodland, or Late Prehistoric groups often occupied many of the same sites. Years of plowing, erosion, and other cultural or natural processes eventually mixed the Paleoindian materials with those of later components, making interpretation of their onsite activities difficult.

The first systematic attempt to study Paleoindian settlement in Kentucky was Martha Rolingson's (1964) analysis of projectile point distributions by physiographic region. Although based on a very small dataset, her analysis indicated that the most intensive area of Paleoindian activity was along the Ohio River, particularly in Kentucky's Jackson Purchase and Western Coalfields. These two regions comprise only 18 percent of Kentucky, but contained 32 percent of Rolingson's Paleoindian point sample (1964:71).

Subsequent work in western Kentucky focused on several sites containing Paleoindian material (Rolingson and Schwartz 1966). The discovery of several Clovis points by artifact collectors at the Parrish Village site (15Hk45) in 1938 inspired William Webb to excavate there in hopes of finding out more about early cultural developments (Webb 1951). WPA archaeologists found four more fluted points, but most of the cultural deposits turned out to be Archaic. Nevertheless, Webb suggested that an earlier component, indicated by the fluted points and certain other artifacts, might be attributable to "an earlier hunting stage of cultural development" (Webb 1951:423). Reanalysis of the Parrish material led Rolingson and Douglas Schwartz to conclude that, "The presence of a tentative Paleo-Indian occupation was limited to seven fluted projectile points which appeared to be unique examples at the site and not culturally associated with any of the chipped stone tool assemblage" (1966:143). Perhaps the most significant contribution of the Parrish Village research was it being the first eastern North American site subjected to large-scale excavations in search of what Webb called an "Early Hunter" component (Freeman et al. 1996:395).

One of the best-documented Lower Ohio Valley Early Paleoindian complexes is located in the Mississippian Plateau region of western Kentucky (Gramly and Yahnig 1991; Sanders 1990). The complex consists of a series of lithic workshop-habitation sites and isolated artifact scatters located in a karstic setting along the Little River, a tributary to the Cumberland River. The four major sites, Adams (15Ch90), Boyd (15Ch236), Roeder (15Ch482), and Ezell (15Ch483), all contain significant Clovis components. These sites appear to have primarily served as workshops where Early Paleoindian people procured and processed the nearby "Hopkinsville chert," a high-quality material found in the Ste. Genevieve limestone formation (Freeman et al. 1996:396). This observation was made earlier by John Witthoft (1952) in his comparison of Parrish and the Shoop site.

The Adams site, situated around a large sinkhole and covering about 1.2 ha, is the most intensively studied of the Little River sites. Adams was the first single component Paleoindian site discovered in Kentucky and has yielded one of the largest assemblages of Paleoindian flaked stone tools in eastern North America. Early Paleoindian people probably repeatedly occupied this locality because of the nearby source of high-quality chert. Surprisingly, despite the proximity of easily accessible chert, Adams inhabitants were very conservative in its use, as indicated by their recycling of broken tools and the presence of use-wear on most of the debitage (Freeman et al. 1996:397–398; Sanders 1990:68).

The Adams assemblage presented archaeologists with the rare opportunity to examine the complete sequence of Clovis projectile point manufacturing, including chert procurement strategies, early and mid-stage biface production, and final shaping and grinding of the finished biface. In addition to making Clovis points, inhabitants manufactured true blades and flakes that served as preforms for other kinds of tools. In fact, artifacts associated with blade technology account for nearly 12 percent of the entire Adams assemblage, suggesting that this industry was an important component of Early Paleoindian technology. Tool types manufactured from blades included gravers, spokeshaves, end- and sidescrapers, "backed" blades, beaks, and numerous multipurpose tools exhibiting attributes of several tool types (Haag 2004:71).

Although the products and by-products of tool production dominated the Adams site assemblage, more than 60 percent of all tools were processing implements like gravers, scrapers, shredders, choppers, planes, and knives. Their presence at Adams suggests that, in addition to tool manufacturing, Early Paleoindian inhabitants performed many domestically oriented tasks there as well (Sanders 1990:62–63).

Additional information on Early Paleoindian blade technology comes from the Joe Priddy site (15Hd583) in Hardin County, Kentucky (Haag 2004). Like Adams, Joe Priddy was a multicomponent site situated around a sinkhole in the west central Kentucky karst region. Probable Early Paleoindian blades account for slightly more than 4 percent of the assemblage, considerably less than the 12 percent at Adams. Like in the Adams assemblage, local Ste. Genevieve chert was selected to make most blade-related artifacts (blades, fragments, and cores) (Haag 2004:71). The entire blade production sequence is represented at Joe Priddy, ranging from initial blade detachment from the core to highly retouched blades and exhausted blade cores. Blade tools included "backed" blades, endscrapers, sidescrapers, and combination end/sidescrapers (Haag 2004:86–95).

Ongoing investigations of Early Paleoindian settlement in Indiana (Dorwin 1966; Smith 1990; Tankersley et al. 1990) indicate that more than one-half of the state's documented fluted points come from the lower (unglaciated) portions of the Wabash and Ohio rivers. The largest and most diversified sites are located on Late Pleistocene terraces along the Ohio River and its main tributaries, particularly downriver from the Falls of the Ohio. Many of these sites are concentrated on specific parts of the landscape, particularly in riparian habitats and overlooks of those areas. Selection of these locations may be attributable to the monitoring, hunting, and processing of migratory game (Tankersley et al. 1990:307–310, Figure 3).

In one study, researchers documented 251 Paleoindian projectile points from 107 localities in the karstic region of south central Indiana (Smith 1990). A variety of point types was represented, indicating that the region was occupied throughout the Paleoindian period; however, Early Paleoindian points accounted for at least 93 (37 percent) of the total.

Early Paleoindian site types in this part of Indiana include base camps, lithic reduction and manufacturing sites, and hunting-processing loci. Locally available cherts dominate the assemblages, with more than 90 percent of identifiable lithic material coming from within 60 km of the Wyandotte source. Most expended projectile points were made from Wyandotte chert, so if Paleoindian hunters were ranging far afield, their discarded artifacts do not reflect this.

Smith (1990:248) suggests that these Early Paleoindian groups were highly mobile, but within a more restricted territory that focused on the Wyandotte chert source. The predominance of Wyandotte chert in Early Paleoindian assemblages located up to 250 km from its source provides some perspective on the size of the territories that were extensively exploited by these groups (Tankersley 1990b:283).

Middle Paleoindian

Middle Paleoindian (ca. 11,000 to 10,500 B.P.) artifact assemblages exhibit greater stylistic variability than those of their Early Paleoindian counterparts, probably reflecting the emergence of regionally distinctive populations and adaptive strategies. Middle Paleoindian people made smaller fluted points than during Early Paleoindian times, along with unfluted lanceolate points and fluted or unfluted points with broad blades and constricted hafts (Anderson et al. 1996:11). The Cumberland point, a narrow, deeply fluted lanceolate biface, is often associated with Middle Paleoindian components in the Southeast and Midsouth, including the Lower Ohio Valley (Anderson et al. 1996:11; Rolingson and Schwartz 1966; Tankersley 1996:33). Great Plains Paleoindian Plano points, like the Agate Basin, Scottsbluff, and Midland types, are occasionally found in the western part of the Lower Ohio Valley (Gramly and Funk 1991:30; Rolingson 1964:47–51). The presence of Plano forms in the upper Southeast indicates that some form of interaction or movement took place between the two regions, but the nature of that connection is still unclear (Anderson et al. 1996:11–12).

In addition to greater projectile point stylistic variability, Middle Paleoindian tool kits differed in other ways from those of the Early Paleoindians. For example, bipolar technology became more popular during this time, marking the increased use of lower-quality lithic materials. In addition, spurred endscrapers and limaces, a type of sidescraper, are more common in Middle Paleoindian tool kits (Tankersley 1996:31).

In the Lower Ohio Valley, the geographic distribution of forests and grasslands was in a state of flux during this time, resulting in the reorganization of animal life that included an alteration of ranges or, in some cases, the extinction of species. It is unlikely that all of the Pleistocene megafauna became extinct during this time, but these changes contributed to their substantial reduction. In Kentucky, and probably elsewhere in the Lower Ohio Valley, Middle Paleoindian groups responded to these dynamic habitat changes by emphasizing a mixed foraging strategy that incorporated both large and small game. The wider distribution of Middle Paleoindian artifacts compared to those of earlier Clovis groups may be attributable to this shift in hunting strategies (Tankersley 1996:32).

As with Early Paleoindian components, Lower Ohio Valley archaeologists have had a difficult time locating substantial Middle Paleoindian occupations. An exception is the Henderson site (15Ly27) situated on a terrace of Eddy Creek near its confluence with the Cumberland River (Rolingson and Schwartz 1966). The 1961 excavation of the shallow, disturbed midden revealed a variety of probable Middle Paleoindian artifacts including a Cumberland fluted point and several probable Cumberland point fragments, 63 scrapers, and 63 gravers. While coming from a culturally mixed deposit, many of the tools resemble those found at Paleoindian sites in other parts of eastern North America. The predominance of projectile points, scrapers, and gravers suggests that Middle Paleoindian inhabitants focused their onsite activities on hunting and animal processing tasks (Rolingson and Schwartz 1966:19–26).

Late Paleoindian

Late Paleoindian (ca. 10,500 to 10,000 B.P.) hunters did not make true fluted points. Instead, a variety of stylistically diverse projectile points marks their presence. Most Plano Cluster points are long and narrow, while Quad and Beaver Lake points are short and wide. Some types have wide flaring basal ears. Others have steeply beveled lateral margins, often with sharp, serrated edges, suggesting that they were used as knives, not projectile points. Despite their stylistic diversity, most Late Paleoindian hafted bifaces, like their fluted predecessors, were designed to kill and butcher game (Tankersley 1996:34, Figure 2.9). The increase in site frequency during the Late Paleoindian subperiod suggests that the regional population was on the rise (Figure 4.4).

The Dalton horizon, represented by the Dalton point, is perhaps the best-known Lower Ohio Valley Late Paleoindian cultural manifestation. Dalton material more commonly occurs in southern Illinois and western Kentucky, being less prevalent in southern Indiana (Justice 1987:41–42). The assignment of the Dalton horizon to the Late Paleoindian subperiod is based on Albert Goodyear's (1982) research in northeast Arkansas and Missouri, where he examined the stratigraphic placement of Dalton artifacts and associated radiocarbon dates. The dating of Dalton to the 10,500 to 9900 B.P. interval places it in a time of rapid environmental change when glacially influenced species were being replaced by those of the Holocene. This transition of Lower Ohio Valley lifeforms is reflected by the replacement of earlier Paleoindian strategies and technologies with new ones that formed the core of Holocene hunter-gatherer adaptations over the next 7,000 years.

Classic Dalton points have a lanceolate blade, grinding on the lateral and basal margins, and a concave base that was thinned to the extent that it sometimes resembles a flute. Lateral margins are often beveled and/or serrated (Anderson et al. 1996:12; Justice 1987:40–42; Morse 1997b:14–30). This distinctive resharpening technique differs from those used by earlier Paleoindian hunter-gatherers and was employed by Lower Ohio Valley hunter-gatherers well into the Early Holocene. Use-wear studies suggest that Dalton people used these implements to cut soft materials like meat and skin (Goodyear 1982:391). Some Dalton points were altered to make awls, drills, and burins, and broken points were often recycled into endscrapers (Morse 1997b:Figure 3.5).

Other than the distinctive Dalton projectile point, the majority of implements comprising the Dalton tool kit can be equated with those found in earlier fluted point assemblages (Goodyear 1974, 1982; Morse 1997b). Shared tool forms include unifacial endscrapers and sidescrapers, gravers on flakes, gravers on endscrapers, true blades, bipolar cores, and wedges (*pieces esquillees*). Of significance, the Dalton tool kit contained one important new tool form, the Dalton adze (Morse 1997b:22–30, Figure 3.10; Morse and Goodyear 1973). This hafted woodworking implement was manufactured using a chipped-stone technology, not by grinding like some later adzes (Goodyear 1982:384).

Archaeologists working in northeastern Arkansas have conducted extensive studies of Dalton settlement strategies (Goodyear 1974; Morse 1973, 1975, 1997a; Morse and Morse 1983; Price and Krakker 1975; Schiffer 1975a, 1975b). In the 1970s, Morse proposed that Dalton bands occupied territories along the major watersheds of the Central Mississippi River valley. Permanently or semipermanently occupied base camps, like the Lace site, were centrally located within each territory, and logistical camps, such as the Brand site, were scattered over the surrounding landscape. The Sloan site represented a probable Dalton cemetery. In contrast, Michael Schiffer (1975a, 1975b) and others disagreed with Morse's interpretation, countering with an alternative model in which Dalton band territories crosscut major ecological zones. Instead of permanently occupied base camps, small family groups inhabited scattered, seasonally occupied residential camps. These scattered groups reunited during certain times of the year (Anderson and Sassaman 1996:25–26).

Recent investigations of Dalton settlement-subsistence practices in the Central Mississippi Valley are providing new insights on the organization of hunter-gatherer societies at the Pleistocene-Holocene transition (Walthall 1998b; Walthall and Holley 1997; Walthall and Koldehoff 1998). The proximity of the Lower Ohio Valley makes it possible that Dalton groups living there employed similar strategies.

Compared to those of earlier Paleoindian groups, Central Mississippi Valley Dalton groups appear to have significantly reorganized their mobility strategies as a way of adapting to the northward spread of deciduous forests and the disappearance of migratory animal species. Highly mobile Dalton hunter-gatherers followed an annual round shaped by seasonal and topographical changes in resource distribution. Walthall (1998a:15–16; 1998b:232) has proposed that Dalton groups spent the warm season at floodplain camps, aggregated for a short time at fall encampments, then dispersed into upland, mast-rich sheltered valleys at the onset of cold weather. Part of the cold season, upland adaptation involved the first widespread use of rockshelters (Walthall 1998b:232). The "encounter" hunting strategy used by Dalton groups to kill deer, bear, squirrel, turkey, and other game required hunting parties to spend more time in widely spaced and diverse settings. The repeated use of a circumscribed hunting territory resulted in the identification and annual reuse of particularly favorable camping areas like rockshelters (Walthall 1998b:233–234).

Rockshelters that yielded a low diversity of Dalton material probably represent camps occupied by small hunting parties. Activities conducted at hunting camps included tool manufacturing and maintenance, and deer carcass processing. When rockshelters were unavailable, open areas having good views of hunting areas served as hunting stations. In contrast, big rockshelters containing larger, more diverse artifact assemblages probably represent residential camps regularly inhabited by larger band segments, perhaps family-based hunting groups. Occupations were frequent, but of short duration, during the cold season (Walthall 1998b:233–234).

Dalton people usually collected suitable stone for toolmaking from local lithic sources they visited during their seasonal round (Walthall 1998b:234). Prior to moving out of the river valleys in the fall, Dalton groups made task-specific tool kits using locally available lithic material found along the valley margins. They carried these tool kits, containing finished, highly transportable implements such as adzes, projectile points, endscrapers, and large flakes, with them into the uplands, some of which they cached there for future use (Walthall and Koldhoff 1998:261).

Archaeologists have found caches of large Dalton bifaces, sometimes referred to as Sloan points (Morse 1997b), at a number of Central Mississippi Valley sites (Walthall and Koldehoff 1998). The scarcity of Sloan points, combined with their large size, the consistent use of high-quality Crescent chert, the skill required to make them, and their presence in caches and burials, supports their interpretation as ritual objects (Walthall and Koldehoff 1998:266). The exchange of Sloan points may have occurred during episodes of interband aggregation and interaction, which took place seasonally, probably at strategically located and favored places on the landscape. Alliances established between neighboring Dalton bands could have served as buffering mechanisms to stabilize intergroup social relationships (Walthall and Koldehoff 1998:266–269).

Recent investigations of southern Illinois Dalton components suggest that groups inhabited much of the southern Illinois landscape including fluvial and terrace settings, the till plain, and upland ridges and rockshelters. SIUC's recent investigation of the Hills Branch Rock Shelter (11Pp508) has yielded important new insights on Dalton upland settlements and rockshelter use (Wagner and Butler 2000).

The Hills Branch Rock Shelter, a small half-dome shelter, is approximately

16 km northwest of the Ohio River along the southern margins of the Shawnee Hills (Wagner and Butler 2000). The east-facing shelter has approximately 45 to 50 m^2 of useable floor area, suggesting that it could accommodate from five to ten people, depending on individual space requirements (Wagner and DelCastello 2000:6). Excavations revealed a modest amount of Late Paleoindian material, including Dalton, Beaver Lake, and Greenbriar projectile points, and a possible Dalton adze. The small number of diagnostic artifacts suggests that the shelter's Late Paleoindian occupation was sporadic and ephemeral, conforming to Walthall's (1998b) description of a Dalton overnight hunting station (Butler and Wagner 2000:171).

The Dalton inhabitants made most of their tools from Mounds Gravel chert that was probably collected south of the rockshelter along or near the Ohio River. Both Dalton points, the Greenbriar point, the possible Dalton adze, and several blades were all made from Mounds Gravel. The strong association between Mounds Gravel and Dalton artifacts suggests that the Hills Branch Dalton occupation reflects forays into the Shawnee Hills by groups that spent much of the year living close to the Ohio River's main channel where Mounds Gravel is abundant (DelCastello 2000:122). The fact that they did not use equally good chert found near the rockshelter suggests either that Dalton people spent little time in the area and they did not need to refurbish their tool kits, or that they were not very familiar with the area's lithic resources (Butler and Wagner 2000:171). These findings support Walthall and Koldehoff's (1998) idea that Dalton groups were tethered to major river valleys, like the Mississippi and Ohio, and used rockshelters like Hills Branch on an intermittent, short-term basis.

Dalton sites containing stratified, undisturbed cultural deposits are rare in the Lower Ohio Valley. The Olive Branch site (11Ax267), a probable Dalton residential or base camp situated approximately 40 km northwest of the Mississippi-Ohio river confluence in far southern Illinois, is one such site (Gramly 2002; Gramly and Funk 1991). Olive Branch is located on the edge of the Mississippi River floodplain where the river passes through a narrow stretch known as Thebes Gap. Prior to historic navigation improvements, it was possible to walk across the Mississippi at this point during periods of low water. Such a place would have served as a ford, a portage, and a fishing place for regional hunter-gatherer groups (Gramly and Funk 1991:25).

More than a decade of archaeological investigations at Olive Branch has yielded a diverse collection of more than 4,000 Dalton artifacts, including projectile points, end- and sidescrapers, drills, gravers, denticulates, adzes, and debitage, along with at least one pit feature, numerous artifact caches, and a human cremation (Gramly 2002). The site's complex stratigraphy is a consequence of over-bank alluvial sedimentation, colluvial deposition from nearby bluffs, and debris (sandstone, chert, etc.) introduced by site inhabitants (Gramly 2002:36).

Much of the activity at Olive Branch appears to have centered on a large limestone outcrop that formed a rock platform (ca. 17 x 13 m). Removal of the soil that filled deep crevices in the limestone revealed abundant Dalton material. Adzes were particularly abundant, suggesting that woodworking was a common activity (Gramly 2002:45).

Radiocarbon dates associated with the Dalton component ranged from 9975 +/– 125 to 9115 +/– 100 B.P. (Gramly 2002:Table 4), indicating a relatively unbroken cultural sequence starting about 10,000 years ago and continuing for nearly 900 years. However, the terminal date of 9115 B.P. is nearly 900 years later than the commonly recognized end of the Dalton era (9900 B.P.). A relatively continuous period of occupation is supported by the absence of major disconformities or temporal disruptions in the midden. An Agate Basin point base (ca. 10,000 B.P.) from below the Dalton deposits represents the oldest artifact from an undisturbed context at the Olive Branch site (Gramly 2002:46).

Researchers identified two Dalton phases at Olive Branch based on the differential distribution of artifacts and changes in projectile point morphology. The cultural marker for the Sirkin phase (ca. 9975 +/– 125 B.P.), representing the earliest Dalton occupation, is the Olive Branch point, an auriculate-shaped biface having a concave, ground base and "delicate ears and a long, edge-ground shank or hafting area" (Gramly 2002:178). Utilized flakes, gravers, projectile points, knives, sidescrapers, and endscrapers dominated the Sirkin phase assemblage, consisting of artifacts found ca. 1.7 m below surface in nine excavation units near the rock platform. The spatial patterning and low artifact frequency suggest that the assemblage is attributable to a single occupation. Recycling of broken or expended flaked stone tools indicates that early Dalton inhabitants practiced an "economizing" behavior when it came to their flaked stone technology (Gramly 2002:73). The presence of fire-cracked rock, not commonly found on earlier Paleoindian sites, suggests the development of a new cooking technology associated with changes in hunter-gatherer diet at the Pleistocene-Holocene transition (Gramly 2002:74).

The subsequent Hill phase (ca. 9100 B.P.) represents Olive Branch's terminal Dalton occupation. Hill phase projectile points are morphologically distinct from Sirkin points, exhibiting different resharpening techniques and basal thinning by flake removal. In addition, v-shaped drills were manufactured for the first time. Other types of flaked stone artifacts resemble those associated with the Sirkin phase. Debitage analysis indicated that the Dalton inhabitants used local cherts; however, a few flakes of Pitkin chert from Arkansas were noted. The presence of this material may indicate the range of movement for Late Dalton hunter-gatherers or, more likely, the extent of their social connections (Gramly 2002:75).

Although site investigators identified only one pit feature, its contents included the cremated remains of a woman. The large, shallow pit, measuring more than 1 m in diameter, was filled with dark soil and scattered small pieces of calcined bone. The burned skeletal material represented only a small part of the woman's skeleton, suggesting that only "token" elements were cremated or that some of the remains had been buried elsewhere (Gramly 2002:91).

The Olive Branch investigations revealed that Dalton hunter-gatherers regularly occupied the site for nearly 1,000 years. The accumulation of cultural material, its diversity, and the site's strategic location suggest that Olive Branch served as a residential or base camp for Late Paleoindian hunter-gatherers who maintained contacts with groups living in other parts of the Lower Ohio and Central Mississippi valleys. The scant faunal evidence indicates that the Dalton inhabitants had a diet that included deer, rabbit, fox, fish, and turtle (Gramly 2002:258–259), closely matching the subsistence resources exploited by hundreds of generations of subsequent Archaic hunter-gatherers.

Information from several other Dalton sites in the Thebes Gap vicinity helps to clarify the nature of the Late Paleoindian regional settlement strategies (Webb et al. 1989). Two upland sites, Rose Hill (11Ax116) and Sutton (11Ax251), were situated from 500 to 1,200 m back from the edge of the Mississippi Valley bluffs. These moderately dense lithic scatters largely consisted of unfinished and finished bifaces and associated debitage.

Artifacts from the two sites indicated a broad tool-manufacturing spectrum and a variety of onsite activities, generally conforming to what would be expected for a Dalton residential camp. Neither of the sites resembled what Dan Morse (1973) called a base camp, but the range of represented activities was too high for a short-term camp. Dalton chert knappers used both local and nonlocal (Crescent Hills) resources (Webb 1989:138–141), conforming to the pattern seen at the Olive Branch site. Two other Dalton sites situated on a nearby terrace suggest less intensive activity when compared to the upland sites.

The concentration of Dalton sites in the Thebes Gap vicinity, only 50 km northwest of the Ohio-Mississippi confluence, may be attributable to several factors. First, the Thebes Gap sites could be part of a larger concentration of Dalton settlements that covers much of the Central Mississippi Valley, extending from Crowley's Ridge in northeast Arkansas to southeast Missouri. Second, the geological and hydrological characteristics of Thebes Gap created one of the better river crossings in the Central Mississippi Valley, making access from southeast Missouri to southern Illinois relatively easy (Webb 1989: 141–142).

Late Paleoindian components are also recognized south of the Ohio River. At the Roach site (15Tr10), located on the Tennessee River near its confluence with the Ohio River, excavations conducted prior to the impoundment of Kentucky Lake yielded a Paleoindian assemblage containing probable Dalton points (Tankersley 1996:35), scrapers, flake knives, and gravers (Rolingson and Schwartz 1966:57–58). Like the Henderson site located 20 km to the northwest, the Roach Paleoindian assemblage is probably associated with processing animal hides and meat. The Dalton points' stylistic variability indicated to Tankersley that Dalton people occasionally occupied the site during Late Paleoindian times (1996:25).

The Morris site (15Hk49), situated on a small knoll near the divide between the Green and Tradewater rivers, once overlooked a large swampy floodplain (Rolingson and Schwartz 1966:64). Its typically Late Paleoindian assemblage included 22 Dalton and three Quad points, more than 200 unifacial scrapers, gravers, and lamellar bladelets (Rolingson and Schwartz 1966:116–119; Tankersley 1996:35–36).

Morris's Dalton inhabitants were probably attracted to the site because of several nearby natural features including high ground, wetlands, and a game trail. The wetlands offered a variety of aquatic plants and animals, while Dalton hunters could easily kill the large animals that used the game trail. The many artifacts found at Morris suggest Late Paleoindian groups repeatedly occupied this location (Tankersley 1996:36).

Several Quad points from the Howe Valley Rock Shelter (15Hd12) in Hardin County indicate that Kentucky's Late Paleoindian hunter-gatherers, like their southern Illinois counterparts, also inhabited rockshelters (Rolingson 1964:67). Additional information on Late Paleoindian rockshelter use in western Kentucky comes from excavations at the George Branch site complex (Mocas n.d., cited in Smith 1995b:175). This series of rockshelters located along the Ohio River in Breckinridge County yielded a variety of Late Paleoindian artifacts including Redstone, Beaver Lake, Quad, and Dalton projectile points

The Magnet site (12Pe171), located on the second terrace of the Ohio River in southern Indiana's Crawford Uplands, is perhaps the largest Paleoindian site in this part of the Lower Ohio Valley. The site has produced over 100 diagnostic Late Paleoindian artifacts, including examples of Beaver Lake, Quad, Plainview, Agate Basin, and Dalton projectile points (Smith 1995c). The mixture of "intrusive" Plains projectile points, like the Agate Basin and Plainview specimens, with types that are more characteristic of the Southeast, adds to the site's significance. Unfortunately, much of the material from the Magnet site is from surface contexts, limiting its interpretative potential.

In addition to projectile points, the Magnet site's diverse artifact inventory includes unifacial tools, debitage, and a substantial number of long, narrow lamellar blades that functioned as side- and endscraper preforms. Similar blades were associated with the Early Archaic component at the nearby Swans Landing site (Smith 1995c:171). Most of the lithic assemblage was manufactured from locally available high-quality Wyandotte chert (Smith 1995c).

Many of the Magnet site's Late Paleoindian activities appear to have focused on the production and refurbishing of tool kits, as indicated by a high frequency of projectile point base fragments typically associated with rehafting activities. The diversity of the tool assemblage suggests that the Magnet site served as a quarry-related base camp (Smith 1995c:176).

The Magnet site's topographic setting provided Late Paleoindian inhabitants with easy access to the abundant, diverse natural resources of both the riverine lowlands and the Crawford Uplands. Nearby high-quality chert sources added to the attractiveness of this location, as did several mineral springs that would have attracted game. Surprisingly, the site does not appear to have been located near a ford in the river where both humans and game animals could easily cross the Ohio (Smith 1995c:177).

Summary

The end of the Pleistocene witnessed the migration of the first humans into the Lower Ohio Valley. Precisely when the first people arrived is uncertain, largely because the material evidence of their presence is very limited and widely scattered. In addition, more than 10,000 years of dynamic natural and cultural processes have buried or destroyed much of the evidence that once existed. Although they probably were not the earliest inhabitants, by 11,000 years ago, small groups of Clovis-toting hunter-gatherers lived in the Lower Ohio Valley. These small, highly mobile groups hunted the remaining Pleistocene megafauna, but they exploited modern plant and animal species as well.

Increased projectile point stylistic diversity throughout eastern North America during the Middle Paleoindian period reflects a decrease in group mobility and the emergence of regionally distinct social units that were adapting to different social and physical environments. Like their Clovis predecessors, some of these people continued to make fluted projectile points.

In contrast, Late Paleoindian Dalton groups no longer manufactured fluted projectile points, reflecting a significant shift in animal procurement strategies associated with the extinction of Pleistocene megafauna. Settlement-subsistence data suggest that these groups gradually established a seasonal round of movement within smaller, better-defined territories. Although Dalton material culture indicates the continued significance of hunting, plant foods appear to have become more important in their diet. As succinctly stated by Stephen Williams and James Stoltman (1965:678) more than 40 years ago, "With the depletion of the herds of big game animals, new sources of food were sought, or more likely, old supplementary subsistence patterns were intensified in favorable localities. As a result, the Dalton peoples laid the foundations for the subsistence pattern we are to recognize later as 'Archaic' in northern Alabama and western Tennessee and Kentucky."

5 / Early Holocene Foragers

Introduction

Although the times of massive glacial expansion had ended, Lower Ohio Valley hunter-gatherers continued to feel the lingering effects of the Pleistocene for thousands of years. Archaeologists refer to the segment of prehistory extending from 10,000 to 3,000 years ago, representing more than one-half of eastern North America's known past, as the Archaic period. The beginning of the Archaic period, as currently defined, roughly coincides with recent dates for the Pleistocene/Holocene "boundary"; however, it is becoming increasingly clear that natural conditions and cultural adaptations did not greatly differ on either side of this boundary (Anderson et al. 1996:14). For the most part, it appears as if the foraging strategies used by Early Archaic hunter-gatherers did not differ markedly from those employed by their Late Paleoindian predecessors. As their territories became more bounded during the Holocene, Archaic groups diversified their subsistence strategies, exploiting a wider variety of plant and animal resources through time (Meltzer and Smith 1986:19–20).

The Archaic was a time marked by continuing changes in the cultural and physical landscapes. The forests and rivers, along with the plant and animal communities that inhabited them, slowly evolved in response to the constantly changing post-Pleistocene climate. Eventually, the landscape assumed a character resembling that seen by the Ohio Valley's first European inhabitants (Jefferies 1996b:39).

Throughout the Archaic, hunter-gatherers exploited the Ohio Valley's plant and animal resources using highly flexible hunting, collecting, and fishing technologies. By the late Middle Holocene, approximately 6,000 years ago, some groups were becoming increasingly sedentary and were starting to experiment with plant cultivation as an alternative food source. By the Late Holocene, thousands of years of interaction between the region's inhabitants and their physical and social surroundings resulted in the creation of a very diverse cultural landscape occupied by groups of variable size and social, economic, and political complexity.

The Early Holocene

Although small, highly mobile Paleoindian groups had lived in the Lower Ohio Valley for thousands of years, it was not until the Early Holocene, some 10,000 to 8,000 years ago, that we see evidence for a general population increase and more distinct regional cultural traditions. This two-thousand-year segment of prehistory is known as the Early Archaic subperiod (Jefferies 1996b:40; Muller 1986:56).

For much of this time, the Lower Ohio Valley's physical landscape was in the process of adjusting to the milder Early Holocene climate. This transition to cool, moist conditions was reflected by the northward expansion of Mixed Hardwood forests. The Southeast's boreal forests, common in many areas during the Late Pleistocene, were reduced to remnant fir and spruce populations that continued to grow in the higher elevations of the southern Appalachians. Oak-hickory forests developed in southern Illinois, mixed hardwood forests covered areas to the south and east of the Ohio Valley, and mixed conifer–northern hardwood forests grew across the Midwest (Delcourt and Delcourt 1981:147; Muller 1986:50).

The Early Holocene marks the first time in many millennia that the Lower Ohio Valley's physical landscape was actually favorable for human habitation (Muller 1986:50). Not only did the developing oak-hickory forests provide abundant nuts and nut oils that Early Archaic people could consume, but they also supported a rich population of game animals that provided food, clothing, shelter, and materials to make tools and ornamental items. Throughout the Early Holocene, small, mobile groups of Early Archaic hunter-gatherers inhabited this environmentally rich portion of the North American Midcontinent, continually adapting to the region's changing landscape. If we can equate the number of recorded sites with regional population levels, comparison of the number of recorded Lower Ohio Valley Paleoindian and Early Archaic sites suggests a substantial Early Holocene population increase (Figure 5.1). Figure 5.2 shows the locations of sites that have contributed important information about Early Archaic hunter-gatherers.

In their seminal volume on the Paleoindian and Early Archaic Southeast, David Anderson and Kenneth Sassaman (1996) provide a historical perspective on the development of theoretical models for Early Archaic settlement/subsistence in eastern North America, particularly the Southeast (Anderson 1996b:29–57). While none of these models was developed specifically for the Lower Ohio Valley, they strongly influenced how archaeologists working in the region conceptualized the region's Early Archaic societies. The following heavily draws on their work.

The first good information on Early Holocene hunter-gatherer chronology and material culture dates to the 1950s and 1960s with the excavation of deeply stratified rockshelters like Graham Cave in Missouri (Logan 1952), Russell Cave in Alabama (Miller 1956, 1958), and Modoc Rock Shelter in Illinois (Fowler 1959). Although rare, a few open-air sites, as exemplified by Joffre Coe's excavations at the Doershuck and Hardaway sites in North Carolina (Coe 1964), also contributed important information on Early Archaic societies. Temporally diagnostic projectile points from radiocarbon-dated contexts at these and other sites provided the first reliable Early Archaic cultural sequences (Anderson and Sassaman 1996:18).

In the early 1950s, James B. Griffin (1952a:354–355) described a typical Early Archaic society as consisting of small, highly mobile groups that scheduled their movements to exploit seasonally available food resources. These groups occupied specific hunting territories and maintained ties with similar groups, creating broad social networks that stretched hundreds of kilometers across the Early Holocene landscape. Social networks served as a means for exchanging information about the landscape and its resources, obtaining marriage partners, and initiating and maintaining individual partnerships and intergroup alliances. On occasion, network participants would meet to conduct ceremonies that helped to solidify their social, economic, and ideological ties. Griffin also suggested that Early Archaic people became increasingly familiar with the local natural resources, a concept that Joseph Caldwell (1958) would expand on a few years later (Anderson and Sassaman 1996:18–19).

In 1958, Caldwell (1958:vii–viii) introduced the concept of "primary forest efficiency." This adaptive strategy, hypothetically based on greater hunting efficiency and a growing knowledge of locally available food sources, was

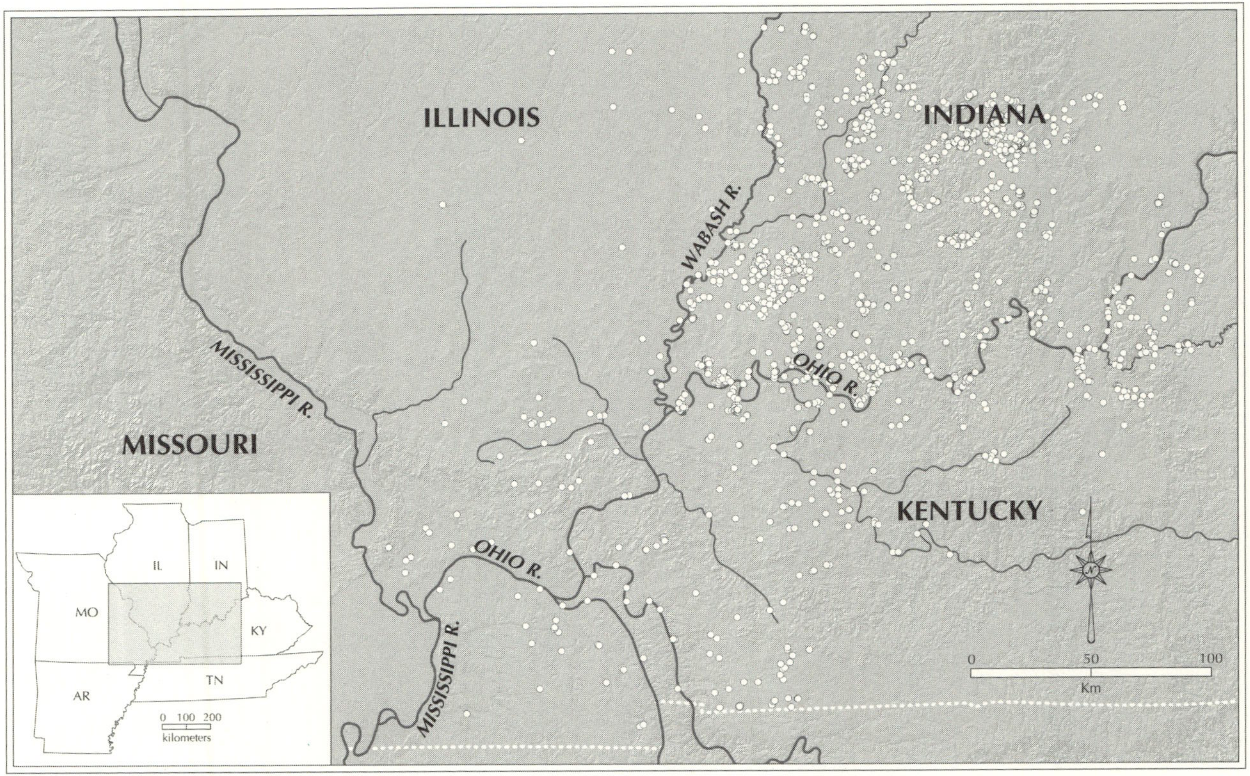

Figure 5.1. Distribution of Early Archaic sites.

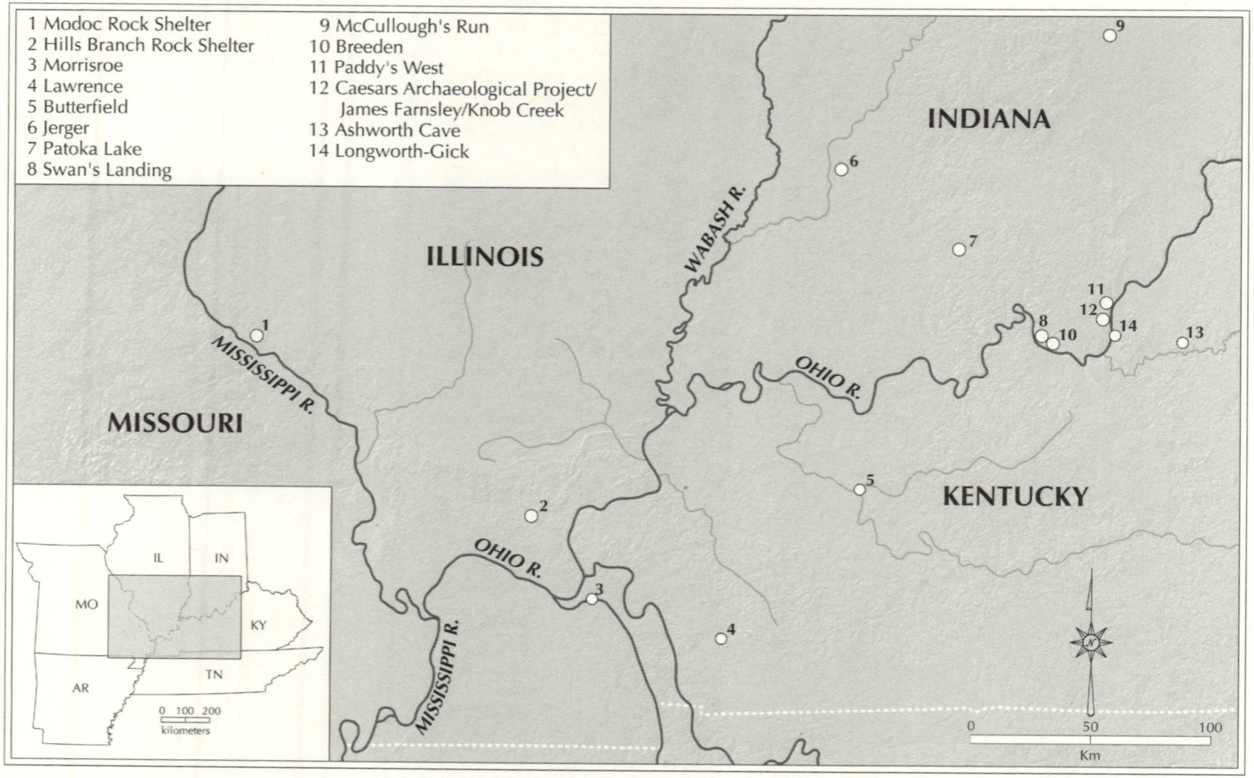

Figure 5.2. Important Lower Ohio Valley Early Archaic sites.

supposedly achieved during the Archaic stage in eastern North America (1958: 18). This was thought to have been a time-progressive process, so Early Archaic hunter-gatherers would have still been on the low end of the learning curve, maintaining a highly mobile, predominantly hunting adaptation (Anderson and Sassaman 1996:20). Caldwell contended that by the Late Holocene, hunter-gatherers had acquired sufficient environmental knowledge to allow them to establish residential stability, maintain a highly diverse diet, and accumulate material wealth, some of which was disposed of through increasingly complex mortuary rituals (1958:vii). More recently, some scholars have argued that the term *primary forest efficiency* is a misnomer because dietary diversification probably resulted from efforts to adapt to a decreasing resource base brought about by a more restricted hunting territory, not a more successful adaptation (Meltzer and Smith 1986:20).

These views of Early Archaic life set the tone for many years to come, particularly the emphasis on the hunting of animals over plant food utilization. Only recently have archaeologists started to recognize the importance of plant foods in the Early Archaic diet (Anderson and Sassaman 1996:20; Meltzer and Smith 1986; Smith 1986:10). Part of the reason for the underestimation of Early Archaic plant utilization may have been the overemphasis on well-manufactured, aesthetically pleasing flaked stone projectile points and a failure to recognize plant processing tools in Early Archaic assemblages (Anderson and Sassaman 1996:20). For example, manos used to grind plant foods were often identified as hammerstones because they were not expected to be there.

The years following 1958 produced abundant data on Early Archaic sites, assemblages, and artifacts, but relatively little in the way of theory with which to evaluate this information. This situation changed in the 1970s and 1980s with the publication of a number of theoretically oriented works containing models of Paleoindian/Early Archaic settlement/subsistence and land use (see Anderson 1996b:29–57 for details). Anderson and Sassaman (1996:22–28) characterize some of these models as the "interriverine deer procurement model" (Luchterhand 1970), "the Flint Run Lithic Determinism model" (Gardner 1983), "the Little Tennessee River model" (J. Chapman 1975, 1978), "the riverine-interriverine model" (Goodyear et al. 1979; House and Ballenger 1976; House and Wogaman 1978); "Morse and Goodyear's Dalton settlement hypothesis" (Morse 1971, 1973, 1975; Goodyear 1974), "Schiffer, Price, and Krakker's Dalton settlement hypothesis" (Price and Krakker 1975; Schiffer 1975a, 1975b), and the "effective temperature/technological organization model" (Cable 1982; Claggett and Cable 1982).

For the most part, researchers developed these models incorporating a combination of regional site distribution data and temporally diagnostic projectile points. Their theoretical frameworks were largely based on variations of Lewis Binford and Sally Binford's (1966) base camp/extraction camp dichotomy and an emphasis on a high level of group mobility (Anderson and Sassaman 1996:28).

More recently, some archaeologists have drawn on ecological and technological organization theory to help formulate their models of Early Archaic cultural dynamics (Anderson 1996b; Daniel 1996). At the regional scale, these models can provide important insights on settlement range and group mobility, population growth, the emergence and extent of distinct Early Archaic subregional traditions (Sassaman and Anderson 1996), and possibly eastern North America's first distinct cultures (Anderson 1996b:54; Wright 1996:431).

Material Culture and Technology

Insights on Early Holocene hunter-gatherer material culture and technology largely come from the analysis of flaked stone and ground stone artifacts, with primary emphasis on the former. Isolated examples of bone, antler, and shell artifacts exist, but implements made from these materials, along with other organic materials (wood, fiber, skin), are not usually preserved in Early Archaic cultural deposits. Understanding Early Archaic technology is further complicated because most artifacts occur in surface contexts and are usually mixed with more recent materials. Other than diagnostic bifaces, it is hard to distinguish most Early Archaic artifacts from those of later site inhabitants.

Important exceptions come from deeply stratified rockshelters or open-air sites in topographic settings promoting the deposition of colluvial, alluvial, or aeolian deposits. Careful excavation and dating of discrete cultural strata at these sites has enabled archaeologists to define temporally distinct projectile point types and artifact assemblages. The stylistic attributes of these artifacts provide a way for cross-dating Early Archaic components throughout much of eastern North America. Sites that have contributed to the clarification of Lower Ohio Valley Early Archaic chronology include St. Albans in West Virginia (Broyles 1971), Modoc Rock Shelter in Illinois (Fowler 1959; Styles et al. 1983), Icehouse Bottom and Rose Island in Tennessee (Chapman 1973, 1975, 1976), the Doershuck and Hardaway sites in North Carolina (Coe 1964), the Morrisroe, Whalen, and Longworth-Gick sites in Kentucky (Collins 1979a;

Nance 1986a, 1986b, 1988, 2001), and sites in the Haw River valley in North Carolina (Claggett and Cable 1982).

Compared to Late Paleoindian projectile points, the biggest change was the introduction of side, corner, and basal notching as a means of attaching, or "hafting," a biface to a spear or javelin shaft. The reasons that hunter-gatherers adopted these new hafting techniques are unclear, but they may have to do with changes in the kinds of game they hunted, the hunting strategies they employed, and/or the kinds of weapons they used to kill game. For the most part, Lower Ohio Valley Early Archaic projectile points consist of an assortment of side- and corner-notched and Bifurcate base types including the Hardin Barbed; Graham Cave; St. Charles; Pine Tree; Thebes; Kirk Corner Notched, Stemmed, and Serrated; MacCorkle; and LeCroy types (Figure 5.3; Table 5.1).

Archaeologists first identified the Hardin Barbed (10,000–7500 B.P.) point, with its lanceolate blade and expanding stemmed form, in the Central Mississippi Valley. Many of its morphological characteristics are similar to those of Late Paleoindian Scottsbluff points, reflecting the technological continuity during this period of prehistory. Hardin Barbed points are most common in the relic prairie areas of central Illinois and west central Indiana, but occur in small numbers in the Lower Ohio Valley (Justice 1987:51–53).

Graham Cave Side Notched (10,000–7500 B.P.) points have deeply concave bases and well-formed side notches. This point type is generally restricted to the Midwest, particularly Missouri and Illinois, but examples have been found in southern Indiana. Examples from Zone I at Modoc Rock Shelter date from 10,000 to 8000 B.P. (Justice 1987:65–66).

Noel Justice (1987:54) includes Thebes and St. Charles points (ca. 10,000–8000 B.P.) in his Thebes cluster. Thebes points, commonly found across Illinois, Indiana, and Kentucky, are generally large and have pronounced diagonal or side notches. Howard Winters (1967) called this type of point a "Cache Diagonal Notched" projectile point (Justice 1987:54–57).

The Kirk Corner Notched cluster includes a variety of large and small corner-notched points such as Kirk Corner Notched, Palmer Corner Notched (Small Kirks), Charleston Corner Notched, Pine Tree Corner Notched, and Decatur (Justice 1987:71–82). The Kirk Corner Notched type has a large triangular blade with a slightly rounded or straight base and serrated, sometimes beveled, lateral margins. Bettye Broyles (1971) distinguished large and small variety Kirks.

Large Kirks (9500–8900 B.P.) occur throughout much of the eastern United

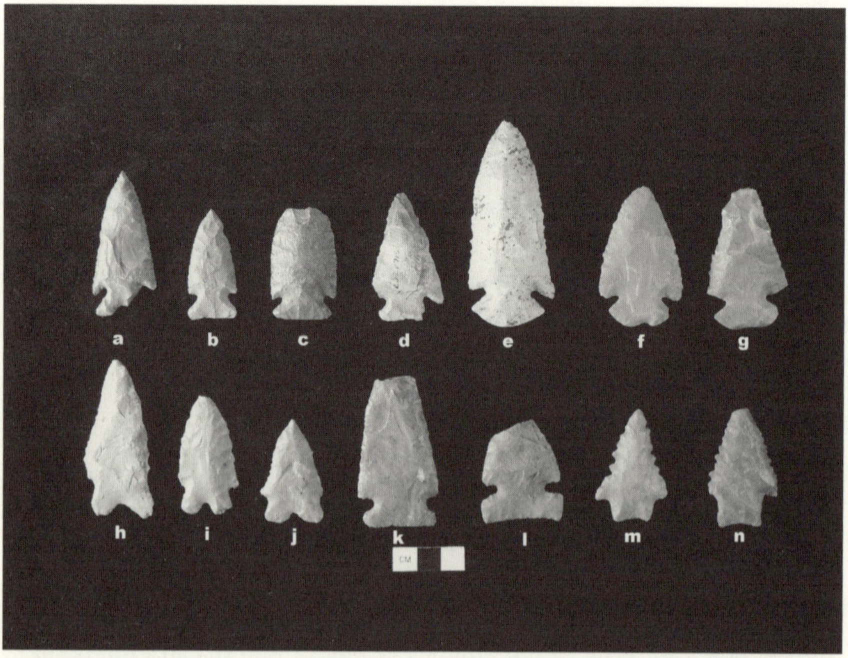

Figure 5.3. Examples of Lower Ohio Valley Early Archaic hafted bifaces—Kirk Corner Notched (a–c), Pine Tree (d), St. Charles (e–f), Thebes (g), MacCorkle (h–j), Graham Cave/Big Sandy I (k–1), Kirk Stemmed (m–n) (William S. Webb Museum of Anthropology, University of Kentucky).

States, including the Lower Ohio Valley. They are common in Indiana, Illinois, and Kentucky (Justice 1987:71–71). Palmer Corner Notched points, also known as Kirk Corner Notched (small variety), predate the Large Kirk type at the St. Albans site in West Virginia (Broyles 1971), the Icehouse Bottom site in Tennessee (Chapman 1977), and the Hardaway site in North Carolina (Coe 1964). The Pine Tree Corner Notched type, a resharpened variant of a Kirk Corner Notched point, and the Decatur point date to the same general time as other members of the Kirk Corner Notched cluster (Justice 1987:78–82).

The Kirk Stemmed and Kirk Serrated types (8900–8000 B.P.) comprise Justice's (1987:82–85) Kirk Stemmed cluster. These points generally have a long, thick, narrow blade that is often serrated and curved. Kirk Serrated points are similar to the stemmed variety, except that the haft element is straight-sided or slightly contracting. The distribution of both the Serrated and Stemmed types conforms to that of the Corner Notched varieties (Justice 1987:82).

MacCorkle Stemmed points (9000–8500 B.P.), characterized by large basal

Table 5.1. Lower Ohio Valley Early Archaic Projectile Point Types.

Projectile Point Type	Date Range	Sources
Hardin Barbed	10,000 to 7500 B.P.	Justice 1987:51–53; Scully 1951:3
Thebes	10,000 to 8000 B.P.	Justice 1987:54–57; Winters 1967
St. Charles	10,000 to 8000 B.P.	Justice 1987:57–58; Scully 1951:4
Graham Cave Side Notched	10,000 to 7500 B.P.	Justice 1987:62–66; Scully 1951:8
Decatur	9500 to 9000 B.P.	Cambron 1957:17; Justice 1987:81–82
Pine Tree Corner Notched	9500 to 8900 B.P.	Cambron and Hulse 1969: 95–96; Justice 1987:79–80
Large Kirk Corner Notched	9500 to 8900 B.P.	Justice 1987:71–72; Coe 1964: 69–70
Small Kirk Corner Notched (Palmer)	9500 to 8900 B.P.	Coe 1964:67; Justice 1987: 78–79
MacCorkle Stemmed	9000 to 8500 B.P.	Broyles 1971:71; Justice 1987: 86–89
Kirk Stemmed	8900 to 8000 B.P.	Coe 1964:70; Justice 1987: 82–84
Kirk Serrated	8900 to 8000 B.P.	Coe 1964:70; Justice 1987: 82–84
LeCroy Bifurcate Stem	8500 to 7800 B.P.	Justice 1987:91–92; Kneberg 1956:27–28

notches, are often referred to as "bifurcate based" points. The large, rounded basal ears give the points a lobed appearance. Basal grinding is common in the hafting area (Justice 1987:86). MacCorkle Stemmed points are common throughout the Midwest, including the Lower Ohio Valley.

LeCroy Bifurcate Stem points (8500–7800 B.P.) also have bifurcated bases, but unlike the MacCorkle Stemmed points, they have pointed rather than rounded basal ears. Blades are often serrated. LeCroys occur throughout most of the eastern United States (Justice 1987:91–92).

In many ways, Early Archaic tool kits resemble those of earlier Late Paleoindian times, suggesting that many aspects of Paleoindian life only gradually changed during the several thousand years marking the Pleistocene-Holocene transition (Jefferies 1996b:40). In addition to projectile points, Early Archaic assemblages include bifaces representing various stages of projectile point manufacturing (early stage blanks and late stage preforms) and an assortment of multipurpose and task-specific tools like drills, endscrapers, sidescrapers, knives, gravers, perforators, denticulates, adzes, and retouched flakes.

A number of deeply stratified Lower Ohio Valley sites have contributed significant information on Early Archaic material culture and technological organization. At these sites, most of which are located on the Ohio River floodplain, Early Archaic cultural deposits are deeply buried in alluvial sediments deposited by the river over the past 10,000 years (Stafford 2004). In these situations, it is possible to identify temporally discrete artifact assemblages and to compare changes in technology through time.

Archaeological investigations at several sites or site complexes near the Falls of the Ohio River, including Longworth-Gick, Swan's Landing (12Hr304), Paddy's West, and the Caesars Palace Casino, have been particularly informative with respect to Early Archaic technology. Collectively, these sites reflect the heavy Early Archaic seasonal use of the Ohio River floodplain and provide important new insights concerning the content and variability of Early Archaic artifact assemblages and lithic technological organization (Smith and Mocas 1995:347).

At Longworth-Gick, situated on a low floodplain ridge near Louisville, excavations revealed a series of Early Archaic cultural zones, the deepest of which was buried below more than seven meters of alluvium (Collins 1979b). During the Early Archaic, this part of the floodplain was highly susceptible to flooding, which probably affected its use by humans. This view is supported by geomorphological and archaeological data suggesting that many of the Early Archaic cultural zones were deposited rather rapidly and that the occupations represented by those zones were brief (Collins and Driskell 1979:1025–1026).

Archaeologists found the earliest evidence of Early Holocene human activity at Longworth-Gick in Zone XXXVII at a depth of more than seven meters below modern ground surface. The deposit contained no diagnostic artifacts, but nearly 400 small chert flakes representing the final stages of bifacial manufacture and trimming were associated with a date of approximately 9800 B.P. (Collins 1979b:545).

The earliest definitive Early Archaic components (Zones XIII and VII), dating from approximately 9800 to 8800 B.P., yielded small Kirk projectile points. A second Early Archaic component (Zone V) dating to ca. 8700 B.P. was positioned approximately one meter above Zone VII. Diagnostic artifacts consisted of large Kirk Corner Notched points with pronounced barbs and deep serrations. Bifaces dominated the remainder of the Zone V assemblage, many of which appear to represent different stages of projectile point manufacturing. The assemblage also contained drills, unifacial endscrapers, marginally modified flakes, and some ground stone items (Collins 1979b:Table 5.17). A third Early Archaic component (Zone IV) yielded a heterogeneous collection of projectile points, but they generally resembled those in the Zone V assemblage. A fourth Early Archaic component (Zone III), also dating to approximately 8700 B.P., was characterized by very small bifurcated stemmed LeCroy and Kanawah projectile points, marking a major break from the Kirk material that preceded it. Changes in biface size (large to small) were attributed to differences in tool function (Collins 1979b:581).

Longworth-Gick inhabitants exploited three general raw material types to make flaked stone tools—local Kentucky Muldraugh chert, Wyandotte (a.k.a. Harrison County) chert from southern Indiana, and a variety of other chert types occurring in low frequencies. The percentage represented by the three categories remained relatively constant throughout the Early Archaic occupation—Muldraugh cherts from 55 to 77 percent, Wyandotte from 22 to 36 percent, with the miscellaneous category accounting for the remainder.

Flaked stone technological attributes, primarily the presence/absence of cortex, suggested that Early Archaic chert knappers acquired Wyandotte chert at the southern Indiana source area and that they removed much of the exterior cortex prior to bringing it to Longworth-Gick. Approximately 30 percent of the Muldraugh chert consisted of water-worn stream gravel, with the remainder coming from ledge and colluvial slope sources. They probably collected the other chert types, some of which appear to be from more distant sources, from gravel beds in the nearby Ohio River. Michael Collins suggested that the acquisition of Wyandotte chert "involved intermediate people and exchange, or journeys to the outcrops" (Collins 1979b:588). Regardless, significant reduction of Wyandotte chert occurred sometime prior to its arrival at Longworth-Gick. In contrast, Early Archaic knappers brought Muldraugh chert to the site in large pieces bearing a significant amount of cortex, reflecting the proximity of its source to Longworth-Gick (Collins 1979b:587–588).

Comparison of the functional and technological attributes of Longworth-Gick's Early Archaic assemblages suggests that the diversity of flaked stone artifacts, along with the number of tasks for which they were used, increased over time. In general, there were more tools associated with plant processing/

use (pitted stones, digging tools) and fewer scrapers, often associated with hide preparation, suggesting an increasing importance of plant food relative to hunting over time. As far as flaked stone tool manufacturing is concerned, there was a general increase in the rate of early stage biface production; evidence for later manufacturing stages decreased or stayed relatively constant (Collins 1979b:582–583).

Recent restudy of the Longworth-Gick Early Archaic lithic artifacts generally supported Collins's (1979b) results. Michael French (1998) concluded that the role of the site in the regional settlement system changed during the Early Archaic, going from a limited activity field camp during the Zone VII occupation to a more intensively occupied base camp during the Zone V and III occupations. In addition, changes in lithic technology appeared to coincide with apparent changes in site function. For example, earlier Palmer groups employed an exclusively "curated" technology, while later groups used more expediently produced tools (French 1998:191–192).

In the mid-1980s, excavations at Swan's Landing, located approximately 90 km downriver from the Falls of the Ohio in southern Indiana, provided additional information on Early Archaic technology and assemblage variability (Smith 1995a). Excavation revealed three distinct Early Archaic zones deeply buried in up to six meters of alluvium. The significance of Swan's Landing is based on three factors. First, Early Archaic hunter-gatherers appear to have occupied the site only during the Kirk horizon (ca. 9500 to 8900 B.P.). Second, the site is located near outcrops of high-quality Wyandotte chert used to make the majority of Early Archaic flaked stone tools. Third, the large number of finished Kirk points suggests that residents were involved in the large-scale production, caching, retrieval, and distribution of Early Archaic bifaces (Smith 1995a:193).

Excavation of portions of the three Early Archaic zones yielded 29 projectile points, most of which resembled Kirk Corner Notched cluster points (Justice 1987). Included were small varieties of Kirk Corner Notched points like those found at Longworth-Gick, as well as larger Charleston Corner Notched and Pine Tree Corner Notched types. Based on the few projectile points made from nonlocal cherts (Holland, Muldraugh, and Derby), it appears that hunter-gatherers using the Swan's Landing site had a minimal range of movement of about 60 km. The restricted variety of the Swan's Landing cherts suggests that if inhabitants exchanged chert with distant groups or exploited an extensive hunting territory, these activities are not reflected in the artifact assemblages (Smith 1995a:203–205, 209).

Unlike many excavated Early Archaic deposits, the Swan's Landing assemblages appear to be functionally specific in that they are more closely associated with flaked stone tool manufacturing than with typical base camp activities (Smith 1995a:213). The intensity of flaked stone tool production is indicated by the many bifaces representing different stages of the manufacturing sequence. In addition, Early Archaic tool kits included unifacially and bifacially flaked endscrapers, often made on bladelike flakes; unifacial sidescrapers; retouched flakes; chipped stone adzes; polished bifacial implements; pitted cobbles; hammerstones; and anvils. With the exception of changes in projectile point morphology, the Swan's Landing flaked stone tool assemblage closely resembles those of earlier Paleoindian hunter-gatherers (Smith 1995a:207).

As Edward Smith proposed (1995a:211), the high-quality Wyandotte chert deposits may have served as a rendezvous point or multiband encampment loci for several Early Archaic hunter-gatherer groups. Easily located, high-quality lithic raw material sources, like those associated with Wyandotte chert, probably served as aggregation points for Paleoindian and Early Archaic bands from over a wide area. The size, artifact density, and probable contemporaneity of the Swan's Landing cultural deposits support such an interpretation; however, the relative homogeneity of the chert types found there suggests more spatially restricted band territories than during the Paleoindian period.

The investigation of Early Holocene hunter-gatherer chert use in southwestern Indiana suggests that earlier Thebes "groups" made over 80 percent of their projectile points from high-quality Wyandotte and Holland cherts. In contrast, these materials account for only 64 percent of the slightly later Kirk points. The frequency of Kirk points made from medium-quality Allen's Creek, Indian Creek, and Attica cherts was more than twice that of Thebes implements (32 percent vs. 14 percent, respectively). Low-quality chert was seldom used to make either point type (Cantin 1989, 2000).

Some of these differences in chert exploitation and use are attributable to variation in the range of group movement and the locations of chert sources. For example, 76 percent of Thebes cluster points were made from "foreign" cherts (mostly Wyandotte) originating more than 30 km from where they were found. In contrast, foreign cherts account for only 65 percent of Kirk cluster points. Considerably more Kirk cluster points were made from "semilocal" (3.0–30.0 km) and "local" (.0–3.0 km) cherts. Collectively, these data suggest that high-quality Wyandotte chert occurred within the larger home range of the Thebes groups and that they used the best available cherts (Wyandotte). In contrast, although Kirk group hunting activities were more spatially restricted, they also selected the best available cherts (mostly medium quality) in their home ranges (Cantin 2000). Mark Cantin (2000) proposed that the larger Thebes territories corresponded to major sections of primary drainages like the lower Wabash and lower Ohio rivers, while smaller secondary drainages served as Kirk group home ranges.

Recent investigations at the James Farnsley site (12Hr520), conducted as part of the Caesars Archaeological Project in southern Indiana (Stafford and Cantin 2009), revealed deeply stratified Archaic cultural deposits extending as much as five meters below the modern ground surface. An Early Archaic Side Notched component dating from 10,000 to 9500 B.P., a poorly documented period in this part of the Lower Ohio Valley, contained large hearths and light scatters of lithic debris. Four associated radiocarbon dates had a pooled mean of 10,060 +/– 52 B.P. Surprisingly, the few diagnostic artifacts from this zone exhibited considerable morphological variability. One large surface hearth, dated to 9954 +/– 86 B.P., yielded two bifaces having deep diagonal notches and a concave base. For lack of a better term, Stafford and Cantin (2009) placed these bifaces in the Early Side Notched cluster.

Although the Farnsley points are contemporary with Big Sandy I Side Notched forms found in the Southeast, the technology used to make them and their shape suggest that they are precursors to Thebes tradition bifaces (Stafford and Cantin 2009). Other bifaces associated with this Early Archaic component included some that resemble small variety Kirk Corner Notched points, except that they exhibit different flaking patterns than found on most Kirk points (Stafford and Cantin 2009).

A slightly later Thebes/St. Charles component, dating to ca. 9500 B.P., is present above the Early Side Notched zone. Near the terrace escarpment, excavators found Thebes points associated with small clusters of knapping debris. The deposit sloped up to the west where archaeologists exposed a lithic workshop containing numerous Thebes points and drills (Stafford and Cantin 2009).

The upper portion of the deposit contained three Kirk Corner Notched components having high densities of chert debitage and tools. The 30-m-long refuse deposit yielded over 200 projectile points and nearly 10,000 other flaked stone implements. Three radiocarbon dates from the lower Kirk zone ranged from approximately 9400 to 8800 B.P., while the middle and upper components yielded dates of 8740 +/– 100 and 8320 +/– 80 B.P., respectively. The similarity of the Thebes date and the earliest Kirk dates suggests that the production of Thebes and Kirk projectile points at least partially overlapped in time (Stafford and Cantin 2009).

Seventy-four percent of the Kirk Corner Notched points from Caesars were Pine Tree Corner Notched varieties, with lower percentages of Large (14 percent) and Small (10 percent) Kirks and Stilwell (2 percent) varieties. The predominance of the Pine Tree Corner Notched points, with their distinctive incurvate and serrated blades, suggests that they reflect a style zone in this part of the Lower Ohio Valley (Stafford and Cantin 2008).

Project archaeologists noted relatively little variation in the distribution of the four point types through the more than one meter of Kirk cultural deposits (Stafford and Cantin 2009). Based on radiocarbon dates from the upper Kirk zone at the Farnsley site, combined with a relatively late date of 8360 +/– 80 B.P. from a Kirk component at the nearby Townsend site (12Hr481), it appears that some Kirk varieties were manufactured as much as 700 years longer in the Lower Ohio Valley than in the Southeast (Stafford and Cantin 2009).

Jack Nance's (1986a, 1988) research at the Morrisroe site, located on the north bank of the Tennessee River in western Kentucky, has done much to clarify the Early Archaic cultural sequence in the lower Tennessee–Cumberland-Ohio River region. Nance's work significantly revised Lewis and Lewis's (1961) nearly 60-year-old cultural sequence based on their work at the Eva site in Tennessee (Nance 1986a:12–13).

Like Longworth-Gick, Morrisroe is situated in an aggrading environment where stratified Archaic occupations are sealed and separated by sterile alluvial sediments. The earliest known Archaic occupation at Morrisroe was a Kirk component buried approximately two meters below modern ground surface in cultural Stratum 4. Kirk Stemmed and Serrated points were the most common types found. Older Early Archaic types, like Kirk Corner Notched and Bifurcate Base points (LeCroy, MacCorkle), were rare, leading Nance to suggest that only the "tail end" of the Early Archaic sequence was present. An early Middle Archaic Stanly projectile point (ca. 8000 to 7000 B.P.) and a radiocarbon date of 8220 B.P. support this chronological assessment (Nance 1986a:22–23).

Nance (1988) documented additional evidence for Early Archaic hunter-gatherer activities at the nearby Whalen site situated on the east bank of the lower Cumberland River (Nance 1988:129–135, 2001). Buried by nearly three meters of alluvium, project archaeologists located a Kirk component dating to approximately 8550 B.P. Mussel shell impressions associated with this Early Archaic zone represent the only known evidence for mussel collection on any lower Tennessee-Cumberland Archaic site (Nance 1988:132).

While deeply stratified floodplain sites have yielded important data about Early Archaic material culture and technology, particularly flaked stone, they are not the only sources of information. In rare cases, Early Archaic features, particularly mortuary facilities, comprise hunter-gatherer "time capsules" containing contemporaneously deposited sets of objects. While the objects associated with the burials were probably deposited as part of a mortuary ritual, they also provide insights into the contents of Early Archaic tool kits.

An artifact cache found with two Early Archaic burials at the Lawrence site (15Tr33) in the uplands east of the Cumberland River contained eight flaked stone implements—two Kirk Corner Notched points, a lanceolate point, a sidescraper, a scraper/spokeshave, two triangular stemmed drills, and a spokeshave (Mocas 1985:Figure 4). Clearly, the bearer of this tool kit was well prepared to carry out a variety of tasks. The pristine condition of the tools indicated that they had not been used prior to placement in the mortuary feature (Mocas 1985:76–85), suggesting that the cache was intended for use in the afterlife. Feature charcoal yielded radiocarbon dates of ca. 7500 B.P. and 7300 B.P. (Mocas 1985:84–85); however, the Kirk points suggest an earlier temporal placement.

Primarily because of preservation issues, most of what we know about Early Archaic technological organization comes from the study of flaked and, to a limited degree, ground stone tools and their manufacturing residues. Compared to later times, we know little about the technological significance of objects made from bone, antler, wood, or fibers. Nevertheless, rare exceptions do occur. Excavations at Ashworth Cave (15Bu236) in west central Kentucky revealed a thick, deeply buried Early Archaic midden outside the shelter (DiBlasi 1981). Along with Kirk Corner Notched points, midden excavations also yielded bone needles, a bone bead, and an antler flaker, indicating the technological significance of this usually poorly represented material.

Settlement Patterns and Intrasite Organization

The ways that people and groups distribute themselves across the landscape reflect how societies are socially, economically, politically, and technologically organized. This was as true for Early Archaic hunter-gatherers as it is for modern societies. In contrast to today's highly visible communities and households, those of Early Archaic hunter-gatherers are much more elusive and difficult to reconstruct. Much of the difficulty stems from the nature of Early Archaic society. Early Holocene hunter-gatherers organized themselves into small, highly mobile groups; built flimsy shelters designed to last a short time; maintained small, highly portable tool kits; and had limited inventories of personal goods. Collectively, these characteristics mean that they did not leave much evidence of where they lived or what they did when they were there. In addition, many Early Archaic components occur in plow-disturbed contexts, often mixed with artifacts dating to later times. Stratified sites often yield high-quality Early Archaic data, but excavations at these sites are usually limited in area because of the depth of the cultural deposits.

Some of the best information on Early Archaic settlement organization comes from large-scale regional surveys that examine the distribution of sites across broad portions of the landscape. This kind of survey provides data on the preferences of Early Archaic groups for particular parts of the physical landscape, presumably reflecting the exploitation of natural resources, social connections among regional hunter-gatherer groups, and ideas about the cultural meaning of the landscape.

One of the first large-scale, systematic studies of Lower Ohio Valley Early Archaic regional settlement strategies was based on research conducted by Indiana University archaeologists prior to the construction of Patoka Lake in southern Indiana's dissected Crawford Uplands (Munson 1980a). For this project, archaeologists carried out extensive surveys and excavations within and around the nearly 9,000-acre impoundment area. Using a standardized measure of land use and demography (number of points/1,000 years), researchers determined that there was a peak in the number of Early Archaic sites compared to earlier and later times (Cook 1980b:684–697, Figures 26.1, 26.2). Clearly, making the jump from numbers of projectile points or sites to Early Archaic population estimates involves many assumptions, but this pattern does suggest that there was considerable hunter-gatherer activity in the region during the Early Holocene.

Early Archaic sites were moderately dispersed across the dissected southern Indiana landscape. Sixty-seven percent were situated on terraces, with the remainder occupying low ridge spur, floodplain, or upland settings. Fifty-seven percent of Early Archaic sites were located along fifth-order streams, in this case, the upper Patoka River, indicating that the river had an important subsistence and/or transportation function. Other Early Archaic sites occurred along large (19 percent) and small (24 percent) tributaries of the Patoka River (Munson and Cook 1980b:698–709).

As a way of investigating the spatial variability of Early Archaic activities, researchers assigned artifact types to specific tasks (extractive, manufacturing, maintenance, etc.), then examined patterns of task distribution across the landscape. These distributions suggested that there were two categories of Early Archaic sites—small ones located along tributary streams and large and small ones situated along the upper Patoka River. The first group, characterized by little evidence for manufacturing and maintenance tasks and no features, probably represents intermittently occupied hunting camps. In contrast, the greater diversity of activities represented at sites along the upper Patoka River suggests more extended Early Archaic occupations (Munson and Cook 1980b:714).

Early Archaic sites lacked midden deposits and other indicators of repeated or long-term occupation. Indirect heating facilities, often associated with nut processing, also were rare, as were nut processing implements commonly found on later sites (Munson and Cook 1980b:717).

The lack of evidence for Early Archaic nut exploitation in the upper Patoka River valley is puzzling since it occurs at contemporary sites in other parts of the Midcontinent (Asch et al. 1972; Broyles 1971; Styles et al. 1983). Cheryl Munson and Thomas Cook (1980b:717) proposed two possible explanations for its absence—both based on the realization that sites in the Patoka Lake area were part of a larger settlement system that covered much of southern Indiana. First, nut-producing trees may not have grown or been common in the Early Archaic postglacial environment of the upper Patoka River. In this case, Early Archaic sites reflect brief seasonal forays by small hunter-gatherer groups that hunted and occasionally made and/or repaired their tools. Large Early Archaic sites, if they existed, were located outside the upper Patoka watershed. An alternative possibility is that Early Archaic groups did collect nuts in the upper Patoka region, but instead of processing them near where they collected them, they took the nuts to base camps situated along larger drainages like the lower Patoka, White, or Ohio rivers (Munson and Cook 1980b:717–718).

C. Russell Stafford's (1994) investigation of Archaic landscape use in southwestern Indiana provided the broader regional perspective to further evaluate the upper Patoka results. This study incorporated data from 21 survey locales, totaling 9,018 ha distributed over 16 counties, including the Patoka River watershed. In order to obtain a topographically representative site sample, survey tracts were placed on a variety of landforms including valley and upland settings, interior upland/tributary valleys, river bluff crests, and river terraces and floodplains. Like the upper Patoka research, Stafford used the frequency and distribution of projectile points to investigate diachronic change in Archaic settlement strategies.

Survey data suggested that Early Archaic hunter-gatherers, when compared to later Archaic groups, more frequently used the upper portions of drainage networks near watershed divides or more distant locations relative to the major rivers. However, Early Archaic groups also exploited major river valley habitats, as indicated by substantial percentages of Early Archaic projectile points at sites along a major river (27 percent) and/or their primary tributaries (45 percent) (Stafford 1994:231).

Stafford maintained that the distribution of Early Archaic bifaces reflects a residential mobility strategy (Binford 1980) that incorporated the movement of foraging groups from one resource to the next as they moved between river drainages, procuring resources on an encounter basis. The high percentage of Early Archaic hafted bifaces made from nonlocal cherts also supports a settlement strategy that incorporated frequent residential movement over a large home range (Stafford 1994:232).

George Milner, Victor Thompson, and I observed a similar pattern of Early Archaic land use across the Ohio River in west central Kentucky (Jefferies et al. 2005). Our archaeological survey along Cypress Creek, a tributary to the Green River, revealed that Early Archaic sites usually yielded only a few projectile points, typical of sites attributable to hunter-gatherers who occupied short-duration and widely scattered camps. Early Archaic groups made greater use of upland environmental settings than did later Archaic groups. They favored some locations more than others, but, based on the low frequency of artifacts, they were not tethered to strategic places like some later hunter-gatherer societies. A few places were repeatedly occupied by Early Archaic hunter-gatherers, presumably because nearby resources were sufficiently abundant and diverse to support the periodic aggregation of these normally scattered, mobile bands (Jefferies et al. 2005).

Because of the ephemeral nature of most Early Archaic occupations, some of the best information on intrasite activities occurs in settings where alluvial and/or colluvial deposition buried and protected occupation surfaces shortly after their abandonment. Such conditions are usually restricted to rockshelters and floodplain settings. Living surfaces sealed by these sterile sediments commonly contain artifacts and features associated with a single, or at least short, occupation episode. Even where such ideal situations exist, however, the depth of the deposits often makes exposure of large surface areas difficult and expensive.

The enormous challenge of trying to understand Early Holocene settlement organization in the Lower Ohio Valley is underscored by the archaeological, geomorphic, and chronological complexity of Early Archaic occupation of the Knob Creek Bottom in southern Indiana (Stafford 2004). The Knob Creek Bottom and its deeply buried archaeological deposits were the focus of the Caesars Archaeological Project in the late 1990s (Stafford 2004). Excavations conducted on several different floodplain landforms disclosed a series of deeply buried Early Archaic occupations.

Isolated large surface hearths and associated small lithic scatters found along most of the exposed portion of the levee indicate that Early Archaic hunter-gatherers briefly camped on these elevated floodplain features, as well as on nearby point bars, while they hunted, fished, and collected the abundant wetland resources. Associated projectile points date these occupations to around 10,000 B.P. (Stafford 2004).

Nearby, on the same levee, but at a slightly higher elevation, project archaeologists exposed a Thebes/St. Charles (ca. 10,000 to 8000 B.P.) component marked by several small (ca. 2 m^2) circumscribed debitage scatters and at least three tool reduction areas representing a lithic workshop. The ephemeral nature of these short-term occupations is typical for the region, contributing to our poor understanding of Lower Ohio Valley Early Archaic hunter-gatherers (Stafford 2004).

A thick, high-density Kirk Corner Notched zone lies just 40 m west of the Thebes workshop, closer to a relic bank of the Ohio River. The close horizontal positioning of components, temporally separated by roughly 1,000 years, is indicative of the complex floodplain topography and of the age differences that can occur within small lateral distances (Stafford 2004).

The Kirk component, representing at least two high-density occupations, included more than 160 oxidized and unoxidized pit features, tool caches, more than 10,000 stone tools, and abundant debitage. A bank-side secondary trash deposit covered in excess of 35 m along the Ohio River levee. Cultural materials from this zone indicate that from about 9,400 to 8,700 years ago, the levee was the scene of numerous residential camps where flaked stone tool production and maintenance were very important activities. Of note, project archaeologists found thousands of Kirk artifacts in a very narrow strip along the levee crest that could have been easily missed using more typical site discovery techniques, underscoring the difficulty of finding buried components in these complex topographic settings (Stafford 2004).

Archaeologists investigated similar kinds of Early Archaic occupations, consisting of features and Kirk projectile points, during their excavation of Zone VII at Longworth-Gick. Zone VII feature clusters contained various kinds of burned areas and charcoal filled pits. The absence of a paleosol and the rapid rate of alluvial deposition suggest that these Early Archaic feature clusters are attributable to brief episodes of intensive human activity rather than an extended period of activity on a stable landscape (Collins 1979b:582).

The Longworth-Gick and Knob Creek Bottom examples of Early Archaic floodplain activity suggest that while many sites located adjacent to major drainages hosted only brief occupations characterized by a few features and artifacts, others indicate much more intensive activity. A similar picture comes from the Green River region where most Early Archaic occupations in both upland and wetland settings were characterized by a few artifacts. However, a few river-edge sites like Butterfield contained substantial Early Archaic material, suggesting intensive and/or repeated use of nearby resources (Jefferies et al. 2005).

Some of the most informative data on Early Archaic activities in upland settings comes from stratified rockshelters. Rockshelters are particularly common in parts of the Lower Ohio Valley where there are major outcrops of sandstone or limestone bedrock, like the Shawnee Hills of southern Illinois. Typical of these kinds of sites is the Hills Branch Rock Shelter that formed in a thick sandstone formation in Pope County. Although exceptions do occur, Early Archaic rockshelter use like that at Hills Branch is usually interpreted as consisting of brief, seasonal encampments (Wagner and Butler 2000).

The environmental setting of the Hills Branch Rock Shelter would have been very attractive to Early Archaic hunter-gatherers because it was located close to several distinctive plant communities that included swamps and other wetlands, dry uplands, and heavily dissected hills and valleys. Sources of high-quality cherts were also located nearby (Butler and Wagner 2000:168–170).

Like the floodplain sites just discussed, evidence for Early Archaic use of the Hills Branch Rock Shelter consisted of Kirk Corner Notched projectile points dated to ca. 9200 B.P. Many of the numerous Kirk points were damaged and missing their distal tips. Other associated artifacts included small triangular endscrapers and a t-shaped drill. Chert debitage attributes suggested that the site's Early Archaic inhabitants conducted both bifacial and core reduction activities while they were visiting Hills Branch. Unlike during the rockshelter's earlier Dalton occupation, Early Archaic hunter-gatherers made most of their tools from locally available Kinkaid chert. The absence of tools made from Mounds Gravel chert suggests that Early Archaic hunter-gatherers were spending much more of their time in the local area rather than making frequent trips to the river valley like their Dalton predecessors (Butler and Wagner 2000:171).

As with most other Early Archaic sites, researchers could not identify complete Early Archaic assemblages at Hills Branch because a variety of natural and cultural processes had mixed the shelter's deposits. Nevertheless, the presence of reworked and discarded projectile points, scrapers, and drills suggested that the rockshelter was more than just an overnight stopping point. The high frequency of projectile points relative to other kinds of tools may reflect the efforts of Early Archaic hunters to repair their broken or worn-out hunting/butchering equipment. Based on this evidence, site investigators interpreted the rockshelter as a short-term base of operations from which hunting trips were initiated and where tool kit repair/refurbishing and initial hide processing was done (Butler and Wagner 2000:172).

Additional information on Early Archaic settlement and activities away from the Ohio River floodplain comes from the Ashworth Cave, a rockshelter located in the Knobs region of north central Kentucky. The rockshelter formed in a limestone outcrop along a tributary to the Salt River. Excavation within and outside the shelter disclosed a buried cultural deposit containing Early Archaic Ashworth Corner Notched points, morphological correlates of Kirk Corner Notched points. Based on dates associated with similar projectile points from other Lower Ohio Valley sites, archaeologists estimated that Ashworth Cave's Early Archaic occupation dated from 9900 to 9500 B.P. (DiBlasi 1981:54).

Ashworth Cave's Early Archaic features consisted of two irregularly shaped burned areas surrounded by limestone slabs and the remains of surface fires that would have served as a source of light and heat for shelter inhabitants. The abundance of chert debitage and the presence of chert bifaces, scrapers, denticulates, and a spokeshave; hammerstones; and bone needles, pins, and a bead suggest that Early Archaic hunter-gatherers performed a wide range of extractive and maintenance tasks while visiting the rockshelter. An Early Archaic burial near the shelter's back wall indicates that activities were not restricted to secular tasks (DiBlasi 1981).

Subsistence

The Lower Ohio Valley's Early Holocene people probably never noticed the extinction of the last remaining Pleistocene megafauna about 10,000 years ago. For the previous several thousand years, these mammals had little dietary significance to regional hunter-gatherers. Unfortunately, because of climatic and soil conditions, well-preserved Early Holocene botanical and faunal specimens are seldom found in the Lower Ohio Valley or most other parts of eastern North America, so we continue to have a poor understanding of Early Archaic subsistence practices. The scant data that do exist indicate that these groups practiced a "diffuse" subsistence strategy reflecting continuity, in many respects, with the diets of both earlier Paleoindian Dalton and later Middle and Late Archaic groups (Meltzer and Smith 1986:19).

Early Archaic hunter-gatherers exploited the Lower Ohio Valley's plant and animal communities using a mobile foraging pattern that included base camps located in areas of reliable, abundant, and predictable subsistence resources, as well as smaller extractive camps established near seasonally specific food sources. Nuts, particularly acorn and hickory, became an increasingly important part of the Early Archaic diet (Gremillion 2003:31–32). Early Archaic people probably ate many leafy plant parts, but there is no record of their dietary contributions.

Existing Early Archaic subsistence data suggest a broad-based utilization of both closed canopy climax forest resources, like gray squirrel and nuts, along with species that inhabited recently disturbed edge and open areas, such as cottontail rabbits, white-tailed deer, and pioneering starchy and oily seed-bearing plants (Meltzer and Smith 1986:17–18). Smith indicates that nearly 20 species of seeds, berries, and nuts are represented in archaeobotanical assemblages from nine Southeastern Early Holocene sites including acorn, hickory nut, beechnut, hazelnut, black walnut, chestnut, and hackberry, persimmon, copperleaf, pigweed, peppervine, goosefoot, bedstraw, maygrass, pokeweed, knotweed, purslane, sumac, blackberry/raspberry, and grape seeds (Smith 1986:Table 1.1).

Generalized Early Archaic tool assemblages containing similar tools as those made by Late Paleoindian hunter-gatherers support a model of subsistence continuity during the Early Holocene. Tools directly associated with plant utilization, such as manos, metates, and nutting stones, are sometimes found on Early Archaic sites, reflecting the dietary importance of plant foods (Meltzer and Smith 1986:17–18).

Some of the best information on Early Archaic subsistence in the southern Midwest comes from the deeply stratified Modoc Rock Shelter in southern Illinois (Styles et al. 1983). Although not in the Lower Ohio Valley, Modoc Rock Shelter is located along the eastern edge of the Mississippi River floodplain approximately 180 km north of the Ohio-Mississippi River confluence. Because of Modoc's proximity to the western end of the Lower Ohio Valley, plant exploitation strategies represented there should approximate those pursued by the Early Archaic hunter-gatherers who lived in southern Illinois and western Kentucky, if not the entire study area.

Analysis of Modoc's carbonized plant remains revealed that Early Archaic cultural deposits contained higher percentages of nut remains, particularly black walnut, pecan, and hazelnut, than found in the site's Middle and Late Holocene deposits. The exceptionally high percentage of pecan is replicated at other nearby sites, possibly reflecting the ease with which these nuts could be processed compared to varieties with thicker shells (hickory and walnut). The increased importance of thick-shelled hickory nuts at the end of the Early Holocene may reflect the adoption of more efficient ways to process them (Styles et al. 1983:284–286).

Modoc Rock Shelter's Early Archaic inhabitants hunted a variety of mammals, particularly small varieties like gray squirrels, cottontail rabbits, and muskrats. The white-tailed deer, while important, played a much smaller role in Early Archaic subsistence than it did in the diets of the subsequent Middle and Late Archaic inhabitants. Researchers who have studied Early Archaic faunal collections from across much of the Midsouth have documented a similar preference for small mammals, particularly squirrel, over white-tailed deer relative to later periods (Styles and Klippel 1996:133).

Early Archaic people occasionally fished while they were at Modoc Rock Shelter. However, the relatively low percentage of fish bone suggests that the backwater lakes and swamps that were so intensively exploited by later Middle and Late Archaic hunter-gatherers had not yet fully developed (Styles and Klippel 1996:117–118).

Analysis of subsistence data from the famous Koster site, located in the lower Illinois River valley approximately 275 km northwest of the Ohio-Mississippi River confluence, revealed that Early Archaic inhabitants also used a wide variety of plants and animals. After about 7500 B.P., hunter-gatherers began to focus their plant and animal procurement efforts on hunting the white-tailed deer, collecting freshwater mussels, and gathering hickory nuts (Neusius 1986:136).

Early Archaic subsistence data from the Lower Ohio Valley proper is even more limited. At sites having sealed Early Archaic cultural deposits like Longworth-Gick, poor preservation conditions severely reduce the likelihood of finding botanical and faunal remains that can inform us about Early Archaic plant exploitation strategies (Lannie 1979:986). Occasionally, we get little hints at what foods were important, like at Paddy's West in southern Indiana where a feature yielded large quantities of black walnut shells. Black walnut also has been occasionally reported from other Early Archaic sites in eastern North America. One feature at Paddy's West contained a possible maygrass seed (Smith 1995b:331–332).

Early Archaic faunal material from Ashworth Cave suggests that hunter-gatherers were focusing their foraging activities on aquatic and forest communities. Mammals comprised 22 percent of the faunal material, followed by reptiles (7 percent), fish (5 percent), birds (4 percent), and amphibians (1 percent). Animal exploitation strategies included hunting deer, turkey, squirrel, and raccoon; collecting terrestrial and aquatic turtles; and fishing for drum fish (DiBlasi 1981:82). These species served as important sources of dietary protein and fat, as well as providing bone, antler, and teeth for making tools and ornaments; skins and hides for shelter coverings and clothing; and feathers for decorative purposes.

Social Organization and Paleodemography

Like the Paleoindian hunter-gatherers who preceded them, the high level of mobility practiced by Early Archaic hunter-gatherers, the small size of their residential groups, and their great antiquity combine to limit what we know about their social organization, the kinds of trauma they experienced, their life expectancy, or the diseases that afflicted them. What information we do have comes from studying the placement and contents of what are usually isolated burials found at scattered open-air and rockshelter sites. Like the Late Paleoindian Sloan cemetery in northeast Arkansas (Morse 1997a), many Early Archaic burials are marked only by the burial goods placed with them, the skeletal remains having long since decayed.

Lower Ohio Valley Early Archaic mortuary programs included both cremations and inhumation burials. Early Archaic mortuary activity at the Jerger site, located along the West Fork of the White River in southwestern Indiana, focused on cremation (Tomak 1979). Investigations there revealed four oval basin-shaped pits on a sand ridge that overlooked what was once a marsh. The pits, containing cremated human skeletal remains and associated cultural material, ranged from .9 to 1.8 m long and from .5 to 1.2 m wide. Agricultural activities had disturbed the upper portions of the features, leaving only the bottom 10 to 12 cm of the pits intact (Tomak 1979, 1983).

Cultural material associated with the cremated human remains included bifurcate base MacCorkle-like projectile points; bifaces; red ocher chunks and powder; thin, flattened bone segments; and marine shell beads. The plow-disturbed soil surrounding the features contained similar kinds of artifacts, including several flaked stone drills, and drilled or perforated animal canine teeth (Tomak 1979:62–67).

Most of the artifacts associated with the cremations were burned, indicating that they accompanied the deceased when they were cremated. The burial pits showed no evidence of burning, so the individuals were probably cremated somewhere else and then buried on the ridge. Red ocher was part of the mortuary ritual, as indicated by red stains found on some of the bones and in the pits. The perforated teeth and shells may have been parts of ornaments worn by the deceased or attached to their clothing (Tomak 1979:67–68).

In the absence of other kinds of features and artifacts, it appears that Early Archaic activity at Jerger focused on mortuary ritual. The marine shell beads are one of the earliest examples for the acquisition of exotic (nonlocal) materials in this part of the Lower Ohio Valley. Even at this early date, a few hunter-gatherers were participating in social networks that extended across major portions of the North American Midcontinent and beyond.

Indiana archaeologists recently documented a similar form of Early Holocene mortuary behavior at the McCullough's Run site, located along the East Fork of the White River (Cochran et al. 1998). The mortuary area, consisting of 10 shallow pits situated along the edge of a sand dune complex, contained the cremated remains of infants, children, and adults. Some pits held the remains of just one person; others contained up to three. Like at Jerger, many individuals were cremated somewhere else, then buried in the shallow pits. Associated artifacts included St. Albans and MacCorkle bifurcate base projectile points, chert bifaces, red and yellow ocher, a sandstone abrader, and several bone tools, many of which were severely heat damaged during cremation. Acceptable radiocarbon dates associated with the cremations ranged from 8130 +/– 70 to 7590 +/– 90 B.P. (Cochran et al. 1998).

Findings from the Jerger and McCullough's Run sites suggest that some people associated with the Bifurcate tradition preferred to dispose of their dead by first cremating them, then burying their remains atop sand dunes bordering wetlands. However, at McCollough's Run variation in the treatment of individuals and their personal possessions suggests subtle differences between the MacCorkle and St. Albans mortuary programs (Cochran et al. 1998). The selection of areas bordering wetlands for mortuary purposes may indicate the importance of these kinds of places in the lives of Early Holocene hunter-gatherers. This preference becomes much more pronounced during the late Middle and Late Holocene.

Early Archaic people did not always bury their dead at special mortuary sites like Jerger and McCollough's Run. Investigations at Ashworth Cave revealed the skeletal remains of an Early Archaic woman buried in a midden near the back wall of the cave (DiBlasi 1981). The woman, probably in her late twenties, was buried in a loosely flexed position with her face turned down. She apparently met her death as a consequence of the large Early Archaic projectile point that was deeply embedded in the body of her third thoracic vertebra. The projectile entered the woman's back, splitting the vertebra's neural arch between the left superior and inferior articulating surfaces and the spinous process. The weapon entered the woman's body with enough force to split the vertebra in half, resulting in a wound that was sufficiently severe to have caused her immediate death. The exact cause of death could have come from blood loss, shock to the nervous system, paralysis of the muscles used for breathing, or a combination of all three (DiBlasi 1981:74–75). The Ashworth Cave burial demonstrates that Early Archaic hunter-gatherers buried their dead in places that were also used for general domestic activities, and that some of these people met violent deaths at the hands of other hunter-gatherers.

The Ashworth Cave Early Archaic burial also provides a glimpse, albeit a very narrow one, of the health status of an Early Holocene woman. Examination of the woman's dentition revealed that her teeth were in an advanced stage of deterioration. She suffered from extreme dental attrition and large abscesses affected many of her teeth (DiBlasi 1981).

Extreme dental attrition and excessive abscessing like that exhibited by the Ashworth Cave woman are common among Lower Ohio Valley hunter-gatherers (Bassett 1982:1075; Ward 2005). Such extreme rates of wear are often attributed to the use of stone tools for food preparation, resulting in the inclusion of small grit particles in the food that gradually wear away a tooth's chewing surface (Bassett 1982:1075).

In addition to dental problems, the woman also exhibited postcranial abnormalities that included a crushed eleventh thoracic vertebra, resulting in degeneration and lipping of the adjacent vertebrae. She also had a vestigial pair of ribs attached to her twelfth thoracic vertebra (DiBlasi 1981:73–74).

Yet another example of Early Archaic mortuary behavior comes from the previously discussed Lawrence site in western Kentucky (Mocas 1977, 1985). Excavation disclosed an oval pit containing the loosely flexed skeletal remains of two individuals estimated to be between 22 and 28 years old. One individual was associated with a heavily utilized and resharpened Kirk Serrated projectile point and a cache of eight flaked stone tools consisting of Kirk Corner Notched projectile points, drills, spokeshaves, and scrapers. These implements resembled others found at the Lawrence site, except that they had not been extensively resharpened. Artifact placement suggested that implements had been intentionally placed with the individual at the time of its interment.

Both individuals wore necklaces made of perforated dog canines and beaver incisors. Like at Jerger, members of the burial party sprinkled red ocher on the skull of one of the deceased (Mocas 1985). The association of Kirk projectile points with one of the individuals indicates an Early Archaic temporal placement; however, charcoal from the feature yielded a slightly later radiocarbon date of 7325 B.P. (Mocas 1985:84).

The Early Archaic Kirk component (Stratum 4) at the Morrisroe site has also yielded several human burials. One of these, the partial remains of a young adult male who was placed on his back with his head to the west, was discovered during a site inspection in 1984 (Nance 1986a:20).

The general absence of Early Archaic burials severely affects our ability to say much about how small Early Holocene hunter-gatherer groups were organized. Clearly, a variety of options existed as to how to dispose of a deceased person's physical remains. Sometimes people chose to bury their dead in pits. Other times, the dead were first cremated, along with some of their personal possessions, and then buried. Artifacts accompanying these individuals usually consisted of what most people would call *utilitarian* items—that is, those implements used during the course of everyday life. In contrast, the beads and perforated animal teeth were probably parts of ornaments or were attached to clothing. Most of the burial items were made from locally available materials; however, the marine shell beads buried with the Jerger site cremation represent one of the earliest examples for the exchange of "exotic" materials from distant places, a practice that would become much more common later in the Holocene. The Ashworth Cave burial reminds us that Early Holocene hunter-gatherers were sometimes victims of interpersonal violence. It appears that people were victims of such acts, regardless of age or gender, even at this early stage in the peopling of the Americas.

Exchange and Interaction

Evidence for exchange most often occurs as raw materials or finished goods that are found outside of the geographic area where they naturally occur. At first glance, this may seem like a good criterion; however, it does not hold for all instances. For example, Early Archaic hunter-gatherers sometimes made projectile points from high-quality lithic materials, much like those used by earlier Paleoindian people. In many cases, the source of this high-quality stone is hundreds of kilometers from where the expended biface was finally discarded. Most researchers attribute the presence of these "exotic" cherts far from their place of origin to large hunting territories and a high degree of group mobility, not individual or intergroup exchange (Kelly and Todd 1988:237). In other words, hunter-gatherers collected this high-quality stone when they were near its source during the normal course of their seasonal movements (Goodyear 1979). Nevertheless, the possibility always exists that some trading partners might have exchanged small quantities of high-quality stone and/or finished tools from time to time.

Not surprisingly, convincing evidence for the exchange of nonlocal materials during the Early Archaic is rare. This is partially attributable to low population densities, small group sizes, and high group mobility that makes evidence of most Early Archaic activities difficult to identify. In addition, the greater social and economic independence of Early Holocene hunter-gatherer groups, compared to later groups, meant that there was less need to develop and/or maintain strong intergroup social networks. Consequently, there was less interaction among groups, including the exchange of goods, to reinforce those social connections. The intergroup interaction that did take place probably occurred during regularly scheduled trips to group aggregation camps where people could find mates and take part in various social integration activities (Anderson and Hanson 1988; Sassaman 1995:179).

Evidence of Early Archaic exchange and interaction in the Lower Ohio Valley and nearby areas is restricted to small marine shell beads. For example, investigators found four nearly identical marine shell beads in Early Archaic cultural deposits at Modoc Rock Shelter. The beads were manufactured from the shells of *Marginella apicina,* the common Atlantic marginella, which lives hundreds of kilometers away along the south Atlantic and Gulf coasts (Ahler 1991). An Early Archaic cremation pit at the Jerger site in southern Indiana also yielded marine shell beads (Tomak 1979, 1983:70). The five beads, associated with Early Archaic bifurcate base projectile points, appear to have been made from the shells of *Detracia bullaoides,* a type of gastropod found along Florida's east and west coasts (The Academy of Natural Sciences 2006). The pointed tip of the shells had been removed, suggesting that the beads were once part of a necklace or perhaps were attached to an article of clothing. These incipient Early Archaic networks formed the foundation for the much more intensive and diversified group interaction that developed during the late Middle Holocene and early Late Holocene.

Summary

The Early Holocene marked the transition from the glacial or peri-glacial environment present during the Late Pleistocene to more moderate climatic conditions approaching those of today. Gone were the species of megafauna for which the Pleistocene is well known, along with the cold-adapted plant communities that supported them. Simultaneously, animals and plants that now populate the modern landscape expanded in range and number, filling in the void left by extinct or relocated Pleistocene species.

The actions of hunter-gatherers who were responding to the rapidly changing physical environment shaped the Lower Ohio Valley's Early Holocene cultural landscape. These groups were still highly mobile and occupied extensive home ranges or hunting territories, but the rising regional population meant that their space was more circumscribed than before.

In most cases, people were uniformly distributed across the landscape, with no particular habitat being more extensively utilized than another. Archaeologists have found little to indicate that groups stayed in one place very long. A few places probably had more cultural significance than others, particularly locations where they buried their dead.

Early Holocene technology also reflects a mobile lifestyle, largely consisting of the tools needed to acquire and process food. We know very little about the social organization of these groups, but kinship and marriage probably strongly influenced a person's or group's social relationships and cultural identity.

The Early Holocene hunter-gatherer way of life would continue relatively unchanged for another 1,500 years or so, extending well into the Middle Holocene. As we will see in the next chapter, however, by around 6,500 years ago, the Lower Ohio Valley's physical and cultural landscapes experienced changes that helped shape the way people lived until the arrival of the first European explorers.

6 / The Middle Holocene

Settling Into the Valley

Introduction

Lower Ohio Valley archaeologists traditionally refer to the time from 8,000 to 5,000 years ago as the Middle Archaic subperiod. As with the other Archaic subdivisions, the precise starting and ending dates for the Middle Archaic varies from one locality to another across the Midcontinent (Emerson et al. 2009). Much of this inconsistency is based on the differential timing for the appearance of cultural traits that we use to define the Middle Archaic (Muller 1986:57).

The Middle Archaic subperiod coincided with much of the Middle Holocene (ca. 8000–5000 B.P.). Environmentally, the Middle Holocene is sometimes referred to as a "window of adjustment" when the postglacial environment stabilized, hill and slope sedimentation rates decreased, stream and river channels adjusted to newly forming floodplains, and critical resource zones developed (Schuldenrein 1996:3). Collectively, these conditions contributed to the gradual establishment of backwater swamps, oxbow lakes, and shoal habitats that provided abundant and diverse aquatic plant and animal foods that Lower Ohio Valley hunter-gatherers could easily exploit (Smith 1986:23–24). By the end of the Middle Holocene, the Lower Ohio Valley, as well as the rest of eastern North America, was experiencing essentially modern environmental conditions (Schuldenrein 1996:3). Ultimately, these environmental changes had a major impact on the number, kind, and distribution of the plants, animals, and people that lived on the Lower Ohio Valley's evolving landscape.

Evidence for Middle Holocene environmental change comes from a variety of sources, but the fossil pollen record is the primary contributor. Pollen profiles, like those from Old Field Swamp (King and Allen 1977) in southeastern Missouri and Jackson Pond (Wilkins et al. 1991:236) in Kentucky, indicate that the Lower Ohio Valley experienced major vegetation changes between 9,000 and 3,000 years ago. Old Field Swamp experienced its driest conditions about 7000 B.P., reflected by the transition of dense bottomland forest into savannah-like grasslands. By about 5,000 years ago, more moist conditions developed, resulting in the gradual return of the forests (King and Allen 1977).

Archaeological data indicate that by this time, regionally distinct archaeological cultures were developing throughout the eastern United States. Variation among these cultures, attributable to the different physical and cultural worlds in which they were evolving, is reflected archaeologically by a variety of technological, settlement, subsistence, and social characteristics (Jefferies 1996b:47). Although these cultures developed along generally similar trajectories, their levels of social complexity were quite variable.

The Middle Archaic also was a time of major transition for Lower Ohio Valley hunter-gatherers. For approximately the first 1,500 years, small, highly mobile hunter-gatherer groups, whose material cultures and adaptive strategies closely resembled those of Early Archaic groups, populated the region. In many ways, the division between the Early Archaic and early Middle Archaic subperiods artificially separates societies that were probably organized along very similar lines of social, economic, and technological complexity.

Our knowledge of early Middle Archaic (ca. 8000–6500 B.P.) societies and lifeways has been hampered by relatively few radiocarbon dates, a poor idea of associated projectile point forms, the highly mobile nature of early Middle Archaic hunter-gatherers, and the burial of their sites under thick alluvial and colluvial deposits (Stafford and Cantin 2009). Fortunately, archaeological investigations along the Ohio River in western Kentucky (Morrisroe site) and southern Indiana (Caesars Palace complex) are starting to provide a clearer picture of what these societies were like (Nance 1986a, 1988; Stafford and Cantin 2009). Much of what we know about early Middle Archaic hunter-gatherers is based on the distribution of the few known diagnostic projectile points found during regional-scale surveys and data from a few deeply stratified sites along the Ohio River.

Throughout much of the North American Midcontinent, including the Lower Ohio Valley, early Middle Archaic settlement systems consisted of numerous small sites in different resource zones where a few people lived for a short time. In many respects, this pattern mimics that of the Early Archaic, presumably because early Middle Archaic groups used similar adaptive strategies. Investigation of early Middle Archaic settlement-subsistence strategies to the north in the Lower Illinois River Valley (Brown and Vierra 1983:190) suggests that small hunter-gatherer groups exploited seasonally available resources through high residential mobility. Stafford (1994) documented a similar settlement-subsistence strategy for early Middle Archaic groups that lived in southwestern Indiana. This type of settlement-subsistence organization has traditionally been associated with a "forager" adaptive strategy (Binford 1980). Early Middle Archaic foragers appear to have organized their annual movements so that the group arrived at a resource point or zone as its seasonal resource(s) became available.

Data on late Middle Archaic (ca. 6500–5000 B.P.) societies have dramatically increased over the past 30 years, resulting in a much-improved picture of the development of more complex hunter-gatherer societies in the Lower Ohio Valley. This information comes from numerous research projects conducted throughout the Midcontinent that have included both cultural resource management-related and grant-funded investigations (Bader and Granger 1989; Brown and Vierra 1983; Cook 1976; Granger 1988; Janzen 1977; Jefferies and Butler 1982; Lewis and Lewis 1961; Munson 1980a; Nance 1988; Stafford 2004; Stafford et al. 2000; Styles et al. 1983). Figure 6.1 shows the locations of Lower Ohio Valley sites that have yielded important data on Middle Archaic hunter-gatherers.

Data from these and other sites suggest that by about 6500 B. P. some Lower Ohio Valley hunter-gatherers significantly altered their settlement-subsistence strategies, resulting in decreased mobility, smaller home ranges, and the establishment of multiseasonal base camps (Brown 1985). These changes are expressed on the landscape by the numerous large shell, rock, and earth middens that dot the edges of the Midcontinent's resource-rich wetlands (Milner 2004:37–42). The groups that lived at these sites exhibited characteristics usually associated with more culturally complex societies, such as substantial habitation structures and storage facilities, plant domestication, cemeteries, and interregional exchange networks (Brown 1985; Jefferies 2004). In some cases, decreased mobility is reflected in their lithic assemblages by a smaller proportion of chert from nonlocal sources compared to that found in most Early Archaic tool kits (Cantin 1989; Jefferies 1982a:1508). Collectively, these data suggest that many late Middle Archaic groups were shifting from a residential mobility strategy associated with foraging to a logistical strategy (Binford

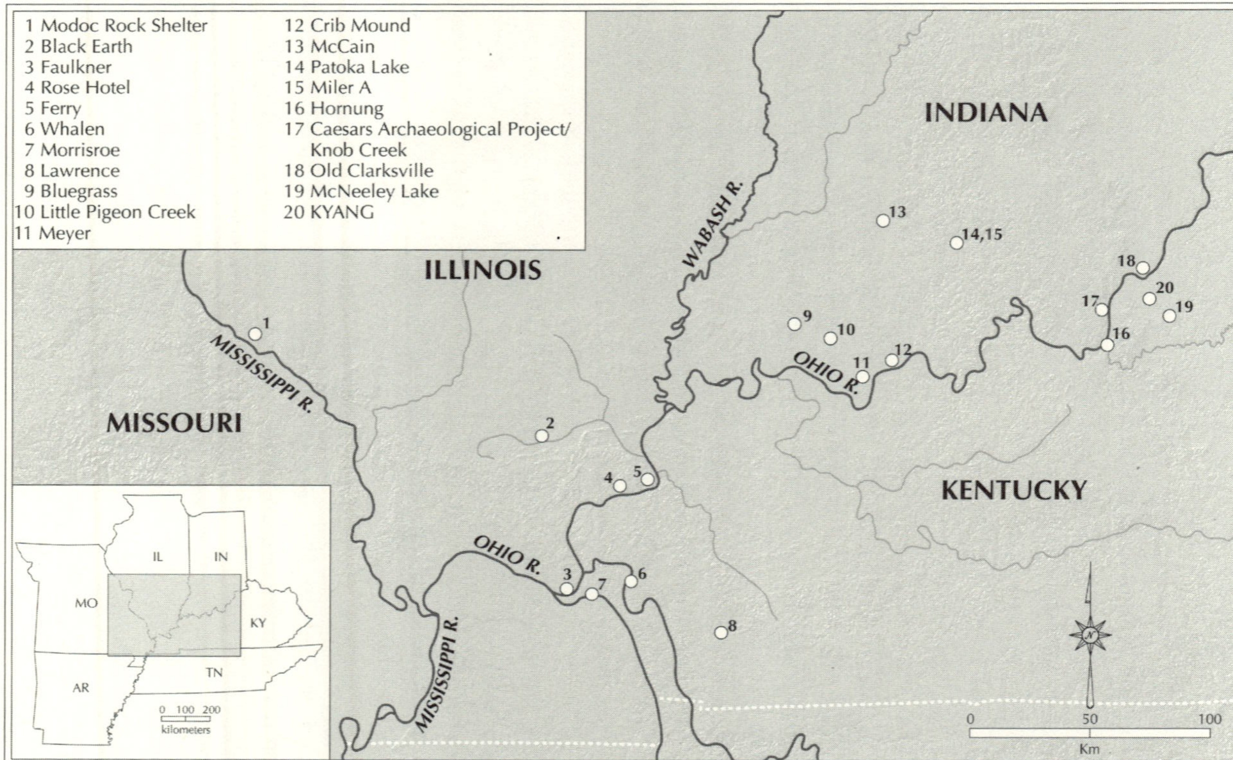

Figure 6.1. Important Lower Ohio Valley Middle Archaic sites.

1980) commonly used by collectors (Stafford 1994:222–223). In pursuing this new strategy, small task groups would have exploited different resource zones from a centrally located base camp.

Explanations for why these changes in mobility occurred at this time vary widely, but usually fall into two categories—environmental or social factors. In reality, it was probably a combination of many factors, including these two, that brought about this major change in lifestyle. The association of the Hypsithermal's drier, warmer climatic conditions, decreased use of the uplands, and the establishment of base camps, many of which were positioned adjacent to wet areas, has led some researchers (Jefferies 1983; Nance 1988) to emphasize environmental factors. Following this scenario, decreased upland productivity caused by drier conditions would have "forced" or "pushed" hunter-gatherer groups out of the uplands and into the wetter river valleys (Brown 1986:323). The establishment of base camps that were strategically placed adjacent to rivers, swamps, and other wetlands would have allowed late Middle Archaic groups to exploit upland and wetland habitats simultaneously, ensuring an abundant, reliable, and diverse food supply (Jefferies 1997:467).

In contrast, James Brown (1985:220; Brown and Vierra 1983) proposed that Middle Holocene environmental changes led to an increase in productivity of aquatic habitats that actually "pulled" late Middle Archaic hunter-gatherers to the Midwest's major river valleys. The increased abundance, diversity, and reliability of aquatic habitat foods contributed toward this shift away from a highly mobile way of life and toward a more sedentary existence. As a cautionary note, however, the presence of base camps in places that lacked aquatic habitats (Lewis 1983) indicates that this process was not solely tied to increased aquatic productivity (Stafford 1994:232).

Brown (1985:222–223) also maintained that social risk factors had a significant role in the reorganization of Middle Archaic settlement-subsistence systems. He suggested that the attractiveness of resource-rich aquatic and circum-aquatic loci promoted sedentism as late Middle Archaic hunter-gatherers discovered that the costs of decreased mobility were compensated for by a lessening of intergroup conflict. Therefore, increased sedentism was a consequence of efforts to maintain intergroup spacing when resource abundance made this a viable option (1985:224).

The gradual shift toward a more sedentary way of life, larger residential groups, and smaller home ranges also brought about changes in the technological, economic, social, and political spheres of hunter-gatherer culture. Among the most important of these new challenges was the maintenance of intergroup social connections. The loss of risk-reducing measures commonly used by mobile hunter-gatherers to maintain social ties was compensated for by the development of broader social networks through which late Middle Archaic people maintained the flow of information, resources, and mates needed to ensure their existence (see Anderson 1996b:52). Over time, these evolving networks appear to have served as the foundation for larger, more complex regional social groups that inhabited extensive territories both north and south of the Ohio River (Jefferies 2004).

Material Culture and Technology

Like Early Holocene hunter-gatherers, most of the evidence for Middle Archaic technology comes from flaked stone artifacts and the by-products of their manufacture, use, and rejuvenation. In some cases, excavations of Middle Archaic sites have yielded sizeable inventories of ground stone artifacts (Jefferies 1982b), and, where good preservation conditions existed, bone and antler artifacts and their production residues (Breitburg 1982). Middle Holocene hunter-gatherers also made objects from copper, shell, and other exotic (nonlocal) materials, but little is known about the specifics of their manufacture. Cultural features, like cooking and storage pits, can provide additional insights on hunter-gatherer technology. I will discuss these kinds of facilities later in this chapter.

Flaked Stone Technology

Over the past 75 years, archaeological investigations at a number of sites in the Midcontinent have contributed valuable data on Middle Archaic hunter-gatherer technology. Just like the Early Archaic, some of these were floodplain sites (J. Chapman 1973; Coe 1964; Nance 1988) or rockshelters (DeJarnette et al. 1962; Fowler 1959) where thousands of years of deposition sealed and separated discrete Middle Archaic occupations. Additional data came from the excavation of extensive shell and earth middens that appear in the archaeological record for the first time during the Middle Archaic (Bader and Granger 1989; Janzen 1977; Jefferies and Butler 1982; Lewis and Lewis 1961; Stafford et al. 2000).

As did Early Archaic flaked stone tool kits, Middle Archaic assemblages include an assortment of notched and stemmed projectile points (Figure 6.2; Table 6.1). There was a general trend toward the increased use of local lithic resources due to regional population increase and decreased group mobility. In

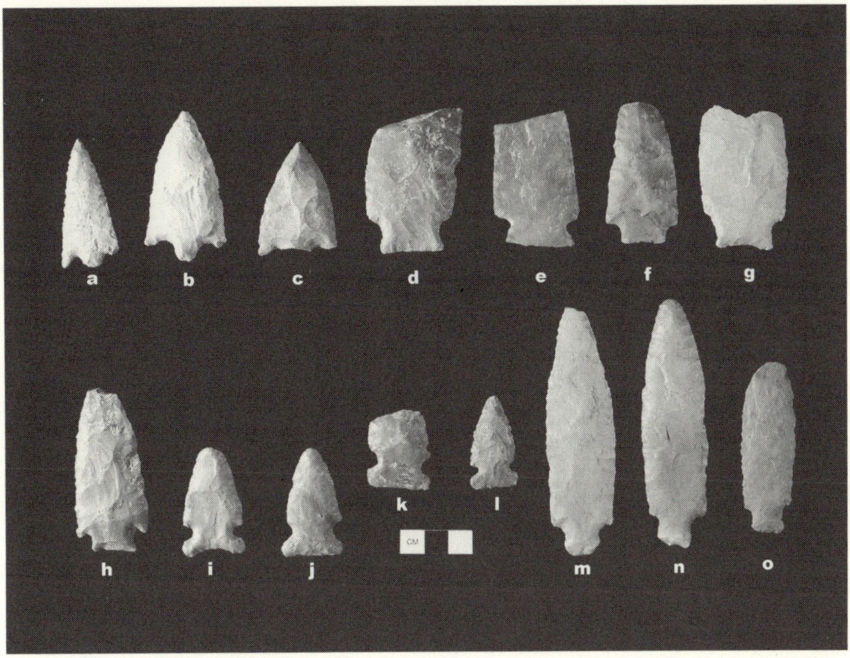

Figure 6.2. Examples of Lower Ohio Valley Middle Archaic hafted bifaces—Eva II (a–c), Benton and Benton/White Springs (d–h), Big Sandy II/Godar (i), Matanzas Side Notched (j), Big Sandy II/Godar (k–1), Elk River (m–o) (William S. Webb Museum of Anthropology, University of Kentucky).

addition, the reliance on heavily curated formal tools decreased, as reflected by the gradual abandonment of strategies, like beveling, that were designed to extend biface use-life (Amick and Carr 1996:44–45). At the same time, there was an increased emphasis on expediently produced tools.

Throughout much of the region, components dating to the first half of the Middle Archaic (8000–6500 B.P.) are very poorly represented and spotty in their distribution. Archaeological surveys of the Cypress Creek watershed in the lower Green River drainage of west central Kentucky revealed only two early Middle Archaic components, compared to 27 dating to the late Middle Archaic (Jefferies et al. 2005). Researchers working in southern Indiana (Stafford 1994:227) also rarely report early Middle Archaic components.

Part of the reason that Lower Ohio Valley early Middle Archaic components have been hard to find is because many are deeply buried in alluvial or colluvial deposits along the river margins. Perhaps more importantly, it is often difficult to assign projectile points associated with those occupations to previously de-

Table 6.1. Lower Ohio Valley Middle Archaic Projectile Point Types

Projectile Point Type	Date Range	Sources
Stanly Stemmed	8000–7000 B.P.	Coe 1964:35–36; Justice 1987:97
Cypress Creek I and II	7800–7000 B.P.	Justice 1987:71; Lewis and Lewis 1961:37,39–40; Nance 1986a:26
Knob Creek Stemmed	7200–6700 B.P.	Stafford and Cantin 2009
Eva Cluster		Lewis and Lewis 1961:40
Eva I	8000–6000 B.P.	Justice 1987:100
Eva II	6000–4000 B.P.	Justice 1987:103
Morrow Mountain Cluster	7000–6000 B.P.	Coe 1964:37–43
Morrow Mountain I		Justice 1987:105
Middle Archaic Side Notched Cluster	10,000–4500 B.P.	May 1982:1358–1361
Graham Cave Notched	10,000–7500 B.P.	C. Chapman 1975:248–249; Justice 1987:62–66
Big Sandy II	8000–4500 B.P.	Justice 1987:68–69; Lewis and Lewis 1961:34–37; May 1982:1359
Faulkner Side Notched	7000–5000 B.P.	May 1982:1360; Winters 1967:18–19, 23
Brannon Side Notched	7000–5000 B.P.	Cook 1976:148–149; May 1982:1360
Godar Side Notched	8000–4500 B.P.	Cook 1976:146; May 1982:1359–1360
Matanzas (Modal and Deep Side Notched)	5700–4500 B.P.	Cook 1976:126, 140–145; May 1982:1360–1362; Munson and Harn 1966
Salt River Side Notched	5700–4500 B.P	Janzen 1977; Munson and Cook 1980c:723; Justice 1987:124
Hidden Valley Stemmed	8900–8000 B.P.	C. Chapman 1948; Justice 1987:84; May 1982:1361–1362
Shouldered (Stemmed) Karnak	5700–5000 B.P.	Cook 1976:138–140; Justice 1987; May 1982:1362; Winters 1967:19,25

Projectile Point Type	Date Range	Sources
Benton Cluster		
Benton Stemmed	6000–4000 B.P.	Hofman 1984; Justice 1987: 111, 113; Kneberg 1956: 25–26; Lewis and Lewis 1961: 34; Meeks 1999
Elk River Stemmed	5500–4000 B.P.	DeJarnette et al. 1962; Justice 1987:111–112, 114
White Springs Cluster		
Sykes	6500–5000 B.P.	Hofman 1984; Justice 1987:108; Lewis and Lewis 1961:40–43; Meeks 1999
White Springs	6500–5000 B.P.	DeJarnette et al. 1962:70; Hofman 1984; Justice 1987: 108–110; Meeks 1999

fined types. C. Russell Stafford (2004) suggests that the stylistic uncertainty of early Middle Archaic projectile points is a significant factor in the underrepresentation of sites dating to the 8000–6500 B.P. era. In southern Indiana some points from deeply buried early Middle Archaic deposits, if found on the surface, would be easily confused with earlier or later types.

In contrast, Nance's (1988) research along the Lower Tennessee-Cumberland rivers located several sites containing thick early Middle Archaic middens reflecting relatively intensive activities. Apparent differences in site distribution and occupational intensity suggest that considerable variation existed in how early Middle Archaic hunter-gatherers exploited different parts of the Lower Ohio Valley.

Early Middle Archaic projectile points include the Stanly, Cypress Creek, Eva, and Morrow Mountain types (Justice 1987). Most of these were defined based on research in Tennessee and North Carolina, and, consequently, the Lower Ohio Valley lies along the northern and western margins of their distributions. In many cases, these types are more frequently found on Kentucky sites than on sites north of the Ohio River in southern Indiana and Illinois (Justice 1987:Figures 42–45).

The excavation of deeply stratified sites in the Lower Tennessee-Cumberland region has shed new light on the temporal placement of many of these point types (Nance 1986a, 1988). At Morrisroe, Jack Nance identified a substantial early Middle Archaic Cypress Creek–Eva component (Stratum 3) dating from approximately 7800 to 7100 B.P. (1986a:25). He noted that Cypress Creek points, morphological correlates of the Kirk Corner Notched type (Justice 1987:71), tended to occur in deeper levels than did the Eva points. Stratum 3 basally notched points most closely resembled the Eva II point type (Lewis and Lewis 1961). Stratum 3 also yielded a few Morrow Mountain and Kirk Corner Notched points, and several stemmed varieties associated with the upper part of the stratum (Nance 1986a:23).

Nance's Morrisroe research demonstrated that basally notched Eva points were manufactured nearly 700 years earlier than previously thought, pushing their first appearance in the Lower Ohio Valley to approximately 7800 B.P. Dates associated with Cypress Creek points (7800–7000 B.P) suggested that this type spanned the typological/temporal gap between earlier Kirk and Bifurcate point types and later Eva types (Nance 1986a:25–26).

At the Lawrence site in western Kentucky, Cypress Creek projectile points were associated with radiocarbon dates ranging from ca. 7500 to 7300 B.P. (Mocas 1977). The occurrence of Kirk Stemmed/Serrated points suggests that they were manufactured well into the early Middle Archaic subperiod.

In southern Indiana, archaeological investigations conducted as part of the Caesars Archaeological Project identified several early Middle Archaic components deeply buried in Holocene alluvial deposits. Excavations placed along a .5-km section of the levee indicated that it started forming about 7500 B.P and became stable by about 2500 B.P. (Stafford and Cantin 2009).

Project archaeologists identified three separate stratified Middle Archaic occupation zones within the levee deposits, characterized by surface hearths and clusters of flaked stone tools and debitage. Two zones yielded artifacts associated with the early Middle Archaic Knob Creek complex, dating from ca. 7200 to 6700 B.P. (Stafford and Cantin 2009). Surprisingly, most of the 66 Knob Creek points did not conform to any extant point types. The most common form, known as Knob Creek Stemmed, was a small, short biface having either a shallow concave or straight base. The concave base variety resembled the Stanly Stemmed type (Coe 1964), but was smaller and some 600 years younger. With respect to size, they more closely resembled Bifurcate cluster points, like Kanawha or St. Albans (Broyles 1971; Justice 1987:246). Other Knob Creek points, if found in a surface context, would most likely be confused with Terminal Archaic Riverton points (Stafford 2004).

Two Kirk Corner Notched-like points, dated to 6942 +/– 60 B.P and resembling Middle Holocene Amos projectile points, further cloud the waters of early Middle Archaic projectile point stylistic variability. Amos points, once thought to be part of the Kirk Corner Notched cluster, are now considered a Middle Archaic type. It seems that, at least in southern Indiana, early Middle Archaic projectile points include the newly recognized Knob Creek Stemmed point, along with corner-notched and Bifurcate specimens (Stafford and Cantin 2009).

Due to the difficulty in finding single component occupations, little is known about the technological organization of early Middle Archaic tool kits. However, exceptions do occur. At Morrisroe, researchers used radiocarbon dates and temporally diagnostic artifacts to define three early Middle Archaic artifact assemblages (H–J) (Conaty 1985:Table 33). Using assemblage attributes, Conaty was able to study the contents of each assemblage and identify patterns of diachronic variability in the activities of the site's early Middle Archaic inhabitants.

Early Middle Archaic assemblages contained an abundant and diverse inventory of extractive, processing, manufacturing, and maintenance tools, along with their production residues (Conaty 1985:Appendix D). Flaked stone artifacts included gravers, notches, unhafted and hafted bifaces, drills, steeply retouched flakes, and reworked bifaces. Production residues consisted of cores, angular fragments, cortical and secondary flakes, bifacial thinning flakes, and bipolar flaking by-products. Associated ground stone items included mortars, pestles, pitted and battered cobbles, and a piece of cannel coal.

Conaty's analysis of Archaic technological organization suggested that assemblages dating to the earlier part of the Middle Archaic contained large amounts of debitage representing the entire biface production trajectory. In contrast, later Middle Archaic assemblages had shorter trajectories suggesting a more restricted range of production and maintenance activities. Earlier groups apparently used more ground stone and pecked-battered artifacts than did later ones.

Early Middle Archaic knappers relied on local St. Louis/Salem cherts to make most of their flaked stone tools. These same sources accounted for nearly 80 percent of all early Middle Archaic debitage (Conaty 1985:Table 24). Cherts from outside the Lower Tennessee-Cumberland region were rare and there was no evidence suggesting that inhabitants selected certain chert types to make specific kinds of tools. Collectively, these findings suggest an expedient use of lithic raw materials. Assemblage attributes indicate that early Middle Archaic hunter-gatherers occupied Morrisroe for relatively long periods and that they conducted a wide range of activities while they were there (Conaty 1985: 339–340).

Late Middle Archaic projectile points are much better documented than those of the early Middle Archaic. Some late Middle Archaic middens have yielded literally hundreds of hafted bifaces reflecting the intensity of occupation at these sites (Jefferies 1982b).

Many late Middle Archaic projectile points fall into what Ernest May (1982: 1358) called the Archaic Side-Notched cluster based on his study of the Black Earth site bifaces. He assigned four point types to this cluster—Graham Cave Notched, Godar, Faulkner, and Matanzas. The Godar and Matanzas types were very common at the Black Earth site, the Faulkner and Graham Cave types much less so. The Graham Cave, Godar, and Faulkner types share common notch, lobe, and stem edge characteristics.

Although these types co-occur on many late Middle Archaic sites, the deeply side notched types with squared ears, like Godar and Big Sandy II, are thought to have been manufactured earlier than the shallow side notched types like Matanzas. For example, at the Knob Creek site, the cultural zone immediately above the Knob Creek component yielded a deeply side notched Big Sandy II point associated with a radiocarbon date of 5830 +/– 90 B.P. Although evidence is limited, it appears that, at least in southern Indiana, points resembling the Big Sandy II type can be used to define a pre–French Lick phase dating before 5500 B.P. (Stafford and Cantin 2009).

The morphology of the Matanzas type, which consists of several variants itself (Cook 1976), is distinctly different from that of the other three types. These points were found in direct association with Godar and Faulkner points in the Black Earth site's Middle Archaic zone (Zone 3). Godar points were numerous in the lower and middle part of Zone 3, while Matanzas and Faulkner points were most common in the middle to upper levels (May 1982:1358). Other contemporary, morphologically similar projectile points are often included in the Archaic Side Notched cluster, including Brannon and Salt River Side Notched.

Late Middle Archaic projectile points also include a variety of stemmed bifaces, most of which date to the later part of the subperiod. Among these are the Karnak type and the Benton and White Springs clusters. May (1982:1362) defined the Shouldered Karnak as a variant of Howard Winters's (1967) Karnak Stemmed type. At Black Earth, the Karnak Stemmed type occurred in the Late Archaic levels, but the shouldered variety was associated with the late Middle Archaic zone.

Although not common at Black Earth, White Springs and Benton projectile points are sometimes associated with Lower Ohio Valley late Middle Archaic occupations. White Springs points, along with the slightly later Sykes point, are common in the Tennessee River valley, but also occur in western Kentucky and, occasionally, in southern Indiana (Justice 1987:110). Examples of the Elk River Stemmed type, a variant of the Benton cluster, also occur throughout this area (Justice 1987:111–112, 114).

Stylistically similar projectile points, including Matanzas, Big Sandy II, Karnak, and a variety of straight and expanding stem types (referred to as M-B-K-S points), are associated with the French Lick phase in southern Indiana (Munson and Cook 1980c). Archaeologists working at other southern Indiana sites like Crib Mound, McCain, Bluegrass, and many sites in Kentucky and Indiana near the Falls of the Ohio, also report similar kinds of late Middle Archaic points (Janzen 1977; Stafford and Cantin 2009). Thomas Cook (1976) identified similar styles at contemporary sites in the Lower Illinois River valley. Based on their examination of numerous radiocarbon dates associated with components yielding M-B-K-S projectile points, Stafford and Mark Cantin (2009) suggest that the French Lick phase dates from ca. 5300 to 4500 B.P.

Perhaps the clearest picture of late Middle Archaic flaked stone technology comes from the Black Earth site's late Middle Archaic component (Zone 3) (Jefferies 1982b; Morrow 1982). Excavation of Zone 3's thick, dark, organically rich midden revealed that about one meter of the deposit formed during an approximately 1,000-year period of very intensive late Middle Archaic activity (Figures 6.3 and 6.4). A suite of eight radiocarbon dates placed the time of most intensive activity from approximately 6000 to 4900 B.P. (Jefferies 1982a:Table 479). The nearly nine months that SIUC archaeologists spent investigating the late Middle Archaic midden yielded approximately 1,100 flaked stone tools, nearly 20,000 pieces of debitage, and hundreds of bone and antler artifacts. As such, it represents the largest collection of late Middle Archaic cultural material in the Lower Ohio Valley and serves as a good indicator for the kinds of activities that Middle Holocene hunter-gatherers performed at large, multiseason base camps like Black Earth.

The Black Earth site's late Middle Archaic flaked stone tool assemblage consisted of projectile points, scrapers, drills, gravers, and perforators that site inhabitants used for many different extractive, processing, manufacturing, and

Figure 6.3. Aerial photograph of the Black Earth site midden (courtesy of the Center for Archaeological Investigations, Southern Illinois University Carbondale).

Figure 6.4. 1978 excavations of Black Earth site—Area A (courtesy of the Center for Archaeological Investigations, Southern Illinois University Carbondale).

maintenance tasks. In addition, it also contained many other kinds of flaked stone artifacts, like cores, bifaces, and flakes associated with the production and maintenance of the flaked stone implements (May 1982; Morrow 1982).

Hafted bifaces (projectile points, hafted endscrapers, and hafted drills) formed the major component of the late Middle Archaic tool kit. These implements exhibited relatively little stylistic variability, most conforming to one of the notched or stemmed projectile point types (Godar, Big Sandy II, Matanzas, etc.) discussed above. Matanzas points comprised about 70 percent of the side notched points; approximately 21 percent were Godar/Big Sandy II types. Smaller Faulkner projectile points accounted only for about 6 percent of the total (Jefferies 1982b:Table 62).

Approximately one-third of all late Middle Archaic flaked stone tools were side- or endscrapers, representing a consistent percentage of the tool inventory throughout Zone 3. Many endscrapers had side notched haft elements suggesting that they were either recycled broken Godar or Matanzas projectile points or that some endscrapers were specifically manufactured to be used as hafted tools. The very steep working edge of many endscrapers (greater than 70 degrees) suggests that they are associated with a variety of heavy-duty scraping tasks (Jefferies 1982a:1470, 1990a). Similar hafted endscrapers, along with similar evidence for recycling, occur at other late Middle Archaic sites throughout the Lower Ohio Valley (Hensley 1996:104).

Microscopic examination of use-wear and damage on the edges and surfaces of 20 hafted endscrapers indicated that the distal end of most specimens, that is the part used for scraping, was rounded and polished (Keeley 1981). This kind of polish is usually caused by scraping dry hide, suggesting that the site's late Middle Archaic inhabitants used these implements to curry deer hides during the later stages of tanning. Once hide processing was completed they would have used the tanned hides to make shelter covers, clothing, bags and pouches, or other items (Jefferies 1982b, 1990a).

Other kinds of edge wear suggested that late Middle Archaic hunter-gatherers used hafted endscrapers for woodworking. Edge wear indicating the cutting of meat might reflect the use of these implements as projectile points and/or knives prior to their being recycled into endscrapers. Hide, bone, and antler polish near the base of the scrapers marks where a wooden or bone handle was once attached (Jefferies 1990b).

Late Middle Archaic people also made end- and sidescrapers that were not intended to be hafted. These implements were usually manufactured from large flakes that could be easily held in the hand. Although no edge wear data are available for the Black Earth specimens, these tools were probably used to do the same kinds of tasks as their hafted counterparts. The similarity of edge angles on hafted and unhafted endscrapers supports this interpretation (Morrow 1982:1317). Based on the high percentage of scrapers in the late Middle Archaic tool assemblage, butchering deer and processing their economically important hides was a major activity at the Black Earth site during the late Middle Holocene.

Other kinds of flaked stone tools (drills, gravers, denticulates, and notches) collectively accounted for just over 10 percent of the late Middle Archaic tool kit. The low frequency of these implements suggests that their associated tasks comprised only a small part of site activities during this time—at least compared to those for which projectile points and scrapers were used.

The percentages of the different tool types found in each of the four late Middle Archaic midden zones (3A–3D) were fairly consistent. Some of this homogeneity may be due to the continual mixing of the midden as site inhabitants excavated numerous storage, processing, and burial pits. A more likely scenario is that the kinds and intensity of activities conducted at the Black Earth site during the late Middle Archaic occupation remained relatively consistent.

As illustrated above, many of the late Middle Archaic flaked stone artifacts were tools—most of them discarded or broken implements that had reached the end of their use-lives. In contrast, many other flaked stone artifacts were the products (bifaces) or by-products (core, debitage) of biface production (Morrow 1982).

The Black Earth site assemblage contained more than 800 unhafted bifaces representing various parts of the biface production trajectory (Morrow 1982). Some bifaces had been further modified for use, but many were just fragments representing failures that broke during production. Thin bifaces (preforms), representing the later stages of biface production, outnumbered thick bifaces (blanks), associated with the early stages of production, by a ratio of almost 9 to 1. This was not unexpected since much of the evidence for early-stage bifaces would have been lost as they progressed through the biface production process.

Carol Morrow's (1982) comprehensive study of the late Middle Archaic flaked stone tool manufacturing trajectory provided more important insights about tool production activities at the Black Earth site. Her investigations showed that finished bifacial tools (projectile points, endscrapers, and drills) outnumbered unfinished bifaces (failed blanks and preforms) by 9 to 1. This ratio again suggests that while some flaked stone tool manufacturing was done at Black Earth, its material correlates accounted for only a small portion of the late Middle Archaic artifacts (1982:1299).

The technological analysis of thousands of pieces of chert debitage provided additional insights into late Middle Archaic stone tool manufacturing at Black Earth (Morrow 1982:1298–1301). Cores, angular fragments, and percussion flakes indicated that late Middle Archaic hunter-gatherers did do some primary stage lithic reduction; however, secondary stage shaping activities, represented by bifacial chert preforms and bifacial thinning flakes, accounted for the majority of the late Middle Archaic debitage. Morrow contended that late Middle Archaic chert knappers brought chert preforms to the site where they further shaped them to produce tools on an "as needed" basis. Overall, the production of flaked stone tools appears to have been a relatively minor late Middle Archaic activity at Black Earth (1982:1300). Instead, site inhabitants apparently focused most of their efforts on the use and revitalization of tools, as indicated by the large number of finished, broken, discarded, and recycled implements (Morrow 1982:1317).

Information resulting from the flaked stone technological analysis, combined with data on the chert sources exploited by the Black Earth site's late Middle Archaic inhabitants, helped to develop a clearer understanding of hunter-gatherer mobility and economic organization (Morrow 1982). Specifically, the restricted geological distribution of some southern Illinois cherts made it possible to evaluate how different chert types were procured and used in the technological system (Morrow and Jefferies 1989).

Late Middle Archaic inhabitants made most of their flaked stone tools from chert that originated in the eastern Shawnee Hills (St. Louis, Ft. Payne, Kinkaid, and Mounds Gravel) approximately 30 km southeast of the Black Earth site. They also used other chert types (Cobden-Dongola, Kaolin, Mill Creek) that came from more distant sources located 60 to 70 km to the southwest in the western Shawnee Hills.

The high-quality western Shawnee Hills cherts found at the Black Earth site raised questions concerning how they were obtained and their technological role(s) (Morrow and Jefferies 1989). If the western cherts were obtained by someone traveling to the source area specifically to collect chert, the time and energy expended to do this might have made western Shawnee Hills cherts more "valuable" to Black Earth site inhabitants. Hypothetically, because of their greater "cost," they may have been used for specialized tasks or used more conservatively than local cherts. Yet, if acquiring western Shawnee Hills chert was "embedded" (Binford 1979:260–261) in the normal movement of hunter-gatherers across the southern Illinois landscape, then there should have been no additional "costs" and the chert probably would have been used and conserved in the same manner as cherts found closer to the site.

Analysis of tools and debitage demonstrated that the Black Earth site inhabitants used local and nonlocal cherts in a technologically similar manner. Western Shawnee Hills chert was not used to make any specific kinds of tools, and, once on site, it was not conserved to any greater extent than chert from more local sources. These findings suggested the "exotic" cherts were collected by individuals or groups during their normal movement throughout southern Illinois. The acquisition of western Shawnee Hills cherts as part of an "embedded" procurement system would explain why local and nonlocal cherts were used in the same technological capacity; that is, the costs of procurement for all cherts were essentially the same (Morrow and Jefferies 1989).

The Black Earth site's late Middle Archaic flaked stone industry closely resembles those associated with other nearby late Middle Archaic and early Late Archaic cultural manifestations including the Faulkner "focus" or "complex" (MacNeish 1948; Muller 1986:59–60) in southern Illinois, the French Lick phase (Munson and Cook 1980c) in southern Indiana, and the Old Clarksville phase at the Falls of the Ohio (Granger 1988; Janzen 1977). The Helton phase (Cook 1976) occupation at the Koster site (Horizon 6) in the lower Illinois valley and Strata 4–5 at Modoc Rock Shelter in southwestern Illinois (Styles et al. 1983) also yielded comparable materials.

Ground Stone Technology

Late Middle Archaic hunter-gatherers used their flaked stone technology to make many of the implements in their tool kits, most of which they used to kill animals or to process their meat, hide, fur, and bones. They produced other important technological items, along with some that functioned in the social realm, using ground stone technology.

The production of ground stone implements is thought by many to mark the transition from the more focused Late Pleistocene/Early Holocene hunting economies to the more generalized economies of mid-Holocene hunter-gatherers. Apparent increases in the number and diversity of ground stone tools is considered good supporting evidence for an increasingly diversified economy, including a greater importance of plant foods (see discussion by Sassaman 1996:57–58).

Bruce Smith (1986:18) has argued, however, that in the Southeast, there was no significant increase in the frequency of informal ground stone tools associated with plant food processing (pitted stones, grinding stones, manos, metates, etc.) from the Early to the Middle Holocene. Formal ground stone tools, like grooved axes and pestles, also had Early Holocene antecedents. Overall, Smith (1986:21) sees little evidence suggesting that Middle Holocene hunter-gatherers possessed increasingly diverse tool kits or that any "revolutionary" technological innovations took place.

Kenneth Sassaman (1996) has recently proposed that while the antecedents of many technological innovations attributed to Middle Holocene hunter-gatherers originated thousands of years earlier, and that mid-Holocene economies were similar to those of the Early Holocene, Middle Holocene technology did differ in several significant aspects. First, Middle Holocene hunter-gatherers improved the production efficiency and use of many innovations that Early Holocene hunter-gatherers first developed. Second, some of these technologies held conspicuous roles in the organization or control of labor. Third, the patchy distributions of some of these innovations, both temporally and spatially, reflect patterns in the social integration of regional populations, as well as basic differences in economic pressure (1996:57–58). Among these innovations were items like spearthrowers or atlatls, grooved axes, and soapstone cooking vessels. Importantly, the manufacture and distribution of these three items were closely tied to the social networks through which distant groups and individuals interacted to reinforce their long-term social and economic relationships (1996:73–74).

Although few studies of ground stone tool function have been carried out in the eastern United States, the morphological attributes of many "rough" or "informal" ground stone tools, like pitted cobbles, grinding stones and slabs, manos and metates, and mortars, suggest that they were used to process seeds, nuts, and other plant parts and products (Smith 1986:18). They also could have been used to grind minerals or other hard materials. Others kinds of ground stone tools were associated with a variety of abrading, polishing, pulverizing, percussion, and perforating tasks. Hunter-gatherers also fashioned an assortment of "paraphernalia," like atlatl weights or bannerstones, plummets, and personal ornaments, using ground stone technology (Adams 2002).

Like flaked stone tools, some ground stone items were shaped for use by flaking, pecking, and/or grinding (formal tools), while others were shaped as a consequence of being used (informal tools). At the Black Earth site, late Middle Archaic hunter-gatherers made many formal tools from hard metamorphic rock that they collected from nearby glacial till deposits. They manufactured most of their informal tools from sandstone that was locally available along the banks of the nearby Saline River (Jefferies 1982b:231, 1987:50).

The Black Earth site's formal ground stone tool inventory included grooved axes, atlatl weights/bannerstones, and plummets. Although detailed use-wear studies are lacking (Sassaman 1996:66), grooved axes were probably used to perform a variety of heavy-duty woodworking tasks. However, they certainly were not the first tools that Southeastern hunter-gatherers used to do this kind of work (Sassaman 1996:64; Smith 1986:18). Examples of much earlier Dalton and Early Archaic chipped stone axes and adzes abound in the archaeological record. Fully grooved ground stone axes first appeared in the Midwest at the Koster site about 7000 B.P., followed by three-quarter grooved varieties about 1,000 years later (Helton phase) (Brown and Vierra 1983:183).

The Black Earth hunter-gatherers made and used only fully grooved axes. While not common, they occurred throughout the late Middle Archaic midden suggesting that they were a consistent component of late Middle Archaic technology (Jefferies 1982b:234). The relatively small number of grooved axes is not surprising, however, since their manufacture was a long, involved, and labor-intensive process, meaning that they were probably highly curated. Little of the bit portion remained on most specimens, indicating that many of the axe heads represented discarded tools that had reached the end of their use-life. A miniature grooved axe, one of many objects found in what was probably a late Middle Archaic "medicine bundle" (Jefferies 1987:Plate 3-21), suggests that these implements were occasionally used for ritual purposes.

Atlatls or spearthrowers, indicated by a stylistically and lithologically diverse array of ground stone atlatl weights or "bannerstones," had become an integral part of Lower Ohio Valley hunting technology by the Middle Archaic (Burdin 2004; Lutz 2000) (Figure 6.5). This temporal placement was confirmed at the Morrisroe site where an atlatl weight associated with Eva, Morrow Mountain, and Kirk points was found in a deeply buried Middle Archaic stratum (ca. 7500 B.P.) (Conaty 1985:Appendix D, Table 5). Artifact hunters have reported dozens of additional fragments eroding from the same cultural zone at Morrisroe (Lutz 2000:113, 165).

Ground stone atlatl weights occur throughout the Lower Ohio Valley, particularly at large midden-bearing sites where they were often placed with human burials (Lutz 2000). During the early days of Ohio Valley archaeology, the function of atlatl weights, as well as other atlatl parts, was the subject of considerable debate (Moore 1916). It was not until William Webb's comprehensive study (1957) of atlatl parts (handles, weights, and hooks) found in many

Figure 6.5. Artist's interpretation of a Middle or Late Archaic hunter using an atlatl (adapted from Webb 1957:Figure 7, William S. Webb Museum of Anthropology, University of Kentucky).

of the Green River Archaic middens that the argument was finally put to rest. Webb observed that the arrangement of handles, weights, and hooks found in some burials conclusively demonstrated that they were parts of spearthrowers (Webb 1946:159–168, Figure 18A).

Ground stone atlatl weights were made from a variety of local (sandstone) and exotic (marble, slate, quartzite) materials. Their shapes were quite variable, ranging from relatively plain to highly stylized (Lutz 2000), but determining the age of specific atlatl weight styles has been somewhat problematic. Dating problems are largely attributable to the scarcity of atlatl weights and the lack of examples from well-dated contexts. Despite these problems, some general trends have been detected (Burdin 2004:64–69; Kwas 1982; Lutz 2000).

In his analysis of Lower Ohio Valley atlatl weights, Rick Burdin (2004:64) noted that the early Middle Archaic atlatl weight sample was small and consisted of only five types (crescent, reel, shuttle, paneled, winged) (Figure 6.6), with "winged" varieties accounting for nearly 60 percent (Lutz 2000:Figure 142). The hunter-gatherers who made these atlatl weights showed a preference for softer stone, like slate and claystone, only occasionally using granite, quartzite, and other kinds of hard stone (Burdin 2004:65). Examples of the Slate

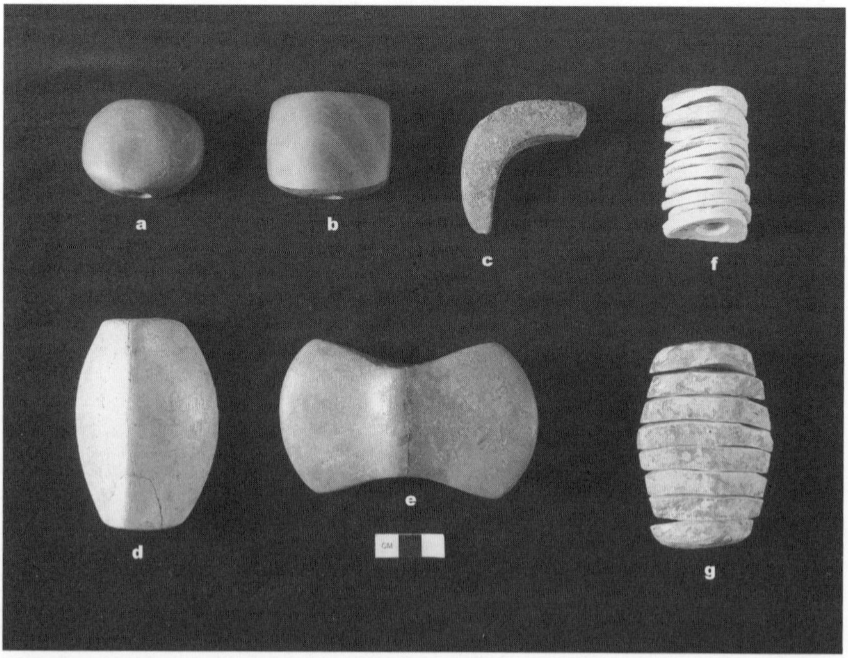

Figure 6.6. Probable Middle Archaic bannerstone/atlatl weights—groundstone (a–e) and marine shell (f–g) (William S. Webb Museum of Anthropology, University of Kentucky).

Butterfly and Crescent types were associated with the Morrisroe early Middle Archaic occupation (Lutz 2000:165).

Both the number of atlatl weights and the types represented increased during the late Middle Archaic. The "humped" and "tube" types account for nearly 75 percent of the 119 specimens analyzed. Winged varieties decreased to only 4 percent, and new types that became popular during the Late Archaic, like the "bottle" form, made their first appearance. Like earlier varieties, late Middle Archaic atlatl weights were most commonly made from slate, but new materials like limestone and marine shell started to appear (Burdin 2004:63–69).

Apparently, the ground stone atlatl weight was not a particularly critical part of the Black Earth site's late Middle Archaic tool kit since project archaeologists found just nine fragments. Several were of the "tubular" style; others were "winged" or "prismatic." Some of the Black Earth specimens resembled examples from the Ferry site (Fowler 1959) in Hardin County, Illinois (Jefferies 1982b:236).

An unfinished atlatl weight from the Black Earth site provided some insights as to how they were made. First, a piece of stone was roughly shaped by grinding to establish the atlatl weight's general form. Once shaping was completed, a hole was drilled through the center using a reed in combination with sand. Holes in other weights appear to have been made using a stone drill (Webb 1974:268). Drilling of the hole prior to completion of the atlatl weight suggests that this was a risky job that was done before expending more time and energy doing the final shaping of the weight.

The Black Earth site ground stone tool inventory also included many informal tools like grinding slabs, grinding stones, pitted cobbles, abraders, and hammerstones. Many of these implements (grinding slabs and stones, pitted cobbles) could have been used for plant food processing, although other functions cannot be excluded. Site inhabitants probably used sandstone abraders for shaping and sharpening bone and antler items, like awls and pins. Hammerstones could have been employed in a variety of capacities that required the direct application of force, such as manufacturing flaked stone tools.

Ground stone artifacts associated with extractive tasks (grooved axes, axe parts, and atlatl weights) represented approximately 9 percent of the Black Earth site's late Middle Archaic ground stone tool inventory. Tools used to manufacture or maintain other tools (hammerstones, grooved abraders) represented 31 percent. Finally, tools that were probably associated with food processing, particularly seeds and nuts, account for 60 percent of all functionally diagnostic ground stone implements (Jefferies 1982a:1472).

Comparison of the Black Earth site's late Middle Archaic ground stone tool inventory with those from roughly contemporaneous Lower Ohio Valley sites indicates a similar range of tool diversity and associated activities. For example, the University of Chicago's 1944 excavations at the nearby Faulkner site yielded a sizeable ground stone tool assemblage that included what Richard MacNeish called grinding stones, grinders, mortars, pestles, anvils, and atlatl weights (Cole et al.1951:219; MacNeish 1948:238–239, Figure 47). Although the names that MacNeish assigned to these tool types are slightly different, their descriptions generally match those of many of the Black Earth tools. Similar types of ground stone tools (hammerstones, manos, nutting stones, axes) from the Miler A site in southern Indiana, associated with the slightly later French Lick phase, suggest that Lower Ohio Valley hunter-gatherers relied on a generally similar ground stone tool inventory once a pattern was established during the Early to Middle Archaic (Munson and Cook 1980c:727).

Bone and Antler Technology

Stone tools represented a very important part of the Middle Archaic hunter-gatherer tool kit; however, due to issues of preservation, their relative importance may be highly inflated. Where ideal preservation conditions exist, such as in anaerobic environments, very alkaline soils, or extremely dry places, evidence for objects made from much more perishable organic materials can often be found.

Although much more durable than wood, leather, or plant fibers, artifacts made from animal bone, antler/horn, and teeth are seldom preserved in the Lower Ohio Valley's acidic soils. Where soil conditions are suitable, however, like in the very alkaline soils found in many late Middle Archaic middens (Woods 1982), the importance of bone and antler tools in hunter-gatherer technologies is made abundantly clear. For example, of the 56,487 cataloged artifacts from the five largest Green River sites, 23,614 (ca. 42 percent) are bone, antler, or dental tools (C. R. Moore, personal communication 2008). Analyses of antler and bone tool inventories from Lower Ohio Valley Middle Archaic sites show that the range of activities with which they were associated replicate many, if not most, of the tasks for which flaked stone tools were used. In addition, Middle Archaic hunter-gatherers used bone and antler to make many kinds of utilitarian, ornamental, or ceremonial objects usually not made from stone, like cups, beads, and rattles.

The largest, most diverse collections of late Middle Archaic bone and antler artifacts come from deep midden deposits like those found at the Black Earth (Breitburg 1982), Kentucky Air National Guard (KYANG) (Bader 1992), Baker (McBride 2000), and McCain sites (Miller 1941). These sites are characterized by very alkaline soils (7.0 to 8.5 pH) that acted to preserve bone that would have quickly decayed at sites having more acidic soils (Woods 1982:1394).

Black Earth site inhabitants made implements and ornaments from the skeletal elements of many species of mammals, birds, and reptiles, but deer bone and antler account for almost one-third of all late Middle Archaic bone artifacts (Breitburg 1982). Site inhabitants modified most of the bone objects so extensively that the animal of origin could not be determined. Numerous sandstone abraders and abundant production residues indicate that many of the bone and antler items were manufactured onsite.

Deer antler was cut using the score-and-snap technique, which involved incising a groove around the antler's circumference, then breaking it along the incision. The resulting tines and shafts were then scraped, abraded, chiseled, and shaved to make a variety of chert-working tools, gouges, projectile points, atlatl parts, cutting and scraping tools, beads, tubes, and cups. Bones from virtually every part of the deer skeleton, particularly forequarter and hindquarter elements, were used to make a variety of awls, needles, pins, and fishhooks (Breitburg 1982).

Middle Archaic hunter-gatherers also made tools from the bones of various small mammals. They made awls from bobcat and dog longbones. They used beaver, woodchuck, and squirrel mandibles for engraving, chiseling, or abrading, and the longbones of aquatic and terrestrial birds, especially turkey, to make awls and fishhooks. They occasionally made bowls, cups, and rattles from turtle shells (Breitburg 1982).

Approximately 70 percent of all functionally diagnostic Black Earth late Middle Archaic bone tools are awls, underscoring the importance of these implements. Awls, made from deer bone or the bones of other large mammals or birds, varied in size and shape, depending on the particular bone from which they were made and the extent to which the awl had been used. Awls were probably used for many different tasks, particularly hide working, basket making, and sewing, which produced a highly polished surface. Other pointed bone objects, sometimes described as awls, may have actually been projectile points (Bader 1992; Webb 1946).

Bone pins, many of which had elaborately carved heads and engraved shafts, represented about 13 percent of the bone tool assemblage. Most of the Black Earth site bone pins were made from deer metapodial longbone shafts (Breitburg 1982:925). Many of the Black Earth site pins resemble those made at contemporary sites throughout the Lower Ohio Valley and adjacent parts of the Midcontinent, suggesting some degree of interaction among the people who lived at these sites (Jefferies 1997, 2004).

Fishhooks and needles, while accounting for only a small percentage of the bone tool assemblage, also were important parts of the late Middle Archaic tool kit. Fishhooks, largely represented by their production debris, were commonly made from deer ulnae. Needles had distinctive "eyes" drilled in one end of the shaft (Breitburg 1982:922).

Additional information on Middle Archaic bone and antler technology comes from Anne Bader's comprehensive analysis of more than 250 tools from the KYANG site (Bader 1992). While it was difficult to distinguish tools made by the site's late Middle Archaic inhabitants from those of later groups, most of them appeared to originate in the earlier Old Clarksville stratum (1992:67). Bader (1992) identified many of the same kinds of bone and antler implements at KYANG (carved bone pins, awls, chert-working implements, projectile points, etc.) that Breitburg did at the Black Earth site.

Mid-twentieth-century archaeological investigations at the McCain site (Miller 1941), located along the East Fork of the White River in Dubois County, Indiana, also yielded bone artifacts closely resembling those from Black Earth, KYANG, Miler A, and other late Middle Archaic midden sites

(Munson and Cook 1980c:727, Table 28.1). Antler artifacts included tine and butt flakers, and projectile points. Awls, both splinter bone and tubular, were the most common bone tool. Longbones of deer and other mammals were split and shaped to make splinter bone awls; tubular awls were made from bird leg and wing bones, particularly from the wild turkey and heron (Miller 1941:38).

McCain inhabitants also manufactured needles from bone splinters, similar to the ones they used to make awls. Needles, in contrast, were carefully shaped so that they had a round or elliptical cross section and a highly polished surface. The proximal ends of most needles were either eyed or notched (Miller 1941:38). Like at Black Earth, fishhooks made from mammal or bird bone were part of the late Middle Archaic tool kit. Inhabitants also made elaborately carved and engraved bone pins, many resembling those from Black Earth, KYANG, and even Koster in the lower Illinois valley (Cook 1976).

Settlement Patterns and Intrasite Organization

Intersite Patterning

In much of eastern North America, the Middle Holocene was a time of significant change in the way people distributed themselves over the landscape (Anderson 1996a; Stafford 1994:221), and the Lower Ohio Valley was no exception (Jefferies et al. 2005; Stafford 1994). In most areas, early Middle Archaic hunter-gatherers, like their Early Archaic predecessors, pursued a highly mobile way of life that required residential groups to make frequent moves to take advantage of seasonally available food sources, acquire other needed goods, and maintain social ties with other people. This adaptive strategy is reflected in the archaeological record by many small sites consisting of light lithic scatters that are widely distributed throughout most topographic settings. Figure 6.7 shows the distribution of recorded Middle Archaic sites across the Lower Ohio Valley landscape.

In many parts of the Lower Ohio Valley, Middle Archaic sites in general are poorly represented. For example, Kentucky site file data indicate that Middle Archaic components are found in comparatively few areas. Where they do occur, they are often clustered in the vicinity of resource-rich riverine areas, like along the Green River and near the mouths of the Tennessee and Cumberland rivers (Anderson 1996a:170).

Sites dating specifically to the early part of the Middle Archaic are very poorly documented (Jefferies et al. 2005; Stafford 1994:227). The low fre-

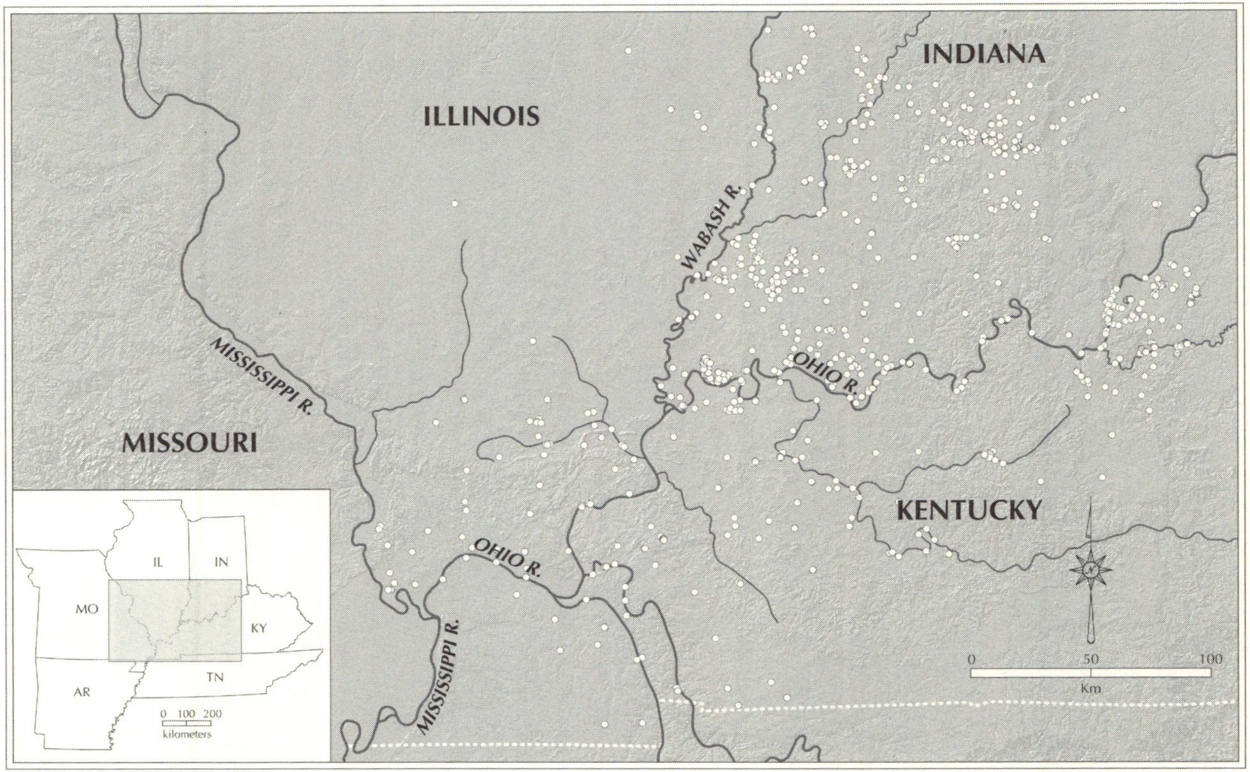

Figure 6.7. Distribution of Middle Archaic sites.

quency of early Middle Archaic sites may be attributable to a generally low population throughout much of the valley at this time, but exceptions to this trend clearly exist (Nance 1986a). Another possible explanation is tied to our inability to identify diagnostic artifacts dating to the ca. 8000–6500 B.P. era (Stafford 1994:227). Importantly, Stafford (2004) has recently identified a number of previously unrecognized early Middle Archaic projectile point types from sealed and well-dated cultural deposits that, when found in surface contexts, might be confused with older or younger point types.

A variety of archaeological indicators suggest that by approximately 6,000 years ago, during the late Middle Archaic or early Late Archaic (depending on local cultural chronologies), some Lower Ohio Valley hunter-gatherer groups were starting to alter the way they moved about the landscape, obtained their food, and exploited the uplands and river valleys. These groups appear to have been shifting from a residential mobility strategy to one emphasizing logistical mobility (Stafford 1994:221). Changes in group mobility strategies appear to be, at least in part, tied to environmental shifts associated with the Hypsithermal interval.

This shift in mobility is reflected by the establishment of multiseason base camps characterized by deep middens, a high diversity of features and tools, burials, and evidence of interregional exchange (Brown 1985). Other kinds of sites also occur, but their wide distribution and generally low artifact content suggest that late Middle Archaic hunter-gatherers occupied them for short periods when performing task-specific activities. The archaeological signatures associated with these increasingly complex hunter-gatherers marked significant changes in their ways of life that set the foundation for future cultural developments.

Lower Ohio Valley sites exhibiting characteristics associated with the emergence of culturally complex hunter-gatherers include the Black Earth site (Jefferies 1982b) in southern Illinois; the Bluegrass (Stafford and Anslinger 1988) and Miler A (Guendling 1980) sites in southern Indiana; and the KYANG (Bader and Granger 1989), Old Clarksville (Granger 1988; Janzen 1977), and numerous other sites near the Falls of the Ohio (Janzen 1977). Occupations at many of the famous Green River middens were also initiated during this time (Jefferies et al. 2005; Marquardt and Watson 2005a:631).

In other parts of the Lower Ohio Valley, some late Middle Archaic hunter-gatherers appear to have maintained a way of life reminiscent of earlier times, continuing an existence based on small group size, high mobility, and a simple, portable technology. Whether this variation in social complexity was by necessity or choice is still open to question (Sassaman 1995:187).

Survey data from the upper Patoka River valley (Patoka Lake) in southern Indiana indicate significant variation in occupation of the region during the Middle Archaic period (Cook 1980b:689). Using standardized measures of land use and demography (number of projectile points/1,000 years), there appears to have been only marginal use of the Patoka Lake area during the early part of the Middle Archaic (ca. 7500 to 6000 B.P.), representing a sharp decrease in activity from the Early Archaic. In contrast, hunter-gatherers extensively utilized the Patoka Lake area during the late Middle Archaic (ca. 6000 to 5000 B.P.) (Cook 1980b:689). Making the jump from numbers of projectile points to Middle Archaic population estimates involves many assumptions, but at least it does give some indication of hunter-gatherer activity within the study area. In addition, it is difficult to determine if the apparent early Middle Archaic absence is a reflection of reality or simply our inability to identify sites dating to that time.

As a point of clarification, Cheryl Munson and Thomas Cook (1980c) defined many of the sites counted as late Middle Archaic at Patoka Lake as belonging to the Late Archaic French Lick phase (5000–3500 B.P.). They based this decision on the results of their initial research that suggested that points dating to the respective periods often co-occurred on sites. Based on later investigations, side notched hafted bifaces (Godar, Matanzas, Big Sandy II, Faulkner) were determined to be older than the stemmed forms (Saratoga, Karnak). Because of these distinctions, side notched hafted bifaces are considered to be late Middle Archaic, while stemmed types are assigned to the Late Archaic (Stafford 1994:227). In addition, many of the "cultural dimensions" (artifact styles, technology, subsistence, etc.) of the French Lick phase are very similar to those of the late Middle Archaic Helton phase (5700–5000 B.P.) and those identified at the Black Earth site. Therefore, for the sake of comparison here, the French Lick phase will be included in the discussion of late Middle Archaic hunter-gatherers.

As during the Early Archaic, French Lick phase sites in the upper Patoka drainage are widely distributed across the southern Indiana landscape. Sixty-three percent are on terraces, with the remainder scattered along floodplains (18 percent), on ridge spurs (7 percent) and slopes (10 percent), and in the uplands (2 percent) (Munson and Cook 1980b:Table 27.3). Fifty-four percent are situated along the upper Patoka River, 10 percent are along large tributaries, and 31 percent along small tributaries (Munson and Cook 1980b:Table 27.6).

Using attributes such as site size, topographic setting, and artifact types represented, project archaeologists identified three French Lick phase site categories in the upper Patoka valley (Munson 1980b). Small hunting camps, located along the Patoka River's tributary streams, were primarily occupied during the fall and winter. Activities conducted at these sites included hunting and butchering of game animals and the refurbishing of tool kits; maintenance and manufacturing tasks were rarely performed.

Other small camps located along the Patoka River exhibited abundant evidence for maintenance activities, but little for nut processing or heat-related tasks. A spring-summer occupation was proposed for this site category.

The third site category consisted of large base camps located along the Patoka River and tributary streams. The abundance and diversity of artifacts and features suggest that late Middle Archaic hunter-gatherers conducted many different kinds of maintenance and manufacturing tasks at these intensively occupied locations. Nut processing and late stage biface reduction were particularly important activities. In contrast, little evidence was found for hunting. Minimally, these sites were occupied in the fall, but based on evidence from other parts of the Lower Ohio Valley they may represent multiseason or year-round occupations (Jefferies 1982b).

Intensive surveys of the Patoka Lake region revealed no French Lick phase mortuary or ceremonial sites; however, shell middens in the lower Patoka River valley and in the nearby White and Ohio river valleys have yielded burials containing French Lick phase artifacts that link them to the Patoka Lake sites. Other evidence from outside the immediate Patoka Lake area indicates that French Lick phase hunter-gatherers used rockshelters during fall and winter hunting forays, and that they established small fishing camps along streams during the summer (Munson 1980b:677–678).

Based on the Patoka Lake settlement data, Munson (1980b:677–680) proposed that French Lick phase hunter-gatherers seasonally occupied the region while they were pursuing game, gathering plant foods, and collecting stone for tool manufacturing. The area was generally devoid of major settlements, serving only to supply seasonally available food to hunter-gatherer groups that based their long-term habitation in other parts of southern Indiana.

Late Middle Archaic settlement in the upper Patoka River valley was put into a broader context by Stafford's (1994) analysis of Archaic landscape use for the entire southwestern part of Indiana. He discovered that a major shift in Archaic landscape utilization took place between Early Archaic and late Middle Archaic times. As illustrated on a small scale by the Patoka Lake study (Munson 1980b), during the Early Archaic, there was more frequent use of areas away from the major drainages, particularly the upper parts of regional drainage networks near basin divides. Early Archaic hunter-gatherers also used major river valley habitats, but less commonly exploited the middle parts of the tributary drainages (Stafford 1994:231).

Starting during the late Middle Archaic, hunter-gatherers placed a greater emphasis on resources located in the lower parts of the drainage systems, as indicated by increasing frequencies of later Archaic hafted bifaces near the major rivers. The most abrupt change took place between the Early and Middle Archaic, with a more gradual but persistent trend through the Late Archaic. Despite the increased emphasis placed on areas near the rivers, hunter-gatherers continued to exploit upland resources throughout the Archaic period (Stafford 1994:231–232).

Archaeologists working in other parts of the Midcontinent have also noted a decreased emphasis on upland habitats during the late Middle Archaic (Ahler 1984; Cook 1976:118; Jefferies 1983; Nance 1988). Many of the proposed explanations for this shift are tied, at least in part, to climatic changes during the Hypsithermal interval, the underlying theme being that drier conditions brought about changes in upland vegetation that resulted in lower food productivity.

As an alternative explanation, Stafford (1994) suggests that changes in population distribution were related to increased nut production associated with the transition from a mixed deciduous to a xeric oak-hickory forest and a more heterogeneous distribution of nut-bearing trees during the Middle Holocene. For example, during the Early and early Middle Archaic, relatively uniformly distributed food resources could be effectively exploited by small groups of highly mobile foragers that moved through multiple watersheds procuring resources on an encounter basis. In contrast, groups practicing a collecting settlement strategy would have most effectively exploited the more heterogeneous distribution of nut-bearing trees during the Middle Holocene. Small, task specific groups traveling from base camps strategically placed along the rivers would have collected scattered upland nut resources, as well as resources near the rivers.

Decreased residential mobility and the repeated occupation of the same base camps resulted in the accumulation of midden, the construction of many features and facilities, the deposition of numerous artifacts, evidence for food storage and processing, and burial of the dead. Over time, these often-occupied places acquired ritual significance to regional hunter-gatherers, particularly with respect to issues of cultural identity and ancestor veneration (see Crothers 1999:250).

Late Middle Archaic hunter-gatherers who lived along the Green River also placed an increased emphasis on river and bottomland settings (Jefferies et al. 2005). Investigation of hunter-gatherer settlement in the Cypress Creek drainage, a major tributary to the Green River, revealed a sharp increase in the percentage of late Middle Archaic components and hafted bifaces associated with these wetland habitats. For example, 46 percent of Early Archaic sites in the Cypress Creek project area were situated near wetlands, but this figure jumped to 68 percent during the late Middle Archaic. Likewise, the percentage of hafted bifaces found in wetland settings increased from 53 percent during the Early Archaic to 88 percent during the late Middle Archaic. Both of these trends suggest that starting in the late Middle Archaic (and continuing through the remainder of the Archaic period), there was a proportionately greater occupation of bottomland and riverine settings with their abundant, reliable, and predictable food sources (Jefferies et al. 2005).

Intrasite Patterning

Because of the previously discussed problems associated with identifying Lower Ohio Valley early Middle Archaic components, little is known about the variability of sites dating to this time or the distribution of people and activities across the landscape. The little information that does exist suggests that early Middle Archaic hunter-gatherers maintained a way of life that was similar to that of Early Archaic groups. In many ways, the differentiation of Early Archaic societies from those of the early Middle Archaic reflects a lack of knowledge about hunter-gatherer societies when the regional cultural sequence was established decades ago.

Information from the deeply stratified Koster site (Brown and Vierra 1983; Cook 1976) provides some general indications of early Middle Archaic life to the north of the Lower Ohio Valley. Koster appears to have been repeatedly occupied for short periods by small groups of hunter-gatherers who pursued a residential mobility strategy. While there, they hunted a wide variety of animals (large and small mammals, fish, and river mussels) and collected seasonally available wild plants (seeds, nuts, and fruits), much like their Early Archaic predecessors. They prepared plant foods using stone pestles and milling stones (Brown and Vierra 1983).

Information from the few well-documented Lower Ohio Valley early Middle Archaic sites suggests that local hunter-gatherer groups maintained a way of life similar to that seen at Koster. Archaeological investigations associated with the Caesars Archaeological Project identified at least two early Middle Archaic occupations thought to represent short-term encampments. The occupations were marked by surface hearths and associated activity areas having low densities of flaked stone tools, debitage, and charcoal. Thirteen radiocarbon dates placed these occupations between 7200 and 6200 B.P. (Stafford 2004).

Like the early part of the Middle Archaic, many sites dating to the late Middle Archaic (ca. 6500–5000 B.P.) consist of light lithic scatters distributed across the Lower Ohio Valley landscape. Some of these components are in rockshelters; many others are "open-air" sites. At some of these small sites artifact assemblages suggest a relatively narrow range of activities, while others reflect a wider variety of tasks.

In contrast, thick earth and/or shell middens indicating intensive and/or repeated occupation of the same location mark other late Middle Archaic sites. Essentially, it appears that some hunter-gatherer groups were spending more and more time at fewer locations. Some researchers have interpreted this significant organizational change in group movement as marking the transition from a residential to a logistical mobility strategy (Stafford 1994; Stafford et al. 2000). Most of these sites, commonly referred to as base camps, were strategically situated near wetlands or rivers that offered abundant, diverse, and reliable food resources to supplement those found in the uplands (Figure 6.8). It appears that part of these groups, at least the very young, very old, and infirm, stayed at the base camps for much of the year. Other people were more mobile, participating in task groups of varying size and composition that dispersed from the base camps for variable lengths of time throughout the year. Task group activities included acquiring food, collecting lithic materials, and maintaining social and economic ties with nearby and distant groups.

Perhaps the earliest professional investigation of a Lower Ohio Valley late Middle Archaic site was by University of Chicago archaeologists at the Faulkner site in the 1940s (Cole et al. 1951; MacNeish 1948; Muller 1986:59–60). Faulkner, located near the Ohio River in the Black Bottom of far southern Illinois, was strategically situated near both extensive wetlands and drier uplands. This positioning on the landscape provided its inhabitants with access to a wide range of subsistence resources that included deer, turkey, nuts, fish, and aquatic plants and animals (Muller 1986:59).

In the spirit of the times, project archaeologists compared Faulkner site traits

Figure 6.8. Lower Ohio Valley floodplain lake in southern Indiana near the Breeden site.

with trait lists from recently excavated sites in the nearby Green and Tennessee River valleys. Trait similarities between Faulkner and Indian Knoll, Chiggerville, and Ward in the Green River area suggested that these sites were regional variants of the same culture (Cole et al. 1951:221–225).

Comparison of the Faulkner site's side notched and stemmed projectile points with points from more recently excavated sites having good chronological controls (Jefferies 1982b) indicates that the Faulkner component dates from ca. 6500 to 4500 B.P. Based on the presence of a light midden accumulation, storage or cache pits, and at least 11 burials, late Middle Archaic hunter-gatherers appear to have maintained a long-term residence at Faulkner, or, alternatively, returned to the site at different times of the year. Again drawing on data from more recently excavated nearby sites, late Middle Archaic hunter-gatherers appear to have used this part of the Black Bottom as a base camp that served as the focus of their social and economic activities (Muller 1986:59).

In the years since the Faulkner excavations, research at several Lower Ohio Valley sites has provided an even clearer picture of the kinds of organizational changes that took place in the region's hunting and gathering societies. Most of these sites, like Black Earth (Jefferies 1982b) in southern Illinois, Bluegrass (Stafford and Anslinger 1988; Stafford et al. 2000) and Meyer (Bader 2004) in southern Indiana, and KYANG (Bader and Granger 1989) and other Falls area sites in Indiana and Kentucky (Janzen 1977), consist of thick middens containing large artifact assemblages, numerous and diverse features, and numerous burials resulting from the many different activities carried out at these long-occupied spots.

The Black Earth site was located in the southern part of the Mount Vernon Hill Country, part of the Southern Till Plain. Poorly drained rivers and creek bottoms marked this hilly, rolling part of the Lower Ohio Valley. To the south lay the more rugged terrain of the Shawnee Hills. A few kilometers to the east was the Wabash Border with its dense forests, marshes, oxbow lakes, and poorly drained Wabash River floodplain. The site itself was situated on a low upland area that rose above a series of shallow lakes and wetlands that nearly surrounded the site (Figure 6.9). The water covering the lowlands was annually replenished by flooding of the nearby South Fork of the Saline River. Clearly, Black Earth was situated in a place that gave its inhabitants easy access to a variety of environmentally diverse habitats, including upland forests, wet floodplains, grasslands, rivers and creeks, shallow lakes, and swamps (Jefferies 1987:23–24).

Because of the environmental diversity, the forests surrounding the Black Earth site contained many different varieties of trees. Although the southern Illinois forests seen by early Euroamerican settlers probably differed to some extent from those of the late Middle Archaic, vegetation data collected during the early nineteenth century provides a general idea of what the local and regional vegetation was like prior to historical modifications (Lopinot 1982b).

Historical sources, primarily the records of General Land Office (GLO) surveyors, indicate that oaks and hickories dominated the drier uplands, while floodplain forests consisted of oaks, maples, willows, elms, and sycamores. Cypress trees grew around and in floodplain sloughs, swamps, and ponds (Schwegman 1975). Open areas near the site would have hosted persimmon, mulberry, paw paw, sumac, blackberry, grape, marshelder, chenopodium, and giant ragweed (Lopinot 1982b).

Like the plants, the diverse landscape that surrounded the Black Earth site supported numerous animal species. Animals that lived in the forest or along its edges included the white-tailed deer, rabbit, squirrel, opossum, and wild turkey. The abundant wetlands were a prime habitat for beaver, mink, otter, muskrat, goose, duck, and fish (Breitburg 1982). Clearly, the late Middle Ar-

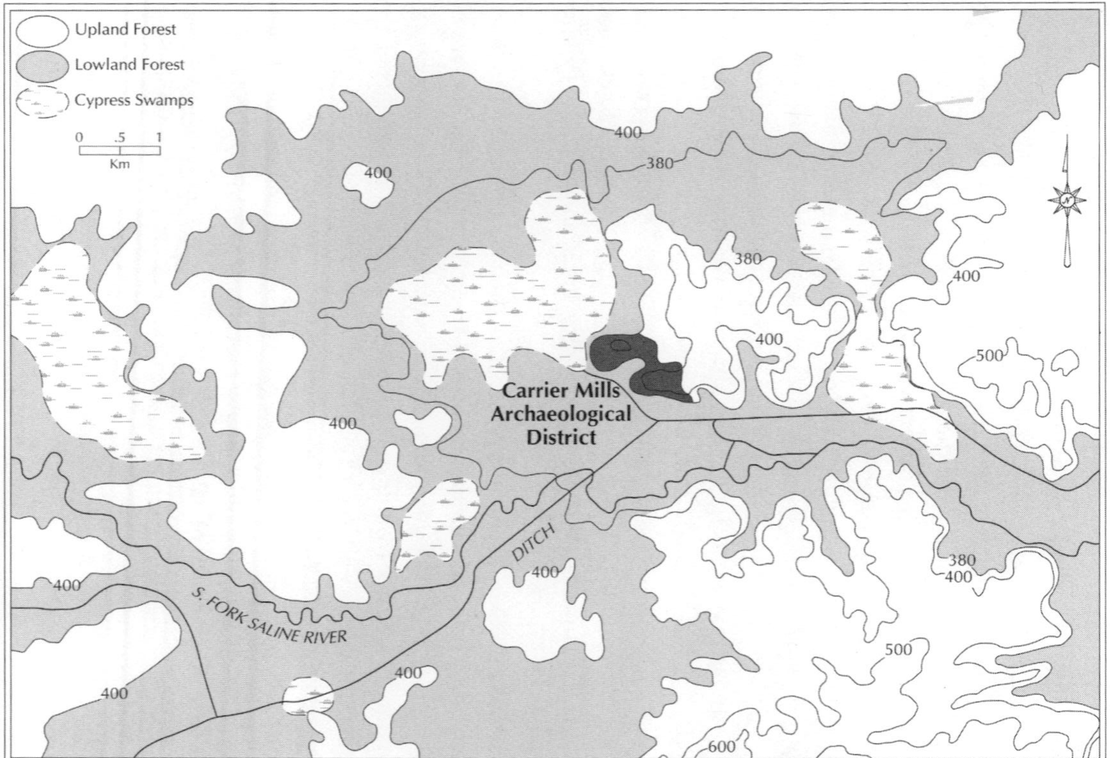

Figure 6.9. Environmental setting of the Black Earth site (redrawn from Jefferies and Butler 1982:Figure 2).

chaic hunter-gatherers who lived at Black Earth had access to many different plant and animal species that could be hunted and collected on a multiseasonal or year-round basis.

The primary late Middle Archaic occupation, covering approximately 1.7 ha, was located at the far western edge of the Black Earth site in an area that once overlooked an expansive, shallow lake. A series of eight charcoal samples collected from different levels of the 1.5-m-thick midden yielded uncalibrated radiocarbon dates ranging from 5905 to 4860 B.P. (Jefferies 1982b:Table 10). Repeated occupation of this spot by late Middle Archaic hunter-gatherers resulted in extensive alteration and enrichment of the natural soil, as indicated by a high sand content and very high levels of calcium, iron, nitrogen, copper, phosphorous, potassium, and zinc. The sand came from the breakdown of sandstone, the only naturally available stone for making features and certain kinds of ground stone implements. Much of the soil's chemical alteration was caused by the deposition of large quantities of wood ash, the by-product of the thousands of fires built by the Black Earth site's Middle Archaic inhabitants (Woods 1982).

Based on the concentrations of phosphorus in the midden and the time over which it was deposited, an estimated 15 to 20 people lived in this part of the Black Earth site during the late Middle Archaic occupation. Since this was a year-round average, the actual number of people living there during periods of peak occupation was probably considerably higher (Woods 1982).

The late Middle Archaic artifact assemblage consisted of a highly diverse tool kit containing flaked stone, ground stone, or bone and antler implements. Most of the flaked stone artifacts were hafted bifaces, consisting of projectile points, endscrapers, and drills. Eighty-six percent of these were side notched bifaces that resembled the Godar, Matanzas (Cook 1976), or Big Sandy II (Lewis and Lewis 1961) projectile point types. Unhafted side- and endscrapers, notches, gravers, drills, perforators, and bifaces in various stages of reduction were also well represented. Ground stone artifacts consisted of grooved axes, abraders, pitted cobbles, hammerstones, and atlatl weights. The exceptionally large, well-preserved bone and antler artifact inventory included awls, pins, needles, fishhooks, turtle shell bowls and rattles, and atlatl hooks and handles (Breitburg 1982; Jefferies 1982b).

Artifact styles, specifically those of the side notched bifaces and the elaborately carved and engraved bone pins, are very similar to those documented at many contemporary sites in the southern Midwest and upper Southeast (Jefferies 1997, 2004). The sharing of artifact styles, and probably many other cultural traits, suggests that late Middle Archaic hunter-gatherers who lived in this part of the North American Midcontinent maintained social connections with groups that lived in other parts of the region for more than 1,000 years. Interaction and exchange with near and distant groups were important for establishing and maintaining alliances, obtaining marriage partners, accessing information about subsistence resources, and gaining knowledge about distant people, places, and things.

Excavation of the Black Earth site revealed nearly 200 features serving a variety of storage, food processing, heating, and lighting functions. The diversity of feature types increased through time, suggesting that the range of activities conducted there also increased. The most common feature type (43 percent) was a shallow, basin-shaped pit probably used for food processing. Small basin-shaped charcoal concentrations, accounting for nearly 30 percent of all features, probably had a similar function. Deep, flat-bottomed circular pits (12 percent) may have been used for storage; however, their small size and relative scarcity suggests that subsurface storage was not a favored way of storing goods at this time (Jefferies and Lynch 1983:311–314). Although researchers found no solid evidence for habitation structures, clusters of post molds in the subsoil at the base of the midden suggest that substantial architecture of some type existed (Jefferies and Lynch 1983:321).

In addition to features, the Black Earth site midden contained the skeletal remains of more than 150 human burials (Lynch 1982). The positioning of individuals was roughly evenly divided between flexed (49 percent) and extended (46 percent). Twenty-seven percent of the people, both children and adults, had associated grave goods consisting of utilitarian, ornamental, and "ritual" or "ceremonial" objects. Overall, the differences in treatment of the dead at time of death appears to have been based on the individual's age, gender, and personal abilities.

Clearly, the interment of the dead was an important activity for the hunter-gatherers who lived at Black Earth, contributing to the long-term ritual significance of that location. However, like at many other Middle and Late Archaic middens in the Lower Ohio Valley, despite the number of burials, there was no evidence suggesting that the site was a formal, organized cemetery or a burial mound (Marquardt and Watson 2005a:636; Milner and Jefferies 1998). The numerous burials are indicative of the number of people who lived (and died) near this spot. This lack of formality, however, did not lessen the cultural and ritual significance of this place on the Middle Archaic (and possibly later) cultural landscape.

Since approximately one-third of the Black Earth site was excavated, a projection of approximately 500 late Middle Archaic burials seems reasonable for the entire site (Woods 1982). Even using this seemingly large number, it represents less than one interment every other year. Using death rate figures from similar kinds of hunter-gatherer societies (Cook and Heizer 1965:8), a burial population of this size could be attributed to a group of roughly 15–25 people, closely conforming to population estimates based on the soil chemistry.

Black Earth site investigations conclusively demonstrated that the late Middle Archaic occupation spanned more than a thousand years. For at least part of that time, it was the scene of intensive activities related to many different aspects of life, including conducting numerous extractive and maintenance tasks, performing rituals associated with burial of the dead, and maintaining social and economic ties with other nearby and distant hunter-gatherer groups.

The Bluegrass site, situated about 30 km from the Ohio River in the forested interior uplands of southwestern Indiana, is another example of a late Middle Archaic base camp/mortuary site (Stafford et al. 2000). The 35-cm-thick midden covered about .2 ha of a knoll overlooking a featureless lake plain. Like the Black Earth site, the dark midden had high densities of fire-cracked rock, nutshell, and well-preserved animal bone, along with lesser quantities of lithic tools and debitage. Excavations exposed more than 130 pits or hearths, 80 human burials, and 12 dog burials. Six uncalibrated radiocarbon dates placed occupation between ca. 5300 and 5000 B.P., well within the Black Earth site date range. Bluegrass also yielded an outlier midden date of ca. 6300 B.P. (Stafford et al. 2000:320; Stafford and Cantin 2009).

Pits were the most common feature type at Bluegrass, some measuring over 1.8 m in diameter (Stafford et al. 2000:320). Some steep-sided pits were up to one meter deep, resembling those interpreted as storage pits at Black Earth (Jefferies 1982b:143).

The Bluegrass site yielded a diverse assemblage of stone and bone tools. Like at Black Earth, hafted bifaces were the most common flaked stone tool types. Projectile points were side notched, resembling Godar and Matanzas points, or stemmed. Other lithic tools included drills, endscrapers, abraders, hammerstones, bifaces, unifaces, gravers, burins, and ground stone axes (Stafford et al. 2000:320).

Elaborately carved and engraved bone pins, similar to those from contemporary late Middle Archaic sites in the Midcontinent (Jefferies 1997), comprised part of the bone tool inventory. Other bone and antler artifacts included awls, antler flakers, atlatl parts, and modified turtle shell representing parts of cups, bowls, or rattles (Stafford et al. 2000:321).

Bluegrass mortuary practices resemble those seen at other late Middle Archaic base camps in the Lower Ohio Valley (Mayes 1997; Stafford et al. 2000). Burials were placed in extended, loosely flexed, or tightly flexed positions. Artifacts accompanied 34 percent of the individuals, but there was no strong association between burial goods and an individual's age or sex. The Bluegrass site's late Middle Archaic inhabitants were in relatively good health, as indicated by low rates of dental caries and abscesses and a lower than expected rate of infectious lesions and arthritis.

More than 100 years of archaeological investigations near the Falls of the Ohio have identified numerous late Middle Archaic occupations that would today be characterized as base camps (Granger 1988; Janzen 1977). Many of these components are now considered representative of the Old Clarksville phase, dating from approximately 6000 to 4400 B.P. (Granger 1988:179).

As in many other parts of the Lower Ohio Valley, these intensively occupied sites consisted of thick middens containing abundant artifacts, features, plant and animal remains, and burials. Some Falls area sites fitting this description include Old Clarksville and Hornung in southern Indiana and KYANG in Kentucky. Like many other Lower Ohio Valley late Middle Archaic base camps, these and most other Old Clarksville phase middens are situated on floodplains or adjacent to expanses of interior wetlands and swamps. In many cases, these sites appear to have been strategically situated near several environmental zones, allowing hunter-gatherers to minimize the need for seasonal movement. The variety of plants and animals supported by these ecologically diverse habitats may have allowed these groups, or at least parts of them, to occupy these base camps on a multiseasonal or year-round basis (Janzen 1977).

Midden contents indicate that the range and diversity of hunter-gatherer activities at these Falls area sites were comparable to those at contemporary sites discussed above. Similarities in projectile point and bone pin styles (Granger 1988:Figures 7 and 8) suggest that Falls area late Middle Archaic groups maintained some level of social contact with contemporary groups living in other parts of the Lower Ohio Valley.

The Kentucky Air National Guard (KYANG) site, located approximately 10 km south of the Ohio River, is one of the best documented Falls area late Middle Archaic middens. The 1.8 ha site is situated on a knoll that rose from a flat, swampy plain. Prior to modern development, this part of the Falls area was covered by marshes and sluggish streams, representing one of the richest seasonal catchment areas in the region. Contemporary sites once existed on many of the other knolls and higher hills that dotted the wetlands, but they have largely been destroyed by the growth of the Louisville metropolitan area (Granger 1988:173, 179).

The Old Clarksville component consisted of a dark midden capped by a layer of mussel shell. Preserved plant and animal remains indicated that late Middle Archaic hunter-gatherers maintained a marsh-oriented subsistence strategy that focused on white-tailed deer, supplemented by freshwater mussels, small mammals, and fish species associated with standing water (Granger 1988:170).

Features included hearths, stone clusters, earth ovens, and bowl- or basin-shaped storage pits. The 32 Old Clarksville phase burials were either single or double inhumations that had been placed in deep bowl-shaped pits. All were interred in a flexed position, many with associated grave goods like engraved bone pins; deer, bear, and wolf tooth necklaces; red ocher; ground stone beads and pendants; adzes; and antler atlatl hooks (Granger 1988:175).

Subsistence data from the Falls areas suggest that sites like KYANG were primarily occupied during the late summer and fall while the inhabitants hunted deer and small marsh animals and collected river mussels. The lack of data supporting a winter occupation may mean that hunter-gatherer groups left the wetland-oriented base camps during the colder part of the year, preferring instead to live farther out in the Outer Bluegrass uplands at sites like McNeeley Lake Cave/Rockshelter (15Jf200) and the Durrett site (15Jf201) (Granger 1988). In contrast, it may simply reflect the lack of specific environmental indicators of late fall through early spring occupation.

Subsistence

More than twenty years ago, Bruce Smith (1986) cautioned the archaeological community that after decades of research, there was little evidence supporting the popular belief that Southeastern Middle Holocene hunter-gatherers gradually adapted to a stable environment through an increasingly developed seasonal round that emphasized plant foods within a broad-scale expansion of utilized plants and animals. Instead, Smith suggested that rather than adopting a single, ultimate adaptive solution, Middle Holocene groups developed regionally distinct adaptive responses for dealing with changes in local food resources.

He supported his contention by comparing Early and Middle Holocene floral and faunal assemblages, emphasizing that there was no temporal increase in the diversity or dietary significance of plants. With one major exception, shellfish, there was no increase in the range of animal species pursued (1986:21).

Recent syntheses of Southeastern Middle Holocene botanical (Gremillion 1996) and faunal (Styles and Klippel 1996) research tend to support Smith's (1986) argument. Clearly, hickory nuts continued to be an important food resource throughout much of the interior Southeast during the mid-Holocene. Acorns and walnuts were important too, but they did not have the same consistent spatial or temporal pattern of use as did hickory nuts (Gremillion 1996:106). Some Southeastern and Midwestern Middle Holocene sites have yielded evidence for the initial stages of indigenous seed plant domestication (Gremillion 1996:110; Smith 1992a:53), but cultivated plant foods did not acquire any dietary significance until the late Middle or Late Holocene (Smith 1992c:282).

Kristen Gremillion (1996:99–100) also argued that apparent local long-term continuity in subsistence resources probably reflects changes in mobility patterns, storage techniques, and processing technology that hunter-gatherers adopted in response to changes in nut abundance and distribution (see Stafford 1994). In these cases, changes in resource abundance and spatial variability would have been masked by corresponding behavioral changes associated with their procurement and processing.

Likewise, comparisons of faunal remains from the Midcontinent indicate generally similar patterns of animal exploitation during the Middle Holocene. At the regional level, there was a trend toward the greater use of aquatic resources and an increased importance of white-tailed deer from the Early to the Middle Holocene (Styles and Klippel 1996:115). Even this trend varied with respect to available local animal populations. In larger river valleys, late Middle Holocene hunter-gatherers added fish and shellfish to their diet in considerable numbers as reflected by the formation of thick middens containing abundant mussel shell (Marquardt and Watson 2005a:631). In upland areas far removed from major rivers, hunter-gatherers substituted small-bodied mammals and reptiles for aquatic resources. In most cases, aquatic foods, like mussels, were still collected where available, but due to their lower frequency in the uplands, they had little dietary significance (Stafford et al. 2000:333).

As Smith (1986), Gremillion (1996), and others have cautioned, generalizations about change and continuity of food resources and hunter-gatherer collecting/hunting/fishing strategies over large areas can hide important spatial and temporal variation. Since the Lower Ohio Valley region lies along the Southeast-Midwest interface, investigations of Middle Archaic plant and animal exploitation strategies will help to clarify the nature of small-scale spatial variation within both of these regions.

Plant Resources

Problems associated with locating Lower Ohio Valley early Middle Holocene (pre-7000 B.P.) archaeological sites make it hard to assess the dietary significance of specific plants. Small early Middle Archaic hunter-gatherer bands that occupied the interior Eastern Woodlands appear to have uniformly exploited the plants and animals that inhabited the region's watersheds (Smith 1992c:282). Most short-term seasonal camps that characterized this period of prehistory have failed to yield much in the way of carbonized plant remains. The meager botanical data "suggest a broad-based utilization of both closed-canopy climax forest plant resources (particularly oak mast and hickory nuts) and early successional species more commonly associated with edge areas and more disturbed situations" (Smith 1992c:282). These small hunter-gatherer groups appear to have maintained a mixed or generalized diet that included many different kinds of nuts collected by foraging groups practicing a residential mobility strategy. Nuts, just one of many plant food resources in the diet, were collected on an "encounter" basis (Stafford 1991; Stafford et al. 2000).

The Morrisroe site is an exception to the general pattern of small, intermittently occupied early Middle Archaic sites (Nance 1986a, 1987, 1988). Excavation of the thick alluvial midden, much of which dates to the early Middle Archaic (Stratum 3) (Nance 1986a:23), yielded hickory nut, black walnut, butternut, and acorn shell, along with grape, hawthorn, black cherry, honey locust, knotweed, and possible sumac seeds (Wymer 1987:Table 2). Hickory nutshell was by far the predominant nut type, but problems of fragmentation and preservation may have masked the dietary significance of acorn and other species with fragile nutshells (Lopinot 1982a:730). Not surprisingly, the Morrisroe paleobotanical record is quite similar to those seen at most Southeastern Early Holocene sites.

During the centuries following 7000 B.P., significant changes occurred in the regional climate, the distribution of people over the landscape, and the significance of plants in the diet of increasingly complex hunter-gatherers. Climatic changes were largely attributable to shifts in the zonal atmospheric flow across the Midwest during the Hypsithermal climatic episode. Long-term changes in rainfall patterns resulted in the transition of Midcontinent river systems from sediment removal and river incising to river aggradation and stabilization. This resulted in the expansion of backwater swamps and oxbow lakes, as well as shoal and shallow-water habitats along many of the Midcontinent's river systems. These new wetland habitats, in turn, supported a diverse, abundant, and easily accessible variety of aquatic and circum-aquatic food resources. Simultaneously, the valleys of these river systems provided disturbed habitats for floodplain species of pioneer plants like goosefoot, marshelder, and gourds to thrive (Smith 1992c:282).

Some Lower Ohio Valley Middle Holocene hunter-gatherers responded to these changes by focusing their attention on strategically situated locations near these developing wetlands. This juxtaposition allowed site inhabitants to exploit abundant wetland food sources as well as those that lived in nearby drier habitats. As we have seen, the time that hunter-gatherer groups spent at these circum-wetland spots, combined with the intensity of their activities, resulted in the accumulation of thick earth, rock, and/or shell middens. The proximity of these midden sites to major river valleys and wetlands has led some archaeologists to suggest that the abundance of food found there tended to "pull" groups to circum-wetland settings where they established their base camps (Brown and Vierra 1983; Marquardt and Watson 2005a:638; Styles et al. 1983).

The creation of continuously disturbed anthropogenic habitats, commonly associated with intensively occupied places, provided excellent opportunities for "colonization" by a variety of floodplain plants that included gourd, goosefoot, sumpweed, and sunflower. Over time, hunter-gatherers domesticated these plants, and they became increasingly important components of late Middle and Late Holocene diets (Smith 1992c:282–283). Of course, people continued to exploit a wide range of wild species for the bulk of their plant foods.

In contrast, others have proposed that instead of being "pulled" by abundant and diverse floodplain resources, subsistence/settlement changes were a consequence of changes in the Middle Holocene climate and forest composition (Stafford 1991; Stafford et al. 2000). The transition from a closed canopy to an open oak-hickory forest resulted in an increased abundance of nut mast and deer. A more heterogeneous distribution of food resources, along with the increased time required for hickory nut bulk processing, resulted in the establishment of base camps associated with a logistical collector system. This model also explains the presence of large midden sites located away from major river valleys (Stafford 1991:228; Stafford et al. 2000:333).

Careful investigations of a number of Lower Ohio Valley late Middle Archaic middens have yielded important information about hunter-gatherer plant use and the dietary significance of wild versus domesticated species. Perhaps the most complete record of Lower Ohio Valley Middle Holocene plant use comes from the study of the Black Earth site's late Middle Archaic midden and features (Lopinot 1982a, 1984). These studies provided valuable insights concerning the kinds of plants that late Middle Archaic hunter-gatherers used, the dietary significance of those plants, the season(s) of site occupation, and local environmental conditions. Since the botanical samples spanned more than 1,000 years, they also helped to clarify how plant use and the local environment changed through time.

Not surprisingly, carbonized nutshell and nutmeat comprised 80 percent of late Middle Archaic plant remains (Lopinot 1984:310). Of this total, hickory nuts alone account for 78.9 percent. Based on differences in shell thickness, Black Earth residents collected many different varieties of hickory nut including shellbark, mockernut, shagbark/pignut/oval, and pecan and bitternut. The economic importance of hickory nuts over other varieties may be attributable to the high proportions of protein and fat that they contain, the relative ease with which they could be processed, their abundance in the surrounding forests, and their long-term storage potential (Lopinot 1984:321). Its abundance may also reflect the durability of hickory nutshell in the archaeological record.

Acorns represent less than 1 percent of all nut remains; however, due to a high rate of shell fragmentation compared to other kinds of nutshell (Lopinot 1982a:730), their true dietary significance is probably masked. Virtually all soil samples contained acorn shell, strongly suggesting that the economic importance of acorns is underrepresented. Based on differences in shell thickness, projected size, and surface markings of cotyledon fragments, Black Earth site inhabitants collected acorns from many different varieties of oak. Acorn probably was a supplemental nut food; that is, they were extensively, but not intensively, exploited (Lopinot 1984:324).

Late Middle Archaic hunter-gatherers also collected other nuts, particularly walnuts and hazelnut, but these species played only a minor role in the diet (Lopinot 1984:319–321). Walnut trees probably occurred as scattered, isolated trees that were collected depending on their local availability. Hazelnuts were probably even more widely scattered (1984:324).

Ninety-four percent ($n = 8,160$) of the Black Earth site seed assemblage consisted of uncarbonized hackberry pits and pit fragments. The economic significance of the fruit of this tree to the late Middle Archaic inhabitants of the Black Earth site is hard to assess, but, as reflected by their highly calcareous pits, they would have been a good source of calcium. Because of the fruit's thin pulp and large stone, some historic Native American groups processed hackberry by finely pounding the fruit, then using it as a flavoring or mixing it with other foods. The high incidence of hackberry pit fragments suggests that this may have been the case during the late Middle Archaic (Lopinot 1984:332).

The small carbonized seed collection was dominated by four seed types—persimmon, wild grape, bedstraw, and wild bean—along with lesser quantities of plum pits and seeds from the legume family. Thirty-nine additional seeds represented 20 other taxa (Lopinot 1984:325–328).

In addition to hackberry, late Middle Archaic residents collected the fleshy fruits from hawthorn, persimmon, honey locust, wild plum, and wild black cherry trees, along with blackberries, haws, and grapes. Persimmons and wild grapes were particularly well represented, accounting for nearly 70 percent of the fleshy fruits. Their presence indicates open woods, forest edge, or other kinds of disturbed habitats near the Black Earth site (Lopinot 1984:333).

Surprisingly, carbonized starchy or oily weed seeds were relatively scarce. The general absence of weed seeds suggests that few abandoned fields, like those associated with old garden plots, existed near the site. All of the represented plants could have lived in or around the Black Earth site, so they may reflect the local vegetation more than dietary preferences (Lopinot 1984:334).

The small carbonized weed seed collection included a few legumes that could only be classified to the family level, a single sunflower seed that appeared to be a wild variety, and a few erect knotweed, goosefoot, and possible giant ragweed seeds. Considering the scarcity of weed seeds, along with their scattered distributions, it is unlikely that the late Middle Archaic people who lived at Black Earth used these plants or their seeds to any great extent, much less cultivated them (Lopinot 1984:334).

Interestingly, in other parts of the Midcontinent some contemporary groups were in the initial stages of domesticating several wild weed species, like marshelder, goosefoot, and sunflower (Smith 1992a:Figure 3.2). Why this was not the case at the Black Earth site is a question of considerable interest.

Late Middle Archaic inhabitants also collected several kinds of wild beans that grew in the moist disturbed ground, damp thickets, and forest edges that surrounded the Black Earth site. The small size of the carbonized beans compared to modern specimens suggests that they were collected before the bean seed matured, possibly indicating that the immature pods were the object of collecting rather than the beans (Lopinot 1984:336–338).

In addition to seeds and nuts, Black Earth site inhabitants also collected plant products that grew underground. One of the most common was groundnut that commonly grows along the edges of swamps, marshes, streams, and lakes. The relatively high protein content of their roots and stems, their abundance around the site, and evidence for use by other late Middle Archaic groups suggests that groundnuts were an important food source at Black Earth (Lopinot 1984:340–342).

The Black Earth plant remains provide a glimpse into the role of plant foods in the late Middle Archaic diet. Although a variety of cultural and natural factors can distort this view, it appears that hickory nuts were the mainstay of a rather specialized plant food collecting strategy. Other nuts also had some dietary and nutritional significance, but based on the carbonized remains, hickory was by far the most important. Likewise, an assortment of fleshy fruits, seeds, and legumes represented supplemental rather than primary foods. The dietary significance of hickory nuts throughout the late Middle Archaic occupation of the Black Earth site suggests a relatively consistent climate and/or that hunter-gatherer population density had not risen to the level requiring a more diversified subsistence strategy (Lopinot 1984:343–344).

Site inhabitants collected these plants across the landscape, but most of them came from along stream bottoms and border zones and in mixed hardwood floodplain forests. Prairie areas were poorly represented, suggesting that there were no nearby grasslands or that their resources were not important to the Black Earth site inhabitants (Lopinot 1984:368).

As is often the case, most of the botanical evidence for season of site occupation pointed to a late summer/fall presence. However, many foods collected during the warm weather seasons could have been stored for later use. In addition, many of the represented plants offered more perishable foods, like greens, saps, bulbs, and tubers, which were available in the early part of the growing season but are seldom preserved at open-air archaeological sites (Lopinot 1984:371).

While the relative importance of different plant species varied during the more than 1,000 years represented in the late Middle Archaic midden, the magnitude of those differences is small. Some variation in resource exploitation may have taken place, but the effects of regional climate change on the Black Earth site area were either slight or nonexistent. The relatively narrow spectrum of the plant food economy pursued by the Black Earth site's late Middle Archaic inhabitants indicates local and regional population pressures were insufficient to bring about major changes, or that the impacts of increased population were dealt with by changing the ways that plant resources were collected and processed (Lopinot 1984:376).

Comparison of the Black Earth site to roughly contemporary sites in adjacent parts of the Midwest and Southeast revealed a generally similar picture of late Middle Archaic plant use (Munson and Cook 1980c; Parker 1999). For example, Helton phase plant remains from the Koster site contain a very high percentage of hickory nutshell (> 90 percent), a high diversity, but low frequency of seeds, and a low percentage of starchy seeds from plants that assumed a much greater dietary significance in the Late Holocene (Asch et al. 1972; Lopinot 1984:391–392; Neusius 1986:130). The emergence of hickory nut as a dominant food source took place slightly earlier (ca. 7600 B.P.) at Modoc Rock Shelter (Styles et al. 1983). Researchers have attributed the increased importance of hickory nut during this time to several cultural factors including the introduction of bulk processing techniques that facilitated nutmeat extraction (P. Munson, cited in Asch and Asch 1985:105) and the adoption of a logistical mobility strategy that facilitated the efficient collection of hickory nuts from widely scattered locations and their bulk processing at centralized base camps (Stafford 1991, 1994).

Regional differences in late Middle Archaic plant use center on variations of minor plant constituents, like weed seeds. Koster inhabitants apparently used more weed seeds, particularly marshelder, than did people at Black Earth, but fruits were more important at Black Earth. Differences in plant remains from these two sites probably reflect differences in availability—weed seeds at Koster are attributable to the site's proximity to active floodplains with their exposed mudflats, shorelines, and other open habitats. With a few exceptions, the late Middle Archaic plant remains suggest that hunter-gatherers were focusing their plant gathering/harvesting on naturally occurring plants from mature plant communities (Lopinot 1984:393–394).

Animal Exploitation

The study of hunter-gatherer plant use and procurement strategies is made possible by the preservation of plant remains through carbonization. In contrast, the moist, acidic soils found throughout the Lower Ohio Valley do little to enhance bone preservation, so our understanding of late Middle Archaic animal exploitation strategies is spotty and incomplete. Most of the information that does exist comes from dry rockshelters and open-air sites having thick middens that dot the Lower Ohio Valley landscape. Bone preservation at these sites is attributable to either extremely dry conditions or very alkaline soil.

The Black Earth site's excellent bone preservation is attributable to alkaline soil conditions created through the cultural deposition of wood ash. Many wood ash constituents, particularly calcium carbonate, iron and magnesium salts, and sodium and potassium carbonates, helped to prevent the formation of acids that would have normally destroyed bone and many other organic materials (Woods 1982:1394).

The analysis of more than 50,000 faunal specimens showed that late Middle Archaic hunter-gatherers hunted, trapped, and fished for numerous large and small animals that inhabited the forests, forest edges, and wetlands that surrounded the site. The high level of habitat diversity found near Black Earth attracted a wide range of animals that, in turn, contributed to an abundant, reliable, and diverse food supply. While many of these animals were pursued to obtain food, site inhabitants also used the bones, skins, sinews, and other body parts to make tools, clothing, containers, and items of personal adornment (Breitburg 1982).

Most of the meat (> 90 percent) eaten by Black Earth inhabitants came from animals that lived along the forest edges—the transition zone between the dense forests and open areas like grasslands and wetlands. The white-tailed deer was the most important of these species, accounting for about 75 percent of the meat from the forest edges. Skeletal indicators for season of death revealed that deer were hunted throughout the year, supporting multiseasonal site occupation. Deer were a particularly important source of bone and antler for making tools and other items. Inhabitants also hunted elk, the second major meat source, as well as woodchuck, rabbit, fox squirrel, and turtles. Wild turkeys provided up to 5 percent of the meat (Breitburg 1982).

Wetland species, like beaver, otter, mink, muskrat, and various birds, fishes, reptiles, and amphibians, contributed approximately 9 percent of the meat to the late Middle Archaic diet. The beaver accounted for most of this total. People at Black Earth hunted many of these animals throughout the year, but others were only seasonally available. Many different kinds of waterfowl were hunted during their spring and fall migrations, while other birds, like most cranes, herons, and rails, were summer residents. Turtles, fish, and various amphibians were probably captured during the warm season (Breitburg 1982:932–933).

The importance of wetland animals in the late Middle Archaic diet is somewhat underestimated because of the small size of many species that lived in and around the shallow swamps and lakes near the site. Careful examination of flotation samples revealed the bones of many small fishes, reptiles, amphibians, and mammals that inhabited aquatic/riparian habitats. The small size of the bones suggests that many of these species were captured using small-mesh seines and nets (Oetelaar 1982). While these species were numerous, they were not major contributors to the late Middle Archaic diet, probably serving as "back up" foods when other more desirable ones were unavailable.

The remaining meat in the late Middle Archaic diet came from woodland species. The raccoon was the most intensively pursued of these animals, but the opossum, gray squirrel, gray fox, and the now extinct passenger pigeon contributed small quantities of meat to the diet. Although little skeletal evidence exists for passenger pigeon hunting, historic Native Americans commonly hunted them for their storable oil (Breitburg 1982:933).

Clearly, Black Earth site residents obtained their animal protein by hunting, fishing, and trapping virtually all of the animal species that lived around them. Meat from animals that were available throughout the year, like the deer, turkey, elk, and raccoon, formed the core of their diet. These species offered the greatest return of meat for the energy expended in their procurement. Seasonally available species, like migratory waterfowl, marsh birds, passenger pigeons, turtles, and fish, served as dietary supplements. Such a pattern of animal exploitation supports multiseason or perhaps year-round occupation. Comparison of faunal material from the different Middle Archaic midden zones indicated that inhabitants obtained approximately the same amount of meat from the same species, indicating very little change in animal exploitation strategies for more than 1,000 years (Breitburg 1982:934).

The plant and animal species represented in the Black Earth site midden, as well as the midden itself, reflect fundamental economic and social adjustments by mid-Holocene hunter-gatherers to changing environmental and cultural factors. Using these newly adopted settlement/subsistence strategies, site inhabitants had sufficient subsistence resources to buffer them against short-term fluctuations in food availability and environmental change. Apparently, these adjustments were sufficiently flexible to enable hunter-gatherers to continue this way of life for the next 1,000 years at this and similar site settings across the North American Midcontinent (see below).

The diversity of natural habitats found across the Lower Ohio Valley landscape produced some variation in late Middle Archaic hunting strategies. For example, a significantly different animal exploitation strategy is indicated at the Bluegrass site (Stafford and Anslinger 1988; Stafford et al. 2000) compared with that seen at Black Earth. Bluegrass is situated in the Wabash Lowland in the upper part of the Pigeon Creek drainage. Like Black Earth, the site overlooks a lacustrine plain containing wetland habitats. In contrast, however, there were no large bodies of open water or permanent swamps near Bluegrass (Stafford et al. 2000:321–322).

The Bluegrass faunal assemblage consisted of three groups of animal taxa based on their abundance. The first group (white-tailed deer, box turtle, squirrel, and snake) was the most numerous. Raccoon, woodchuck, and wild turkey represent medium-sized taxa, while less important animals, like fish, aquatic turtle, cottontail rabbit, opossum, and crayfish, comprised the third group. The Bluegrass faunal assemblage clearly reflects an animal exploitation strategy that focused on terrestrial mammals, reptiles, and birds (Stafford et al. 2000: 324–325).

Deer was the most important animal at both Black Earth and Bluegrass during their late Middle Archaic occupations; however, significant differences existed. At Bluegrass, small mammals comprised nearly 10 percent of the faunal remains, while at Black Earth, they only represented about 1 percent. Birds and turtles accounted for about one-third of the Black Earth faunal assemblage, but only about one-half that at Bluegrass, although turtle still represented over 10 percent of the assemblage. Black Earth inhabitants captured a significant number of ducks and geese, while turkey dominates the Bluegrass avian assemblage (Stafford et al. 2000:329).

The terrestrial focus of the Bluegrass and Black Earth inhabitants' animal procurement strategies sharply contrasted with those pursued by many other late Middle Archaic hunter-gatherers. In many parts of the Midcontinent, particularly along major waterways, the appearance of base camps suggesting decreased mobility is linked to foraging strategies that emphasized deer and aquatic/semiaquatic species (Styles 1986; Styles and Klippel 1996). Outside the valleys, small-bodied mammals and reptiles were largely substituted for the wetland species. The pattern is apparent at both Black Earth and Bluegrass, with the latter being a more extreme example of upland adaptation.

Stafford et al. (2000:333) maintain that, regardless of the surrounding habitat types, processing small-bodied animals required more time, partially accounting for the decrease in time that late Middle Archaic groups spent traveling. This, along with factors associated with vegetation changes caused by

Holocene climatic shifts, contributed to the adoption of a logistical collector system, decreased residential mobility, and the exploitation of the full range of the Midwest's habitats.

Social Organization

Data collected by extensive archaeological surveys and excavations indicate that Lower Ohio Valley hunter-gatherer societies were undergoing significant organizational changes during the Middle Holocene. Early on, most regional hunter-gatherer groups maintained a way of life that was quite reminiscent of Early Holocene hunter-gatherer societies. The regional population density was still quite low and group mobility was relatively high. These characteristics are reflected by numerous small sites having low artifact densities. Early Middle Archaic components are found throughout most of the regional landscape, suggesting that hunter-gatherers were not focusing their collective attention on any particular set of resources and/or that social, economic, or political pressures did not limit their movements to any greater extent than during the Early Archaic. Other than broadly shared projectile point styles, there is little evidence suggesting that much interaction or exchange took place among these scattered groups. That which did take place probably happened at planned or spontaneous multiple band gatherings (Anderson 1996b:44). These trends seen in the Lower Ohio Valley generally parallel those observed throughout most of the North American Midcontinent.

About 6500 B.P. this picture starts to change, at least in some places. As discussed earlier, major changes in the mobility strategies that hunter-gatherers used to deal effectively with the evolving social and physical Middle Holocene landscape resulted in decreased mobility for part or all group members. In some parts of the Lower Ohio Valley, these changes are marked by thick earth and/or shell middens that are thought by many to represent late Middle Archaic base camps. The size and contents of these impressive sites, many strategically situated adjacent to ecologically rich wetlands, suggest that a significant portion of these hunter-gatherer groups was repeatedly visiting these locations. As a consequence, more people were living together for much of the year.

Changes in group mobility may have also been tied to increasing regional hunter-gatherer populations. More people living in larger groups could have led to greater circumscription, ultimately reducing the size of the group's territorial space. A smaller catchment area might mean a reduction in the number and kinds of options available to deal with various kinds of economic and social risks. Increased evidence in the form of shared artifact styles (i.e., carved and engraved bone pins, atlatl weights) and the presence of nonlocal materials (copper and marine shell) suggest a higher frequency and intensity of intergroup exchange and interaction, offsetting the effects of decreased group mobility and fewer episodes of multigroup aggregation.

Collectively, decreased group mobility, smaller territories, and more people living together for longer periods created new social, economic, political, and perhaps even ideological challenges for these groups. Mechanisms that once successfully met the demands of maintaining group order, ensuring an adequate food supply, and conducting various kinds of exchanges and interactions with neighboring and distant groups may have no longer worked. New solutions for dealing with these challenges probably involved significant social changes in late Middle Archaic society. Evidence for increasing hunter-gatherer social complexity is often reflected in the archaeological record by changes in technological organization, subsistence strategies, mortuary practices, settlement organization, architecture, and patterns of exchange and interaction (Brown 1985; Price and Brown 1985).

Despite the obvious changes to the late Middle Archaic cultural landscape in some parts of the Midcontinent, groups that inhabited other parts appear to have pursued a way of life not unlike that of earlier hunter-gatherers (Sassaman 1995). These groups apparently resisted or avoided conditions that brought about increased sedentism in some places. This is indicated by their small sites that usually consisted of a limited inventory of portable stone tools made from local lithic materials. Collectively, it seems that these hunter-gatherers had a high degree of residential mobility within relatively small territories. Late Middle Holocene hunter-gatherers exhibited a wide range of organizational variation that covered the full spectrum of mobility, subsistence, and settlement strategies (Sassaman 1995:181).

Investigations of mortuary practices have provided valuable insights about late Middle Archaic hunter-gatherer social organization (E. Anderson 1991; Buikstra 1981; Buikstra and Charles 1999; Charles and Buikstra 1983; Charles et al. 1988; Hassen and Farnsworth 1987; Magennis 1977). Formal cemeteries, like those found in the Gibson and Elizabeth mounds in west central Illinois, have been interpreted as territorial markers (Charles and Buikstra 1983). The construction of these Helton phase mounds, situated in highly visible spots on the bluffs overlooking the lower Illinois River valley, appear to be associated with increasing regional population and competition between neighboring hunter-gatherer groups. Adults who were incapable (physically incapacitated) of contributing to group well-being were excluded from bluff-top burial, instead being buried in nearby habitation areas (Charles and Buikstra 1983:134). Differential placement of burials based on age also appears to have been practiced during this time at Modoc Rock Shelter (E. Anderson 1991). It seems that where a late Middle Archaic hunter-gatherer was buried depended at least partially on his or her age and ability to participate in group activities.

Very few late Middle Archaic skeletal collections are known for the Lower Ohio Valley. A major exception to this is the large, well-preserved one from the Black Earth site. Because of the midden's excellent bone preservation, all age groups were well represented. Even the fragile remains of infants and children, not commonly found at sites having more acidic soils, were preserved. The comprehensive analysis of the late Middle Archaic mortuary program provided new data on the range of variation represented in mortuary treatment, as well as important new insights on social distinctions within late Middle Holocene hunting and gathering societies (Lynch 1982).

The Black Earth site excavations exposed more than 220 sets of human remains. However, only 124 of these could be confidently associated with the late Middle Archaic occupation and provide the required data for mortuary analysis (Lynch 1982). Infants (0–3 years old) accounted for more than 20 percent of all late Middle Archaic individuals. This relatively high percentage is comparable to figures from known hunter-gatherer groups (Howell 1979:89). The number of deaths assigned to the child and adolescent age group (3–18 years) was very low (9 percent), followed by an increase to 29 percent for young adults (18–35 years) and 36 percent for middle adults (35–50 years), reflecting the aging of those individuals. Old adults, those over 50 years old, accounted for only 3 percent, reflecting the very few individuals who reached advanced age (Lynch 1982:Table 321). Overall, males outnumbered females by a ratio of about 5 to 4.

Nearly 40 percent of the people were buried with their heads toward the west. Fifty-five percent of the males were oriented in that direction. Those oriented in other directions ranged from 4 to 8 percent. No statistical differences were detected between an individual's orientation and the presence or type(s) of grave goods found with them.

Burial position was evenly divided between extended (46 percent) and flexed (49 percent). Infants, adolescents, and young adults tended to be extended, while middle and older adults were usually flexed. Based on these findings, age was apparently an important consideration in determining burial positioning. There was no distinction in orientation based on gender (Lynch 1982).

While Black Earth burials were either extended or flexed, exceptions did occur. A 38-year-old male interred in a deep oval pit, was placed in a seated position with his right leg extended and his left leg bent back at a 90 degree angle. His upper torso was pushed forward over his legs, with his face pressed against his left ankle. The left arm stretched back along the torso, while his right arm extended from the shoulder, then bent back at the elbow. No grave goods accompanied this person (Lynch 1982:Figure 294).

Although the cultural significance of this kind of mortuary treatment is unknown, it clearly does not fit the norm of late Middle Archaic burial practices. It seems unlikely that the body would have stayed in position without the upper torso being held down while being buried. Was this the kind of treatment someone received when they violated group norms? Perhaps he was not a member of the group, but if that was the case, why was he buried at Black Earth in the first place? Perhaps future research will reveal more about this person and the burial treatment he received.

In addition to the treatment of the body, the number, variety, and types of grave goods placed with an individual also helped to identify distinctions within the late Middle Archaic burial population (Lynch 1982). Only 27 percent of the individuals had grave goods of any kind, so the very presence of any utilitarian, ornamental, or ritual/ceremonial object may say something about a person's social standing. Grave goods were more commonly associated with extended burials, but there was no strong correlation between grave goods and a person's gender or burial position.

There were no obvious differences in the kinds of grave goods associated with men and women, but male interments tended to have a wider variety of objects than did female. Utilitarian items were usually associated with adults, of which flaked stone drills, chert cores, projectile points, and grooved axes were exclusively associated with males. Flaked stone scrapers, bone awls, modified pieces of deer antler, chert bifaces, and bone needles were associated with, but not necessarily limited to, females. Ornamental objects restricted to males included bone pins and gorgets, pieces of worked stone, decorated bones, and modified small animal bones (Lynch 1982).

Artifacts used for hunting and other resource procurement tasks were generally associated with males. Many female-associated grave goods could have been used to process these resources. Utilitarian items were found with only one person under the age of 15 years, suggesting that these implements, and the tasks for which they were used, were age restricted. Such distinctions may provide insights into the division of labor within these hunting and gathering societies (Lynch 1982).

Unlike utilitarian items, objects classified as ornamental and ritual/ceremonial occurred with both adults and subadults, including infants. Several adolescent males possessed burial objects, like elaborately carved and engraved bone pins, which possibly served as prestige items (Jefferies 2004). The possession of objects like bone pins suggests that certain members of this society, on the one hand, could attain some degree of individual prestige at an early age. On the other hand, adults over 40 years old did not have access to either ornamental or ritual/ceremonial objects, suggesting that by that age they no longer participated in the full range of adult activities (Lynch 1982:1151).

Several of the Black Earth site burials included collections of objects reminiscent of "medicine bundles" used by historically documented hunter-gatherers. George Catlin (1841) indicates that, among nineteenth-century Native American groups of the upper Missouri River, when a person died, their medicine bag was placed in the grave and allowed to decay with their body (Mooney 1975:114).

At Black Earth, archaeologists found a compact cluster of more than 40 items near the left shoulder of an adult male. The artifacts, which appeared to have once been in a bag or pouch, included bear and eagle phalanges, the beak of a large aquatic bird, modified small animal bones, a deer bone pin and awl, side notched points and endscrapers, quartzite pebbles, pieces of ground limestone, a small grooved axe, hematite, and turtle shell fragments (Lynch 1982: Table 343). Several other Black Earth burials had comparable, but smaller, artifact concentrations buried with them. Webb (1950a:336–343) described similar artifact occurrences at the Carlston Annis shell midden as "medicine bags."

The association of these "medicine bags" or bundles with certain people may signify their involvement in ritual activities. Or, they may simply be items of personal significance to those individuals.

The preferred way for Black Earth inhabitants to inter their dead was to place each person in his or her own burial pit. However, at least nine burial features containing two or more individuals show that this treatment was not always desired or possible. Multiple burials contained individuals of all ages and genders. Males, however, were relatively uncommon, occurring in only three cases. Only 9 percent of individuals in multiple burials had associated grave goods, far less than for the general population (Lynch 1982:1160).

In a few cases, the skeletal remains themselves provide important insights into the Black Earth inhabitants' social world. For example, a 45-year-old male suffered two crushed lumbar vertebrae, resulting in a severe spinal column injury. Skeletal evidence indicates that he was paralyzed below the neck by this injury, making him a quadriplegic. Despite his immobility, other skeletal indicators demonstrate that the man lived for many years after his devastating injury. This man's extended life, despite his debilitating injuries and immobility, supports the concept of a relatively sedentary population at the Black Earth site, at least during this person's lifetime (Bassett 1982).

The late Middle Archaic mortuary program at the Bluegrass site was very similar to that documented at Black Earth (Mayes 1997; Stafford et al. 2000: 321). Analysis of 80 individuals yielded an expected age and sex distribution. Like at Black Earth, infants were well represented, accounting for 13 percent of all individuals. Burials were placed in extended, loosely flexed, and tightly flexed positions. Tightly flexed burials appear to have been bound at death before they were interred in circular burial pits (Mayes 1997:76).

Thirty-four percent of the Bluegrass burials had associated artifacts, slightly higher than at Black Earth. The majority of grave goods was buried with adult males, primarily utilitarian items like projectile points, bone awls, and atlatl parts. Burial objects did not occur with infants (Mayes 1997:78). Ornamental items, such as shell and bone beads, bone pins, drilled canine teeth, and red ocher, were commonly associated with females (1997:88).

One of the Bluegrass burials is of particular interest. This juvenile or young adult of unknown sex was buried in a sitting position with his/her chin resting on the chest area (Mayes 1997:90). Artifacts associated with this person included antler tines, atlatl parts, bone awls, a chert drill, a drilled canine, a limonite bar, a woodchuck ramus, a projectile point, a flake, and a mussel shell. While the significance of this individual's burial position and multiple grave goods is unclear, he/she was one of the few subadults to be given this mortuary treatment (1997:90). Like at Black Earth and Carlston Annis, perhaps these objects were also part of a medicine bundle.

The nearby Little Pigeon Creek cemetery, also located in southern Indiana, represents yet another example of a late Middle Archaic (French Lick phase) midden and mortuary area (Cook et al. 1996). The site, situated at the end of a floodplain ridge that overlooks the Little Pigeon Creek, has up to three meters of midden, the lower one-half of which is attributable to the French Lick phase occupation. Grave goods associated with the 31 French Lick phase burials included engraved bone pins and spatulae, ocher, and unmodified animal bone. The age and gender composition of the site's burial population, along with the kinds and distribution of grave goods, suggest an egalitarian level of social organization (Cook et al. 1996).

Recent investigations at the Meyer site, situated just 2 km north of the Ohio

River in southern Indiana, have yielded new insights about late Middle Archaic mortuary practices (Bader 2004). Like many other contemporary sites of this type, Meyer is located on a low ridge that overlooks extensive wetland/forest habitats. The presence of late Middle Archaic side notched (Godar, Matanzas, Brewerton), as well as a few Late Archaic stemmed projectile points, suggests that occupation spanned several thousand years. A radiocarbon date of 5280 +/– 70 BP was associated with a carved bone pin similar to those found at many other Lower Ohio Valley late Middle Archaic sites.

Site investigations located a minimum of 30 burials, most of which were placed in a flexed position. Interestingly, small circular charcoal stains averaging about 30 cm in diameter found near many burials suggest that the mortuary program involved the use of fire/smoke. One infant had four of these dark, circular features placed around it, one in each of the cardinal directions. Other burials had similar features, most of which were positioned near the knees or the head. Such features may have been used as smudge pits or for burning materials used in the mortuary ritual (Bader 2004:22). Similar features were associated with roughly contemporary burials at the nearby KYANG and Little Pigeon Creek sites (Bader 2005).

The Meyer site midden contained a very low density of habitation debris—even normally abundant debitage was rare. In contrast, features contained a variety of cultural materials like carved and engraved bone pin parts, awls, and beads; modified animal teeth; cut deer antler; turtle shell; flaked stone drills, scrapers, and side notched points; and ground stone axes (Bader 2004:19).

Low midden artifact frequencies, combined with the presence of burials and features containing an unusual assortment of artifacts, suggest that Meyer was not a typical habitation site, instead serving as a special-purpose mortuary area. In many ways, Meyer is reminiscent of what we see at some southern Indiana Early Archaic sites (Jerger, McCollough's Run) that were used almost exclusively for mortuary activities. Bader (2005) has recently proposed that the nearby KYANG site also primarily functioned as a late Middle Archaic mortuary site.

Bader (2005) notes that numerous late Middle Archaic burial areas were often situated to overlook wetlands that hosted critical food resources. The positioning of burial grounds near these important resource areas may have served as a means of identifying territorial prerogatives by local hunter-gatherer groups. If these mortuary areas served as territorial markers (Charles and Buikstra 1983), they had to be visible to other groups. This could have been accomplished by selecting a prominent topographic feature like those occupied by Meyer and KYANG, keeping the area clear of vegetation, and/or erecting markers, perhaps large carved or decorated wooden poles, that would have been visible to those approaching the site (Bader 2005).

The mortuary practices represented at Black Earth, Bluegrass, Little Pigeon Creek, and Meyer are very similar to those documented at contemporary sites in the Midcontinent like Koster and Elizabeth Mounds in west central Illinois (Albertson and Charles 1988; Buikstra 1981; Stafford et al. 2000). It appears that most differentiation in postmortem treatment was based on the gender, age, and personal abilities of each individual. A few people were buried with more and/or a greater variety of grave goods than others, but the association of these objects probably reflects the individual abilities of that person. Good examples of this are the medicine bundles associated with a few late Middle Archaic individuals. Based on the characteristics of the Black Earth and Bluegrass mortuary programs, it appears that most people's status was "achieved" more than "ascribed" in these largely egalitarian hunting and gathering societies (Lynch 1982; Mayes 1997; Stafford et al. 2000).

Bioarchaeology and Paleodemography

The many Euroamerican settlers who came to the Lower Ohio Valley in the early nineteenth century frequently encountered skeletal remains of the region's prehistoric hunter-gatherers. Early cave explorers found desiccated bodies and mummies in some area caves (Meloy and Watson 1969), while farmers' plows often exposed human skeletons in agricultural fields placed atop large riveredge middens. Many regional histories offer descriptions of these skeletons and mummies but provide little explanation (and sometimes a lot of speculation) concerning their origin or age (Haskins and Herrmann 1996:108).

One of the first large hunter-gatherer skeletal collections "excavated" in the Lower Ohio Valley came from C. B. Moore's 1915 investigations at the Indian Knoll site in Kentucky (Moore 1916). In keeping with the times, most burials were not completely excavated, and much of the postcranial skeletal material was reburied at the site. Moore took other skeletal elements, primarily crania, for analysis (Haskins and Herrmann 1996:108). Subsequent archaeological investigations in Kentucky, western Tennessee, and northern Alabama, most of it conducted in the 1930s and 1940s under the auspices of various Depression Era work relief programs, yielded additional large skeletal collections (Webb 1946, 1950a; Webb and DeJarnette 1942; Webb and Haag 1940). In particular, Webb's field investigations along the Green River yielded many burials, most of which (91 percent) were attributable to the region's Archaic period occupation (Haskins and Herrmann 1996:107–110).

Conforming to the existing paradigm, analyses of the skeletal materials from these sites largely focused on craniometrics (Snow 1948). Documentation of trauma and disease was usually limited to describing a few specific cases. While this approach produced data, it provided little insight concerning Archaic populations other than that these hunter-gatherers were robust and had high levels of dental attrition, few caries, and long heads (Smith 1996:134).

During the 1950s and 1960s, some physical anthropologists started to adopt a new analytical strategy for studying human skeletal biology known as the "biocultural approach" (Armelagos et al. 1982). By the mid-1970s, this trend culminated with the emergence of the field of bioarchaeology that employed an archaeological, environmental, and cultural framework to examine osteological data (Haskins and Herrmann 1996:111). The application of bioarchaeological concepts gradually transformed osteological data from something that traditionally appeared as appendices of measurements and cranial types at the back of site reports to culturally significant information (Smith 1996:134).

The Black Earth site's large, well-preserved, and well-dated late Middle Archaic skeletal collection was perfectly suited for such a population-oriented analysis. Once the project osteologist determined the age and gender of each individual, the resulting data were used to address a broad range of questions concerning the adaptation, nutrition, and overall health of the site's inhabitants (Bassett 1982).

Investigators found that, as for many other Lower Ohio Valley hunter-gatherer groups, infant mortality was quite high (Haskins and Herrmann 1996:117; Powell 1996:120). At least 27 percent of the population died within the first year of life. Such a high percentage of infants might also indicate a strong preference for burying newborns in habitation areas like the Black Earth site (Bassett 1982:1052).

Birthing-related deaths for females were not exceptionally high—in fact, the percentage of Middle Archaic females who died during the period of maximum childbearing (15–30 years old) was lower than for males of the same age range (24 vs. 32 percent). Several Black Earth site burials contained the remains of both an infant and a female, suggesting that child-birthing deaths did occur.

At birth, a Black Earth site newborn had an average life expectancy of 24 years. If a child lived through its first year, he or she could expect to live for an additional 32 years. This is about the same life expectancy that people who lived in England had during the Middle Ages (Todd 1927:Figure 7).

Once surviving birth and the first few months of life, chances for a child's survival increased. Only 2 percent of the population died between the ages of two and four years, suggesting low weanling mortality. Mortality averaged about 2 percent per year until the age of 13, then slightly increasing each year until adulthood. Adult males outnumbered adult females by about 1.5 to 1.0 and continued to do so until after the age of 40. If a male survived his first year, he could expect to live for an average of 32 additional years. Females generally outlived their male counterparts by approximately six years. The oldest person buried at the Black Earth site, a female, lived to be about 60 years old (Bassett 1982).

A population's collective growth and development characteristics can provide insights about their general nutrition. Progression of longbone growth is often used as a measure of childhood nutritional stress, while interruptions in bone growth are commonly interpreted as signs of nutritional stress during the major childhood growth period (Bassett 1982:1055). Based on these criteria, Black Earth inhabitants did not display any evidence for long periods of poor diet that could have caused nutritional stress. Undoubtedly, there were times when the food supply ran low or dropped in nutritional quality, but these episodes were too short to affect normal bone growth and development. Stature reconstructions indicate that men were on average about 166.6 cm (5.5 ft) tall; women were slightly shorter, averaging about 158 cm (5.2 ft) (Bassett 1982).

Skeletal evidence also provided information on some of the injuries and traumata experienced by the Black Earth site inhabitants. Although the skeletal series exhibited a variety of pathologies, few appear to have been the cause of death. This is a somewhat biased picture, however, since many diseases and injuries leave little or no skeletal evidence. In many cases, osteological evidence of trauma and disease appears on the bones of those who survived them rather than those who quickly died from them (Bassett 1982).

Many of the late Middle Archaic inhabitants, both men and women and young and old, showed indications of some type of progressive disease or diseases that produced symptoms such as mild inflammation of the boney tissue, deep lesions on the longbones and crania, and severe bowing of the tibiae in severe cases. Although the precise cause of these symptoms could not be narrowed to one disease, they appear to be attributable to a treponemal-like disease such as yaws or pinta (Bassett 1982; Golden 2006). These nonsexually transmitted infections are caused by treponemal spirochetes that are closely related to the bacterium that causes syphilis (Merck & Co. 2006).

Apparently, the severity of the disease increased with age since only 17 percent of the 15–20-year-old age group displayed symptoms. The first severe case occurred in the 30–35 age group, and by the age of 60, almost everyone had symptoms. Despite the widespread occurrence of symptoms, the disease(s) that caused them did not seem to be a significant cause of death.

At least 23, or 15 percent, of Black Earth skeletons had one or more bone fractures. Many of the people buried at the Green River shell middens also had high frequencies of fractures, suggesting that the hunter-gatherer way of life was demanding and full of risks (Powell 1996:125). At the Black Earth site, most fractures occurred in the rib area or on the forearms. Males had more fractures than females, even when their greater numbers were taken into consideration. One of the more common injuries (9 males, 3 females) was a forearm fracture, commonly referred to as a "parry fracture." These kinds of forearm breaks can be caused by warding off blows resulting from interpersonal conflict (Armelagos 1969). The number of men with parry fractures is somewhat puzzling, however, since there were no other signs of conflict (Bassett 1982). Males had the only serious fractures of the cranium, scapula, and pelvis. Several men had multiple fractures of the cranium, rib, and vertebra, suggesting that they experienced a severe fall, perhaps from a tree or cliff, at some point in their lives.

Once a person broke a bone, there was a distinct possibility that the injury would become infected. Some of the Black Earth inhabitants developed infections in the muscle tissue near the damaged bone; these were of relatively minor significance. Other less common, but more serious, infections occurred in the marrow inside the bone. These often-fatal systemic infections, known as osteomyelitis, were sometimes caused by a staphylococcus infection that produced abscesses, pus-draining sinuses, and areas of dead bone (Bassett 1982).

Perhaps the most widespread health issue at Black Earth was osteoarthritis and osteophytosis. Just about every Black Earth man and woman over the age of 35 suffered from arthritis to some extent, and the extent of the illness, just like today, increased with age. Males tended to have a higher incidence in the upper torso, possibly reflecting the use of this part of the body in hunting activities or while carrying heavy objects. Females were most commonly affected in the lumbar region, perhaps from strenuous activities like digging or childbirth (Bassett 1982).

Recent analysis of musculoskeletal stress markers observed on a sample of the Black Earth site late Middle Archaic skeletal collection supports the active use of atlatls by males. In contrast, markers on female skeletons indicate that they commonly performed activities like pounding nuts, grinding food, and processing hides. The large quantity of scrapers found at the Black Earth site also point to hide processing being an important activity (Clapper 2006: 50–52).

The dental health of the population reflects the kinds of food the Black Earth residents ate and the ways they prepared it. Teeth were characterized by a very high rate of apical and alveolar abscessing, a rapid rate of attrition, and relatively few caries. Apical abscessing is often associated with decay of the tooth's pulp cavity, a consequence of rapid attrition. Alveolar abscesses are commonly caused by small pieces of grit that become trapped between the tooth and the gum. Most of the caries were caused by the high attrition rate that often exposed the pulp cavity, as opposed to enamel caries that formed in the fissures (Bassett 1982).

Attrition of the teeth started early in life and proceeded very rapidly. Tooth wear was evident on the teeth of very young children who had not even been weaned. Apparently, the Black Earth site's sandy midden adhered to the objects that children often put in their mouths.

The high rate of tooth wear for older adults was probably caused by the way they prepared their foods. At Black Earth, sandstone was the most plentiful material for making the ground stone tools used to process nuts, seeds, and other hard foods. As people ground these foods, small sand grains detached from the implements, mixed with the food, later to be chewed along with the daily meals. Over time, this process took its toll on the teeth. Virtually all adults exhibited some tooth wear, and several had teeth that had been worn down to the root, below the gum line. While the high attrition rate increased abscessing, it probably reduced the number of caries (Bassett 1982).

Overall, late Middle Archaic hunter-gatherers who lived and died at Black Earth probably led relatively healthy lives. If they survived their first year, they could expect to live for another 35 years or so. Surprisingly, their life expectancy generally matched or exceeded those of other preindustrial societies throughout the world. A variety of skeletal markers (growth arrest lines, dental hypoplasia, fluctuating skeletal asymmetry, porotic hyperostosis) indicate that these people had a pretty healthy diet and experienced very little metabolic stress due to poor nutrition. The average person would probably break a few bones during his/her life, but most would recover in some fashion. Just about everyone had to put up with the pain of arthritis and some kind of systemic disease that affected the skeletal system. Apparently, they got along pretty well with one another and their neighbors since there were few indicators of conflict or aggression (i.e., projectile points imbedded in bones, evidence of scalping, skull bashing, etc.). Overall, it was a pretty good life.

The late Middle Archaic skeletal series from the Bluegrass site, consisting of 80 individuals, closely resembled the Black Earth series. The sex and age distribution appears to be typical of those found in living populations, with fewer females than males. Infants were well represented at the site, accounting for 13 percent of the sample. Female mortality peaked between 20 and 29 years, probably reflecting increased risks associated with childbearing. Male mortality peaked in the 40–49-year age group. If a person survived their first five years, life expectancy was about 29 years (Mayes 1997:93). Bluegrass site men were about the same size (165 cm; 5.4 ft) as those at Black Earth, while women were slightly shorter, averaging 154 cm (5.1 ft) tall (Mayes 1997:Table 5.3). The stature of the Bluegrass inhabitants suggests that they, like those at Black Earth, did not experience prolonged episodes of severe nutritional stress (1997:94).

Overall, it seems that the Bluegrass site inhabitants experienced a relatively healthy, if somewhat short, life. They had relatively few dental caries and abscesses, and lower than expected rates of arthritis and infectious lesions (Mayes 1977:67). Low frequencies of bone pathologies associated with nutritional stress and disease also suggest a generally healthy population. Little evidence for trauma was found, so they probably experienced little interpersonal conflict and lived a life that was not physically stressful enough to cause lots of traumatic injuries (1997:73, 95).

Exchange and Interaction

Lower Ohio Valley hunter-gatherers had probably participated in intergroup exchange and interaction since the first groups arrived in the area at the end of the Pleistocene. Unfortunately, the archaeological record has yielded little evidence for these early transactions. Projectile points made from "exotic" foreign cherts occur, but they are usually interpreted as an indication of a group's mobility and the size of its home territory (Johnson 1994:105). A few marine shell beads are associated with the Early Archaic occupations at Modoc Rock Shelter (Ahler 1991) and Jerger (Tomak 1979), but such finds are rare. It is not until the second half of the Middle Archaic that evidence for more extensive intergroup interaction becomes more common (Jefferies 1995a).

Brown (1985:224) has proposed that increased sedentism during the late Middle Archaic resulted from the efforts of individual groups to maintain intergroup spacing where and when there were sufficient subsistence resources. The loss of risk-averting strategies used by groups that were more mobile might have been partially alleviated by the strengthening of intergroup social networks, including the expansion of economic ties through exchange.

Archaeological evidence for late Middle Archaic exchange in the Lower Ohio Valley and adjacent area occurs in at least two forms. First, items made from "exotic" or nonlocal materials, such as copper and galena, made their first significant appearance in the archaeological record (Brown and Vierra 1983:185). Hunter-gatherers continued to exchange marine shell as well (Jefferies 1982a:1476). Second, the regional-scale distribution of stylistically similar artifacts, other than projectile points, occurs for the first time. Examples of such artifacts include atlatl weights (Burdin 2004) and elaborately carved and engraved bone pins (Jefferies 1997).

Most of the evidence for intergroup interaction comes from the large late Middle Archaic base camps like Black Earth and Bluegrass. Items made from nonlocal materials were occasionally placed with some of the burials interred at these sites. In addition, the preservation of delicately carved and engraved bone and shell items was facilitated by the highly alkaline soils that characterize many of these sites.

Copper artifacts occur at several Lower Ohio Valley late Middle Archaic sites, including Black Earth, Baker (McBride 2000), and Bluegrass (Anslinger, personal communications 1992). Artifacts from the Black Earth site vicinity actually came from a nearby site situated on a low ridge immediately to the south (11Sa88). Artifacts associated with one burial included a late Middle Archaic Matanzas projectile point and a copper wedgelike object that was placed near the person's head (Lynch 1982:Figure 318). The cultural affiliation of a second individual, accompanied by several copper beads, was less certain.

Perhaps the clearest example of late Middle Archaic regional-scale interaction and exchange in the Lower Ohio Valley is the wide distribution of elaborately carved and engraved bone pins (Jefferies 1997, 2004; White 2003a). Although the known distribution of these artifacts is strongly associated with sites having good bone preservation conditions, it is clear that similar pin styles were manufactured throughout much of the southern Midwest, including the Lower Ohio Valley (Figure 6.10).

The people who made the Koster site pins removed long sections of bone,

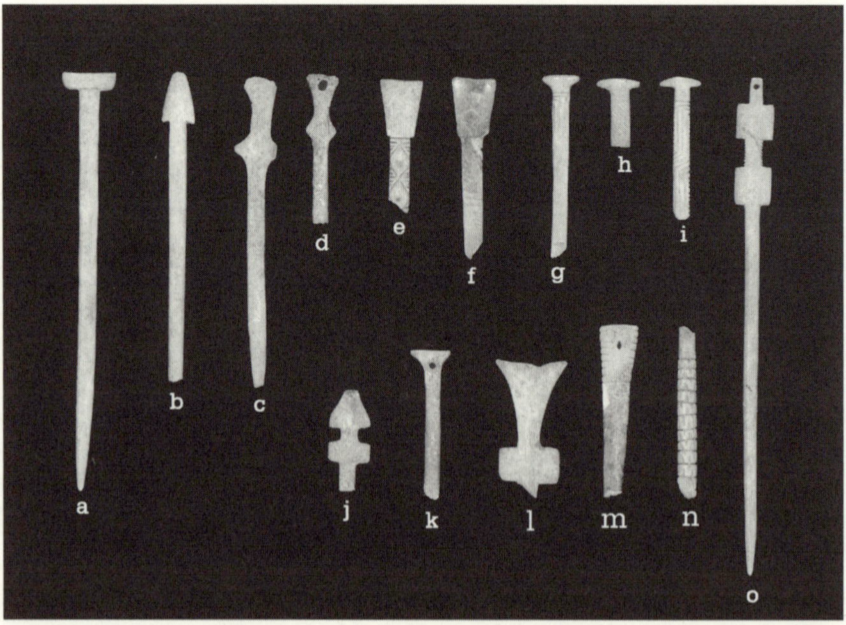

Figure 6.10. Black Earth site (Area A) carved and engraved bone pins (courtesy of the Center for Archaeological Investigations, Southern Illinois University Carbondale).

usually deer bone, by incising and splitting longbones, particularly the metapodials (Cook 1976). Once detached, the sections were ground flat, split into even thinner sections, then carved and ground to achieve their final shape (Bader 1992:280; Cook 1976:38).

Although commonly referred to as "hairpins," the precise way that late Middle Archaic hunter-gatherers used these objects is unclear. The pins showed little evidence for use, so they were not employed for any heavy-duty tasks. The polish present on most of the pin shafts is probably attributable to final shaping of the pin and its subsequent handling and use (Bader 1992:283). Some pins were found in the head area of burials (Webb 1974:167, 170, 172, 222), suggesting that they actually were hairpins. Others appear to have been used to fasten clothing or to hold burial wraps in place (Jefferies 2004:74).

The heads of some late Middle Archaic pins were carved to create one of several common shapes that included "crutch-top," "t-top," "fishtailed-cruciform," and "double-expanded" forms (Figure 6.10) (Jefferies 1997:Figures 3–4). Many pins were engraved with a variety of simple and complex designs such as zoned diamonds, zigzags, concentric squares or steps, vertical or horizontal lines, and zoned zigzags (Jefferies 1997:Figure 7).

Analysis of the spatial distribution of head shapes and decorative motifs revealed that similar styles occurred at a number of late Middle Archaic sites scattered across the southern Midwest. Examination of more than 140 bone pins from eight sites (KYANG in Kentucky; Bluegrass, Crib Mound, and McCain in southern Indiana; Black Earth, Modoc Rock Shelter, and Koster in Illinois, and Arnold Research Cave in Missouri) resulted in the identification of seven pin types. Each type occurred at three or more of these sites.

The "crutch-top" style was the most common and most widely distributed, occurring at six of seven sites extending from the Falls of the Ohio (KYANG) to west central Illinois (Koster). Intricately engraved t-top pins, characterized by their distinctive expanded head shape, occurred at five of seven sites (Jefferies 1997).

Virtually identical pin styles found at widely scattered sites indicates that at least some late Middle Archaic groups participated in regional-scale networks that promoted social interaction and integration. Radiocarbon dates associated with the pins indicate that these groups maintained some level of interaction for at least 1,000 years. Andrew White's (2003a) recent stratigraphic analysis of the Black Earth site bone pins suggests that the popularity of different head shapes and engraving styles varied over time.

Comparison of the stylistic attributes of pins found on contemporary sites north and south of the Ohio River revealed few similarities, suggesting that the river coincided with some type of social boundary demarcating distinctly different regional hunter-gatherer groups (Jefferies 2004).

I think that the distribution of bone pin styles suggests that the emergence of more complex hunting and gathering societies in the southern Midwest, including the Lower Ohio Valley, was accompanied by a restructuring of intergroup relationships and a more broadly defined cultural identity. Trends toward reduced group mobility and smaller home territories toward the end of the Middle Archaic meant that new solutions had to be developed to continue the flow of needed materials, information, and mates, and to ensure that intergroup alliances would continue to operate smoothly (Jefferies 2004:84–85).

Over time, the intensification of interaction among neighboring hunter-gatherer groups led to a broadening of cultural identity and the establishment of larger, more complex social groups. If access to other groups and their resources (information, food, exotic materials, and mates) was restricted to a few privileged individuals, then these intermediaries may have acquired greater status and prestige within their group (Marquardt 1985), ultimately creating a mechanism for amplifying social inequities among group members (Jefferies 2004:85).

The distribution of stylistically similar projectile points, atlatl weights, grooved axes, and other objects suggests that similar social transitions were underway in other parts of the Southeast and Midwest during the late Middle Archaic (Burdin 2004; Johnson and Brookes 1989; Sassaman et al. 1988). In contrast, similar archaeological indicators have not been found in all parts of the region, suggesting that the organizational complexity of late Middle Archaic hunter-gatherer societies widely varied across the Lower Ohio Valley and beyond. Much like today, some people may have preferred to have been left alone so that they could continue living like they did in earlier times (Jefferies 2004:85).

Summary

The Middle Holocene witnessed major changes in the Lower Ohio Valley's physical and cultural landscapes. By this time, an essentially modern climatic regime was established and the distribution of natural habitats and their associated plant and animal communities approximated those of today. This pattern was disrupted during the late Middle Holocene by the influences of the Hypsithermal interval that brought slightly drier and warmer conditions to the region.

Resulting changes in the distribution of food resources appear to have contributed to the reorganization of hunter-gatherer settlement/subsistence strategies. Population growth also played a critical role in shaping the organization of local and regional hunter-gatherer groups and influencing the ways groups maintained their social connections.

By the late Middle Holocene, some Lower Ohio Valley hunter-gatherers were spending more of their time at places that were strategically located on the physical and cultural landscapes, facilitating their access to important natural resources and basic social needs. Thick midden deposits indicative of centuries of intensive cultural activities mark many of these spots. Throughout this time, hunter-gatherers buried their deceased relatives and friends at these places, increasing the ritual significance of these spots for millennia to come.

As we will see in the next chapter, changes in hunter-gatherer social, economic, and ritual life first seen during the late Middle Holocene continued well into the Late Holocene. In some parts of the Lower Ohio Valley, the size and complexity of hunter-gatherer social groups continued to increase, marked by indicators of greater social inequality, more intensive and extensive social connections, and the eventual emergence of horticulture based on the cultivation of domesticated local cultigens.

7 / The Late Holocene

Filling the Landscape

Introduction

Geologists refer to the time from 5,000 years ago until the present as the Late Holocene. Archaeologically, the first 2,000 years of the Late Holocene correspond to the Late Archaic subperiod (5000–3000 B.P.). The drier conditions of the Hypsithermal interval were ending by this time, and sea levels along the Atlantic and Gulf coasts were approaching those of today. The composition of the Southeast's forests, including those in the Lower Ohio Valley, changed very little during the Late Holocene, except for relatively minor spatial and temporal fluctuations attributable to small oscillations in temperature and moisture, as well as the growing impact of more people living on the land (Muller 1986:51; Smith 1986:6–7).

Tall stands of oak-hickory forests covered much of the Lower Ohio Valley landscape. The acorns and nuts that these trees produced were important foods for Late Holocene hunter-gatherers as well as many of the animals they relied on for food, clothing, and raw materials for toolmaking. Southern pine gradually crept into far western Kentucky during this time, along with vegetation commonly associated with cypress-gum forests in the lower Mississippi River valley (Delcourt and Delcourt 1981:Figures 8 and 9). To the north, the prairie grasslands gradually retreated to the west, reflecting the amelioration of the drier Hypsithermal conditions.

Pollen profiles from Old Field swamp in southeastern Missouri near the confluence of the Ohio and Mississippi rivers indicate the replacement of grass and herb plant communities by bottomland arboreal vegetation and an increase in swamp size about 5,000 years ago, reflecting increased precipitation (King and Allen 1977:320–321). At Jackson Pond, in west central Kentucky, an increased percentage of grass pollen and the appearance of prairie vegetation like prairie clover suggest that the mixed prairie-deciduous forests encountered by early Euroamerican travelers were established by this time (Wilkins et al. 1991:236).

Throughout much of the Southeast and Midwest, the Late Holocene was a time of cultural continuity as well as change. In many places cultural traditions established during the late Middle Archaic continued well into the Late Archaic with relatively little technological or social change. In contrast, recent research has demonstrated that several cultural innovations once used to distinguish later Woodland cultures (3000–1000 B.P.) from those of the Archaic actually had their roots in the Late, or in a few cases, the Middle Archaic. These include the establishment of formal mortuary areas, sometimes in the form of low burial mounds (Buikstra and Charles 1999; Charles and Buikstra 1983), and the cultivation of domesticated local plants in small gardens (Smith and Cowan 2003).

In parts of the Southeast, but apparently not the Lower Ohio Valley, Late Archaic groups were experimenting with making clay pots for cooking and storing foods (Sassaman 1993a; Smith 1986). Smith (1986:30), referring to this technological innovation as "the container revolution," noted that early ceramic vessels were technologically simple and aesthetically unassuming. These shallow, hand-molded vessels were made using clay tempered with either plant fibers or sand. In other parts of the Southeast and Northeast, containers sometimes were fashioned from steatite or sandstone (Sassaman 1993a:180–185; Webb and DeJarnette 1942). While Lower Ohio Valley hunter-gatherers undoubtedly used an assortment of wooden, fiber, or hide containers, and possibly dried gourd or squash bottles (Smith 1986), no evidence for ceramic vessel manufacture has been found prior to start of the Woodland period.

Some Late Archaic hunter-gatherers expanded on the tradition of constructing earthworks and mounds started by their Middle Archaic forebears (Russo 1996; Saunders 2004; Saunders et al. 1994). Earthen mounds, often located in prominent settings like on bluff tops above river valleys, served a mortuary function as well as possibly marking a group's home territory (Charles and Buikstra 1983; Klepinger and Henning 1976). The function(s) of other Late Archaic mounds and earthworks, like those constructed at Poverty Point in Louisiana, are more difficult to discern (Gibson 1996, 2001), but they clearly demonstrate the level of cultural complexity achieved by Late Archaic people. So far, no intentionally constructed Late Archaic earthworks or mounds are known for the Lower Ohio Valley, but some researchers have suggested that the Green River shell middens were actually ceremonial mounds similar to later Woodland-period constructions (Claassen 1996a:243).

In parts of the Midcontinent Late Archaic hunter-gatherers continued to move toward the eventual domestication of an assortment of starchy and oily seed-bearing plants like sunflower, sumpweed, chenopod, and maygrass. Changes in seed morphology suggest that domestication of some species predates 4000 B.P., so it seems likely that Late Holocene hunter-gatherers harvested and replanted them somewhat earlier (Gremillion 1996:107). These plants became important food resources during the Late Holocene in some parts of eastern North America (Gremillion 1996; Smith 1989). Gourds, which probably served as containers, rattles, or net floats, were being grown by 7000 B.P. New genetic data suggest that Paleoindian groups brought domesticated bottle gourds with them when they colonized North and South America more that 10,000 years ago (Erickson et al. 2005).

Much of the evidence supporting the early domestication of these native species comes from eastern Kentucky rockshelters and sites in west central and southern Illinois (Gremillion 1996). In contrast, paleoethnobotanists have found little evidence for early plant domestication in areas adjacent to the lower Ohio River (Rossen 2000, 2006), and intensive efforts to find indications for Late Archaic plant domestication at the Green River shell middens have produced few positive results (Marquardt and Watson 2005a:631). Investigations at the nearby upland Ward site yielded nearly 90 chenopod seeds (Bonzani 2002; Jefferies et al. 2008), supporting the contention that experimentation with seed harvesting and plant domestication may have been more common at upland Green River sites rather than along the river (Marquardt and Watson 2005a:631).

Evidence in the form of exotic stone, copper, marine shell, and the bones of nonnative animal species suggests the continuation and probable expansion of interregional interaction/exchange networks during the Late Archaic (Marquardt and Watson 2005a:634). The increased geographic coverage of Late Archaic social networks brought with it a greater opportunity for intergroup cooperation and integration, as well as conflict.

Throughout much of the Lower Ohio Valley, settlement strategies initiated in the late Middle Holocene appear to have extended well into the Late Holocene. Many Green River hunter-gatherers pursued settlement/subsistence practices that were broadly similar to those of their late Middle Archaic predecessors, resulting in the continued accumulation of large shell and earth middens. While the occupation of these impressive river-edge sites continued, a general increase in the number of Late Archaic sites in other environmental zones suggests that regional populations were growing and starting to exploit a wider range of habitats (Jefferies et al. 2005; Marquardt and Watson 2005c).

In contrast, activities at other floodplain locations that were intensively occupied during the Middle Holocene, such as Morrisroe, became more intermittent and less intensive during the Late Holocene (Conaty 1985; Nance 1988). Decreased Late Holocene occupational intensity also characterized the Black Earth site (Jefferies 1983). At the same time, lower Tennessee-Cumberland Late Holocene hunter-gatherers appear to have expanded their use of the uplands where hunting-related activities predominated (Nance 1977:11–13).

In Kentucky, the distribution of Late Archaic components generally corresponds to that of Middle Archaic ones, but Late Archaic sites are more numerous and often occur in clusters. Late Archaic sites are more widely distributed over the landscape, resulting in fewer empty areas or possible buffer zones. These subtle changes are attributable to one or more ongoing processes, including greater intergroup cooperation or competitive interaction, resulting in more sites in intervening areas; greater permeability of social boundaries, promoting increased movement of groups or individuals over the landscape; and/or increased group mobility within larger territories (Anderson 1996a:171). Given the evidence for increased weapon-related trauma during the Late Archaic, areas between group territories may have been where some of this intergroup conflict occurred (Smith 1996:144). Figure 7.1 shows the location of major Lower Ohio Valley Late Archaic sites.

Material Culture and Technology

Late Holocene hunter-gatherers manufactured an assortment of implements to help them cope with the environment, along with an equally diverse collection of objects associated with their social and spiritual worlds. In most ways, however, Late Archaic artifact inventories did not greatly differ from those of the Middle Archaic. In some cases, formal tools replaced more expediently produced ones. For example, Late Holocene hunter-gatherers probably used the many conical, cylindrical, and bell-shaped ground stone pestles found at

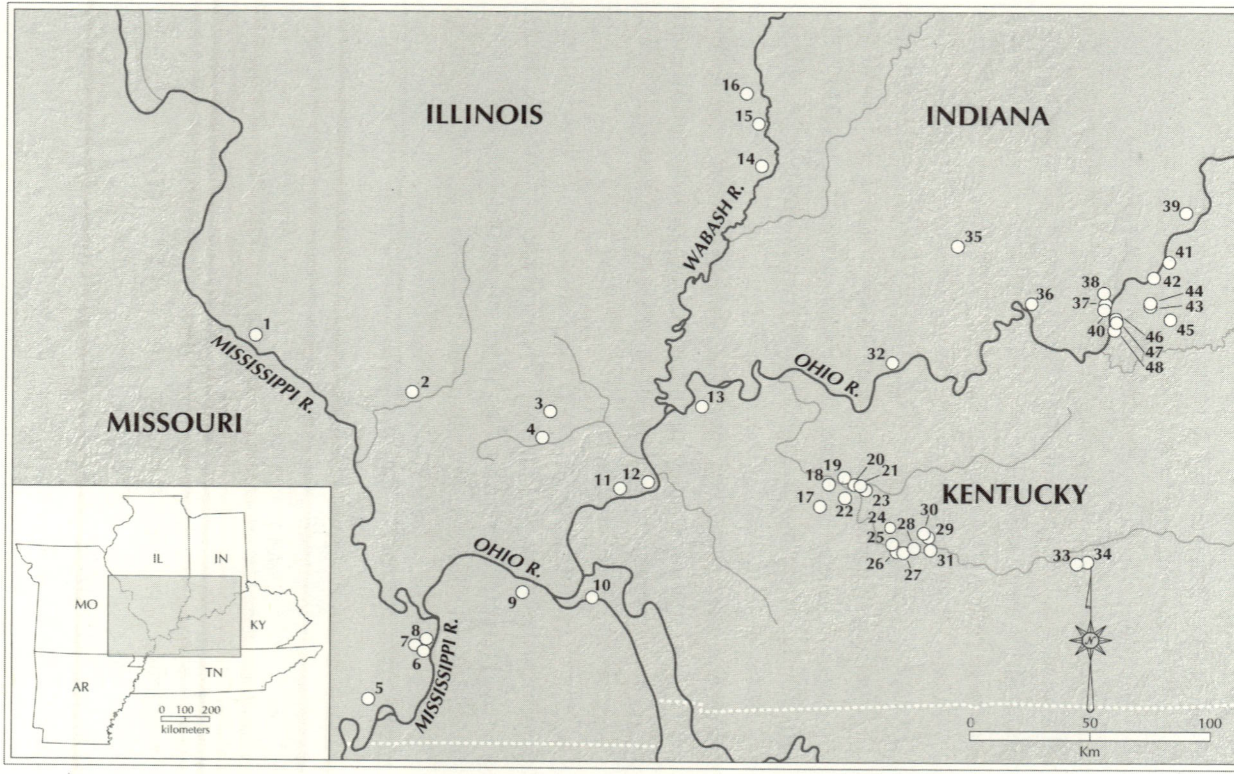
ILLINOIS
INDIANA
MISSOURI
KENTUCKY
MISSISSIPPI R.
WABASH R.
OHIO R.
OHIO R.
MISSISSIPPI R.
1
2
3
4
5
6
7
8
9
10
11
12
13
14
15
16
17
18
19
20
21
22
23
24
25
26
27
28
29
30
31
32
33
34
35
36
37
38
39
40
41
42
43
44
45
46
47
48
0
50
100
Km
IL
IN
MO
KY
AR
TN
0 100 200
kilometers

1 Modoc Rock Shelter	17 Barrett	33 Mammoth Cave
2 Little Muddy Rock Shelter	18 Kirkland	34 Salts Cave
3 Kottmeyer, Funkhouser, Wasson No.1	19 Hollis	35 Patoka Lake
4 Black Earth	20 Kirtley	36 Breeden
5 Clifford LaPlant	21 Butterfield	37 Caesars Archaeological Project/Knob Creek
6 Burkett	22 Ward	38 Reid
7 Weems	23 Jimtown Hill	39 Miles
8 O'Bryan Ridge	24 Jackson Bluff	40 Paddy's West
9 Hedden	25 Indian Knoll	41 Habich
10 Morrisroe	26 Baker	42 Railway Museum
11 Rose Hotel	27 Chiggerville	43 Lone Hill
12 Ferry	28 Bowles	44 KYANG
13 Highland Creek	29 Carlston Annis	45 McNeeley Lake
14 Robeson Hills	30 DeWeese	46 Spadie
15 Swan Island	31 Read	47 Rosenberger
16 Riverton	32 Crib Mound	48 Villier

Figure 7.1. Important Lower Ohio Valley Late Archaic sites.

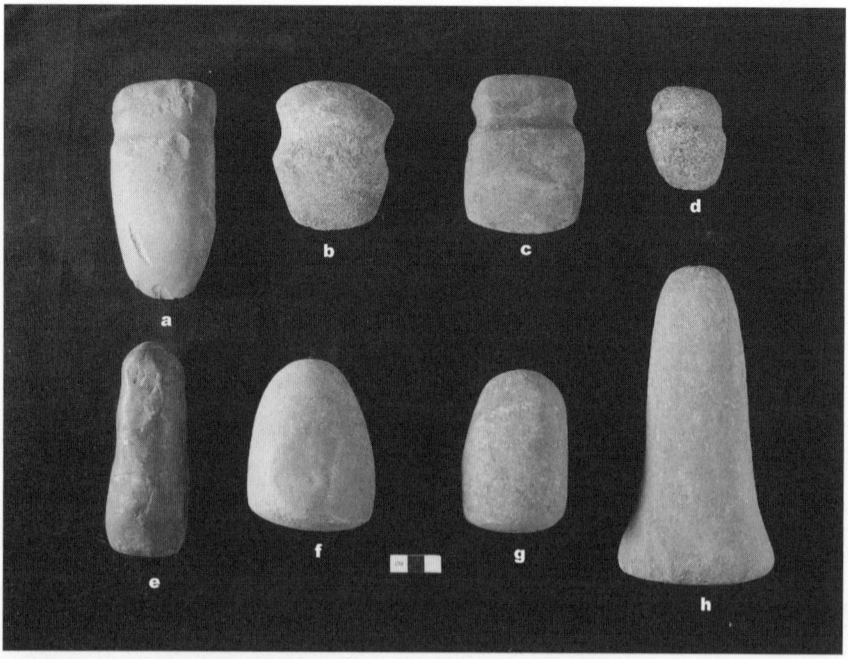

Figure 7.2. Ground stone grooved axes (a–d), cylindrical pestle (e), mullers/manos (f–g), and bell-shaped pestle (h) from the Green River region of Kentucky (William S. Webb Museum of Anthropology, University of Kentucky).

some of the Green River middens to do the same tasks for which Middle Holocene people used informal grinding stones or wooden tools (Figure 7.2; Webb 1946:Figure 39). Most stone tools were made from locally available resources, but more exotic lithic materials were sometimes used to make certain items, such as atlatl weights (Burdin 2004; Webb 1946:267) or plummets (Jefferies 1982b:236).

A continuing trend toward the greater use of local lithic resources is probably linked to decreased group mobility, a trend started during the late Middle Archaic (Amick and Carr 1996). In many areas, Late Archaic groups were "entrenched" in a particular area, systematically reoccupying a series of sites through redundant and infrequent moves. Along with smaller hunting and gathering territories came fewer opportunities to embed resource procurement activities within residential moves. In response, groups practiced a "logistically organized mobility strategy" where small task groups brought needed stone, often in a highly reduced form, back to the base camp. Other organizational responses to "entrenchment" included investment in tools having longer

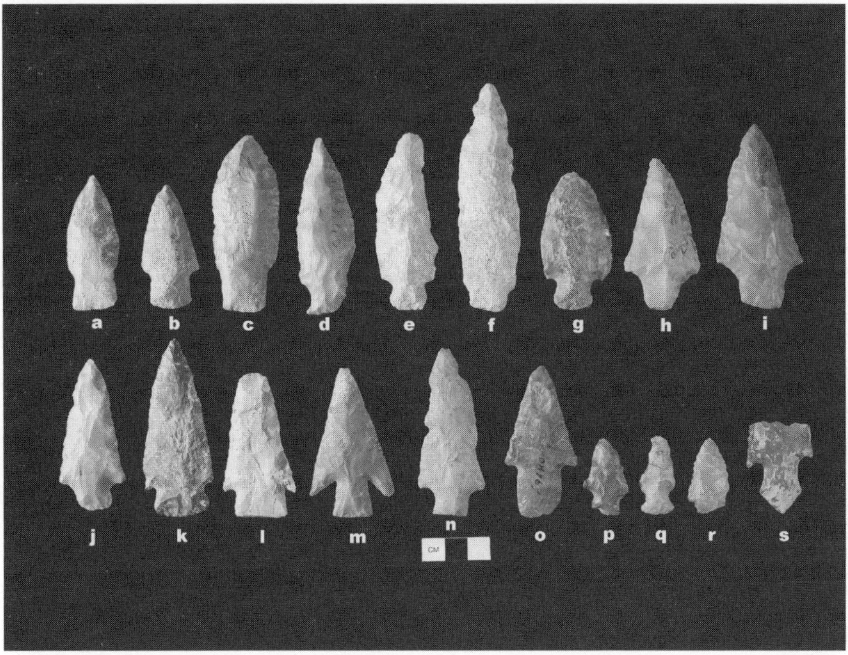

Figure 7.3. Examples of Lower Ohio Valley Late Archaic hafted bifaces—Saratoga Stemmed (a–c), Rowlett/McWhinney (d–g), Pickwick/Ledbetter (h–j), Etley (k–1), Buck Creek Barbed (m), Wade (n), Buck Creek Barbed (o), Merom/Trimble (p–r), Turkey-tail (s) (William S. Webb Museum of Anthropology, University of Kentucky).

use-lives, stockpiling lithic resources, exchanging lithic materials, and adopting new lithic technologies (microdrills, blades). In some parts of the Southeast, this strategy can be traced back to the late Middle Archaic, especially in areas having high population densities relative to available subsistence resources (Amick and Carr 1996:45–46).

Perhaps the most distinguishing characteristic of Late Archaic lithic technology was the gradual replacement of side and corner notched projectile points that were so popular during the Middle Archaic with a variety of straight, expanding, or contracting stemmed points (Justice 1987) (Figure 7.3). The reason(s) for this change is unknown, but I think it had a lot to do with corresponding changes in the hunting strategies used by Late Archaic hunter-gatherers. One possibility is that the stemmed hafting technology was better suited for attaching points to javelins used in conjunction with the atlatl.

Late Archaic groups incorporated a growing list of nonlocal raw materials into their technology, including marine shell and copper obtained through

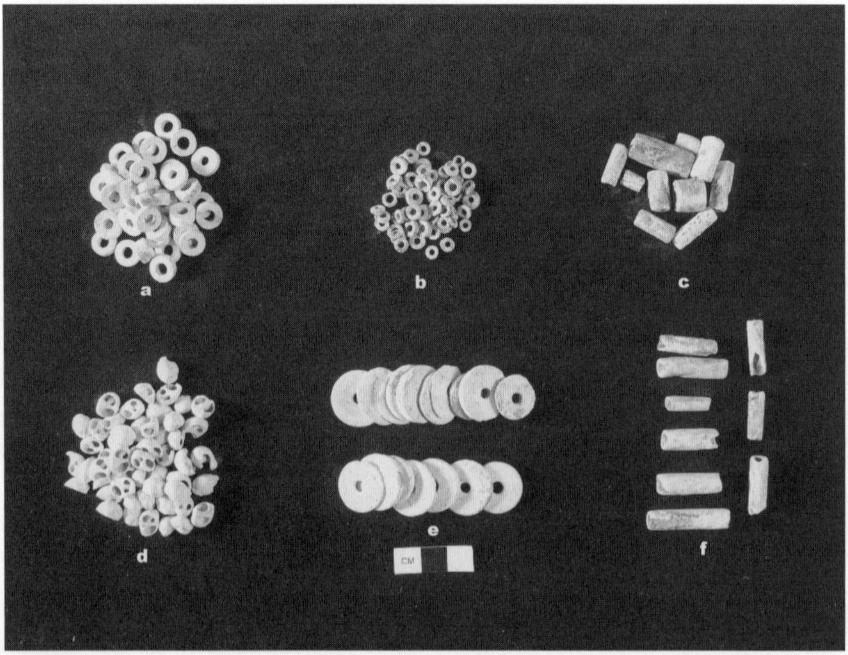

Figure 7.4. Middle Green River shell disk (a, b, e), tubular (c, f), and ground *Leptoxis* (d) beads (William S. Webb Museum of Anthropology, University of Kentucky).

interregional social networks (Marquardt and Watson 2005b:10). While copper objects are rare, marine shell items occur in relatively large numbers, especially in the Green River shell middens. Thousands of shell disk beads of consistent size and form suggest a well-developed technology for their production (Figure 7.4).

The number and diversity of antler and bone items found at some Lower Ohio Valley Late Archaic sites reminds us of just how biased our interpretations are when estimates of activity intensity and diversity are based solely on flaked and ground stone tools. Late Archaic hunter-gatherers used bone and antler implements to do some of the same tasks that they performed with stone tools, but bone and antler tools were specifically required for others.

Flaked Stone Technology

Despite the better representation of Late Archaic sites in the archaeological record, fine-grained information on flaked stone lithic technology and artifact style is still difficult to come by. As with earlier Archaic components, most Late Archaic occupations occur in plowzone contexts where artifacts have been mixed with those of earlier and/or later ones. With the exception of temporally diagnostic artifacts, in most cases we can say little about the organization of flaked stone tool technology.

Large Middle to Late Archaic middens, such as those found along the Green River (Marquardt and Watson 2005), contain abundant lithic materials attributable to the production, maintenance, and rejuvenation of flaked stone tools (Marquardt 2005). In most cases, however, individual occupations were not sealed by sterile alluvium, resulting in the continual mixing of newly deposited materials with those of previous occupations. For example, our spatial analysis of temporally diagnostic projectile points from the Read site (15Bt10) revealed that while the older projectile points tended to be in the lower midden levels, there was considerable mixing of temporally specific types (Milner and Jefferies 1998). Graves, storage pits, and other kinds of pit features constructed by site inhabitants also contributed to component mixing.

Fortunately, archaeological investigations at a number of Lower Ohio Valley rockshelters and floodplain sites, as well as sites in adjacent regions, provide important insights on Late Archaic flaked stone artifacts, technological organization, and tool kit composition. As a result, archaeologists have gradually compiled a list of diagnostic Late Archaic hafted bifaces from well-dated contexts (Figure 7.3; Table 7.1).

Late Archaic hafted bifaces include a variety of large and small stemmed and notched types (Justice 1987). Some of these types are based on research in the Lower Ohio Valley; others were initially defined in other parts of eastern North America (e.g., Pickwick, Ledbetter Stemmed, Brewerton) but occur in at least part of the region (Table 7.1).

The majority of Lower Ohio Valley Late Archaic hafted bifaces, mostly projectile points, are relatively large with straight, expanding, or contracting stems. Justice (1987:133–158) assigns many of these bifaces to the Late Archaic Stemmed cluster, which includes a variety of regionally specific types like Karnak Stemmed (Winters 1967), McWhinney Heavy Stemmed (Vickery 1972), and Rowlett (Duffield 1966; Granger 1988:Figure 5). Other Late Archaic types include Saratoga Broad Bladed, Parallel Stemmed, and Expanding Stemmed (Winters 1967); and Pickwick (DeJarnette et al. 1962:66) and Ledbetter Stemmed (Kneberg 1956). Although regionally distinct stylistic differences do exist, much of the morphological variability within types is related to the quality of lithic materials and the extent of tool resharpening and rejuvenation. Late Archaic hunter-gatherers manufactured many of these stemmed types throughout the early and middle Late Holocene, but a few types are more

Table 7.1. Lower Ohio Valley Late Archaic Projectile Point Types.

Projectile Point Type	Date Range	Sources
Rowlett	4200–3000 B.P.	Duffield 1966:66–69; Justice 1987:139; Mocas 1976:46–50
McWhinney Heavy Stemmed	4200–3000 B.P.	Justice 1987:138–139; Vickery 1972, 1976
Karnak Stemmed	5700–5000 B.P.	Justice 1987:134–138; Winters 1967:19
Matanzas	5700–4000 B.P.	Justice 1987:119–121; Munson and Harn 1966:153–154
Etley	4300–3000 B.P.	Justice 1987:146–148; Scully 1951:2; Phillips and Brown 1983
Brewerton Cluster (Corner Notched, Eared-Notched, and Side Notched)	5000–3700 B.P.	Justice 1987:115–119, 123–124; Ritchie 1961:17,19
Ledbetter Stemmed	4500–3000 B.P.	Cambron and Hulse 1969; Justice 1987:149–153; Kneberg 1956
Pickwick	4500–3000 B.P.	Cambron and Hulse 1975:103; DeJarnette et al. 1962:66; Justice 1987:153–154
Saratoga Cluster (Broad Bladed, Parallel Stemmed, Expanding Stemmed)	4500–2500 B.P.	Justice 1987:154–159; Winters 1967:19, 25; Marquardt and Watson 2005a
Merom Expanding Stemmed	3600–3000 B.P.	Justice 1987:130–132; Winters 1969
Trimble Side Notched	3600–3000 B.P.	Justice 1987:130–132; Winters 1969
Gary	3500–2100 B.P.	Newell and Krieger 1949:164–165; Justice 1987:189–190
Buck Creek Barbed	3500–2600 B.P.	Justice 1987:183–184; Seeman 1975:107
Wade	3000–2500 B.P.	Cambron and Hulse 1969:110; Justice 1987:180–182
Cypress Stemmed	3000–2300 B.P.	Denny 1972; Justice 1987:197–198; May 1982:1366–1367

temporally restricted (Table 7.1). In addition, some point types first manufactured during the later part of the Late Archaic continued to be made well into the Early Woodland (Justice 1987:198).

The distribution of Brewerton Eared Notched points, a Northeastern type associated with the Brewerton phase in New York (Ritchie 1961), extends into west central Kentucky and southern Indiana (Collins and Driskell 1979:1026). These thick, weakly side notched points are morphological correlates of Matanzas cluster projectile points, dating from about 5000 to 3700 B.P.

In southern Indiana, McWhinney and Saratoga points are commonly found at Late Archaic sites along the Ohio River, increasing in frequency after ca. 4500 B.P. In contrast, to the north in the southern Indiana Hill Country, Karnak Stemmed and Unstemmed varieties dominate contemporary assemblages. These geographic distinctions appear to represent two separate southern Indiana Late Archaic phases (Stafford and Cantin 2009).

By about 3500 B. P., most southern Indiana hunter-gatherers shifted their projectile point technology from the production of large stemmed points to manufacturing smaller Riverton types and a variety of barbed points (Stafford and Cantin 2009). Merom Expanding Stemmed and Trimble Side Notched (Winters 1969) points are associated with the Late Archaic Riverton culture (3600–3000 B.P.) defined by Winters (1969) for the lower Wabash River valley. More recent investigations have demonstrated that the distribution of these small Late Archaic points is much wider than that of Winters's Riverton culture and now includes most of southern Indiana, southern and western Illinois, and central and western Kentucky (Tomak 1982, cited in Justice 1987:132).

Field investigations at the Wint site in southern Indiana yielded a large Riverton assemblage (Anslinger 1986). Considerable morphological variability existed among the projectile points, resulting in their being grouped into a Riverton cluster instead of being assigned to the specific types (Merom, Trimble, Robeson) defined by Winters (1969). Radiocarbon dates associated with the Riverton component ranged from ca. 3400 to 2750 B.P. (Anslinger 1986, cited in Stafford and Cantin 2009).

The Caesars Palace excavations at the Knob Creek site (12Hr484) also exposed a significant Riverton occupation. Cultural materials included more than 300 Riverton projectile points, along with 79 features distributed along a .5-km section of the Ohio River levee. Five radiocarbon dates clustered around 3500 B.P., with a sixth falling around 3150 B.P. These dates generally fall at the earlier end of the range of Riverton dates, conforming to Winters's (1969) Wabash valley dates (Stafford and Cantin 2009).

Terminal Archaic (3000–2500 B.P.) straight stemmed and barbed Wade points (Cambron and Hulse 1969) occur at some lower Tennessee River Late Archaic sites; a few specimens have also been found north of the Ohio River in southern Indiana (Justice 1987:180). Terminal Archaic hunter-gatherers who lived in southern Indiana made what are called Buck Creek Barbed points (3500–2600 B.P.). Similar projectile points occur in southeastern Illinois where they are called Cypress Stemmed (Justice 1987:183).

At Knob Creek, the Buck Creek Barbed occupation is mixed with a slightly older Riverton assemblage. A single radiocarbon date placed the occupation at 2980 +/– 70 B.P., several hundred years younger than most of the Riverton radiocarbon dates (Stafford and Cantin 2009).

One of the clearest Late Archaic stratigraphic sequences in southern Illinois comes from the Little Muddy Rock Shelter (11J814) (Moffatt et al. 1992), a deeply stratified site containing cultural deposits ranging in age from Early Archaic to Mississippian. The shelter lies along the lower reaches of the Little Muddy River in the Mt. Vernon Hill Country, just to the north of the deeply dissected Shawnee Hills (Moffat et al. 1992:3). Although not in the Lower Ohio Valley proper, it lies very close to its northern boundary, making it a good analog for the northwestern part of the study area.

The Little Muddy Rock Shelter contained five Late Archaic components (Koldehoff 1992; Moffat 1992), the oldest of which (Late Archaic I) dated to about 4500 B.P. Although excavations yielded only biface tips and midsections, their technological attributes and overall morphology suggest that they are parts of Karnak or Matanzas points (Koldehoff 1992:324). If this date is correct, the Late Archaic I component coincides with the interface between the late Middle and Late Archaic in southwestern Illinois. The Late Archaic II horizon, which yielded one acceptable radiocarbon date of ca. 4000 B.P., produced six Saratoga and/or Karnak points; however, an earlier Matanzas point was also present.

Projectile points from the Late Archaic III component, dating to ca. 3700 B.P., largely consisted of Saratoga types, but an Etley and an older Brannon side notched point were also present. Etley points are seldom found in this part of southern Illinois—they are much more common to the north in the American Bottom and lower Illinois River valley. The few southern Illinois examples are usually made from Burlington chert, a nonlocal material that outcrops along the Mississippi River north of St. Louis, Missouri. The presence of Burlington chert at the Little Muddy Rock Shelter raises the possibility of exchange between southwestern and west central Illinois Late Archaic groups (Koldehoff 1992:327).

The radiocarbon date from the Late Archaic IV zone is virtually identical (ca. 3700 B.P.) to the Zone III date, but the projectile points are quite distinct. Instead of large stemmed points, the zone yielded small stemmed and notched bifaces resembling Merom and Trimble bifaces. The Little Muddy Rock Shelter date is about 100 years earlier than the earliest Riverton date from the lower Wabash valley (Winters 1969), but point morphology falls within the range described for Winters's Merom/Trimble specimens (Koldehoff 1992:329).

Five medium-sized Cypress Stemmed projectile points with contracting, straight, or expanding stems signify a Terminal Archaic occupation of the Little Muddy Rock Shelter (Koldehoff 1992:331–332). Points like these, resembling later Adena Stemmed (2800–2300 B.P.) points, continued to be made well into the Early Woodland (Justice 1987:192) or even later. A radiocarbon date of ca. 3000 B.P. probably marks the beginning of this occupation.

Late Archaic chert exploitation at the Little Muddy Rock Shelter focused on both local glacial till cherts and high-quality cherts from more distant locations. Local cherts are well represented in all Late Archaic horizons. Implements made from Cobden/Dongola and Bailey/Grassy Knob cherts demonstrate that Late Archaic hunter-gatherers traveled to the sources of these cherts some 25 to 35 km southwest of the shelter. In contrast, the occurrence of Crescent chert may simply signify the movement of site inhabitants between the Little Muddy drainage and areas to the northwest near the American Bottoms. Nonlocal cherts usually occur as finished tools and debitage attributable to tool maintenance and recycling (Koldehoff 1992:358–359). Collectively, chert source data suggest that the Late Archaic hunter-gatherers who visited the Little Muddy Rock Shelter exploited a large section of southwestern Illinois that extended at least 75 km from northwest to southeast.

Based on data from the Terminal Late Archaic zone, Koldehoff (1992:332) defined the Au Vase phase, dating from ca. 3000 to 2600 B.P. The Cypress Stemmed point was the diagnostic Au Vase artifact type. The Au Vase phase is closely linked to the American Bottom Prairie Lake phase (Emerson and McElrath 1983) and may encompass all of southern Illinois. Cypress Stemmed points were also associated with a Terminal Late Archaic occupation dating to ca. 2800 B.P. in the Cache River valley of southern Illinois (DelCastello and Butler 2000:34).

Much of what we know about Terminal Archaic hunter-gatherers comes from Winters's (1969) study of the lower Wabash valley Riverton culture (ca. 4000–3000 B.P.). Diagnostic Riverton bifaces consist of a series of small stemmed or notched points that includes the Merom Expanding Stem, Trimble Side Notched, and Robeson Constricted Stem types (Winters 1969:Plates 12–14). Riverton lithic technology was based on the expedient acquisition and reduction of small chert gravels found throughout the Wabash valley. In addition to projectile points, Riverton chert knappers used local "Wabash" chert to make a wide variety of relatively small flaked stone implements like knives, choppers, drills, microperforators, reamers, scrapers, and shredders (1969:24). Winters noted that, "Nodules of Wabash chert are consistently small, making the manufacture of large chert implements possible only with considerable selection of the nodules present in the gravels" (1969:24). Because of the low investment of time, energy, and resources used to make flaked stone implements, Riverton tools show little evidence for extensive resharpening and curation (Anslinger et al. 1994:25–26).

Recent archaeological investigations in southeastern Missouri are beginning to shed light on Late Holocene hunter-gatherers in the Ohio-Mississippi river confluence region (Buchner et al. 2003; Thomas et al. 2004). This is a particularly interesting (and often confusing) part of the Lower Ohio Valley region because of the interface of cultural traditions from several different parts of the Midcontinent. From an archaeological perspective, this has resulted in a jumbled combination of projectile point terminology, phase names, and chronological sequences (Butler 2009).

Late Archaic cultures of Southeast Missouri's Cairo Lowlands are included in the O'Bryan Ridge phase, one of the region's first widespread cultural complexes (Lafferty and Price 1996:Figure 1.2). Distinctive baked clay objects (i.e., Poverty Point Objects) and lapidary items reflect some degree of cultural relationship between O'Bryan Ridge groups and those of the Poverty Point culture (Lafferty and Price 1996:3) of the lower Mississippi River valley (Gibson 1996, 2001). Radiocarbon dates and diagnostic O'Bryan Ridge phase artifacts suggest that this Late Archaic cultural manifestation extended from approximately 4800 to 2500 B.P. (Buchner 2003a:52).

The Burkett site (23Mi20), the type-site for the O'Bryan Ridge phase, is situated in the Cairo Lowlands, just southwest of the Ohio-Mississippi river confluence. Recent investigations at Burkett identified an intact O'Bryan Ridge phase midden buried below later Woodland cultural deposits (Lafferty and Hess 1996).

A suite of radiocarbon dates from Burkett and the nearby Weems site

(23Mi25) indicates that the O'Bryan Ridge phase occupation dates from ca. 4500 to 4000 B.P. Diagnostic O'Bryan Ridge phase projectile points included the Burkett point, a morphological correlate of the Gary type (Justice 1987: 190), along with other "Gary-like" projectile points.

The Burkett site lithic industry appears to have been complex and multifaceted. Statistically significant associations exist between certain chert types and artifact forms. Site inhabitants usually used Cobden chert to make projectile points and Mill Creek chert to manufacture other kinds of bifaces. They often made formal unifaces from Mounds Gravel, an abundant local chert. Core reduction strategies were varied, with examples of bifacial, blade, opposed platform, discoidal, and blocky/amorphous technologies represented, the latter type being most common (Thomas et al. 2004:117).

Lapidary materials, such as finely made jasper beads, were made and/or used by Burkett inhabitants, as were Poverty Point Objects. The morphological diversity of the more than 300 clay objects was less than seen at many "classic" Poverty Point sites, but still included amorphous, biconical, or spheriod forms (Thomas et al. 2004:118–119). Although the precise technological role of baked clay objects is unclear, they were probably used for cooking.

The O'Bryan Ridge phase occupation at the Clifford LaPlant site (23Nm561), located on a low ridge overlooking the Mississippi River approximately 30 km southwest of the Burkett site, was both intensive and extensive (Buchner et al. 2003). The assemblage included a mixture of Midwestern and Southeastern point types like Saratoga Straight Stemmed, Etley, Table Rock Stemmed, Brewerton Eared Notched, Dryoff, Big Creek, and Stone Square Stemmed, as well as several types (Epps, Lambert) associated with the Poverty Point culture. The LaPlant flaked stone tool inventory was manufactured from cherts that originated in many different parts of the region, reflecting inhabitants' mobile lifestyle (Buchner 2003b:467–468).

As at Burkett, LaPlant's O'Bryan Ridge phase technology also included the use of baked clay objects. Over 70 percent of the complete specimens were biconical; other shapes included spherical, cube, waisted cylinder, and grooved forms. The fact that a large earth oven contained more than 50 whole or fragmentary baked clay objects supports their association with food preparation activities. Although the manufacture of baked clay objects continued into the Woodland period, an associated radiocarbon date of ca. 3900 B.P. places the age of this feature and the baked clay objects in the Late Archaic (Buchner 2003b:468).

At the Morrisroe site in western Kentucky, comparison of the technological attributes of Late Archaic assemblages (Assemblages A and B) with those of Middle Archaic assemblages (Assemblages C–J) indicates a significant change in the kinds of activities conducted at the site. Older assemblages were larger and contained a higher diversity of flake sizes and types, suggesting a wide range of activities and a "long" biface reduction trajectory. In contrast, Late Archaic assemblages contained fewer tools and fewer and smaller flakes, suggesting a reduction in activity and a decrease in flaked stone tool manufacturing. These trends point to a substantial shift in the way the site was used, possibly indicating a transition from long-term occupations during the Middle Archaic to short-term, ephemeral occupations by Late Archaic hunter-gatherers (Conaty 1985:340).

A number of sites near the Falls of the Ohio River have provided important insights on Late Holocene hunter-gatherer lithic technology. In this part of the Lower Ohio Valley, archaeologists have defined the Lone Hill phase (Granger 1988) for the early portion of the Late Archaic (ca. 5000–3000 B.P.), while later components are associated with the Riverton phase (ca. 3600–3000 B.P.) (Winters 1969). One of the major distinctions between these two cultural manifestations is their projectile point manufacturing technology, with Riverton technology having a greater emphasis on the expedient acquisition and production of flaked stone implements (Anslinger et al. 1994:25–26).

Excavations at the Railway Museum site, located just upriver from the Falls of the Ohio, identified several Late Archaic occupations (Anslinger et al. 1994: 129). The earliest and most intensive of these was a Lone Hill phase component represented by broad-barbed stemmed points and hafted scrapers resembling Saratoga Broad Bladed, Delhi, Wade, and Buck Creek Barbed points. Production of these stemmed bifaces utilized large blanks and preforms made from locally available Ste. Genevieve (a.k.a. Wyandotte) chert (1994:129–130).

A second potential Late Archaic component may be indicated by McWhinney points that are generally narrower, thicker, and have less pressure retouch than the broad-barbed specimens. Chert selected to make McWhinney points was generally of low quality, a strategy that was widely used by Falls area Late Archaic hunter-gatherers; no McWhinney points were made from the higher-quality Ste. Genevieve chert (Anslinger et al. 1994).

Investigation of an intensive McWhinney component at the Miles site (12C1158) in southern Indiana approximately 50 km upriver from the Railway Museum site provided additional insights on Late Archaic technology near the Falls (White 2003b). The McWhinney bifaces, collected from feature and nonfeature contexts, were relatively thick with haft elements including expanding stemmed, side notched, and lanceolate forms. Charcoal from two probable mortuary features containing McWhinney projectile points yielded dates of ca. 4200 B.P., consistent with other Ohio Valley dates for McWhinney projectile points (White 2003b:Table 3, 124–125).

At the Miles site McWhinney lithic technology centered on an amorphous core technology using locally available Laurel/Marble Hill cherts. Flake tools were manufactured from amorphous cores, as were bifaces. Similar types of cores have been associated with other Late Archaic components in southern Indiana (White 2003b:118–119).

Archaeological investigations at the Villier site, located on the first terrace of the Ohio River a few kilometers downriver from the Falls, revealed an extensive midden containing several Late Archaic components (Robinson and Smith 1979). Although no observable internal stratification was evident, the vertical distribution of temporally diagnostic artifacts suggested a relatively undisturbed and temporally ordered deposit.

Projectile points associated with Villier's Late Archaic occupation included early Late Archaic (ca. 5000–4000 B.P.) Brewerton, Lamoka, McWhinney, and Rowlett types. Many of these types were manufactured as late as 3000 B.P, as indicated by the association of small Brewerton-like points with a date of ca. 3100 B.P. at the nearby Spadie site. The most intensive period of site use dates to the end of the Late Archaic (ca. post-4000 B.P.), as reflected by the high frequency of Merom/Trimble points (Robinson and Smith 1979:648).

Most of the Villier Late Archaic flaked stone tools were made from locally available Muldraugh or Wyandotte chert, both of which were available from nearby outcrops or stream beds. Wyandotte chert was preferred to make bifaces, perhaps because of its excellent flaking characteristics, while Muldraugh was used for other purposes. Interestingly, while Wyandotte was the chert of choice for early Late Archaic bifaces, later groups that made Merom/Trimble points showed a strong preference for other cherts, primarily Muldraugh. The reason(s) for this major shift in raw material selection is unclear, but it may have to do with a shift in social boundaries or changes in the technological requirements of biface production (Robinson and Smith 1979:634).

Perhaps the best information on Late Holocene hunter-gatherer technology and material culture comes from Kentucky's Green River region where more than 100 years of archaeological investigations have identified an impressive inventory of flaked and ground stone, bone, antler, and shell tools, items of personal adornment, and even musical instruments (Marquardt and Watson 2005; Moore 1916; Webb 1946). Most of this information comes from the investigation of the many well-known river-edge and upland middens, many of which were excavated during the late 1930s and early 1940s (Webb 1946, 1950a, 1950b; Webb and Haag 1939, 1940). More recent investigations at Carlston Annis and other sites in the Green River's Big Bend region by the Shell Mound Archaeological Project (SMAP) have provided additional "fine-grained" information on Late Archaic technologies (Marquardt 2005; White 2005). Although earlier radiocarbon dates and older artifacts exist, the period of most intensive occupation at most of these sites extends from about 5800 to 3500 B.P. (Marquardt and Watson 2005a:631).

The Late Holocene hunter-gatherers who inhabited the Green River valley had an efficient lithic technology that they used to make implements to do everyday cutting, grinding, and food-processing tasks. Their flaked stone industry focused on biface manufacture, modification, and recycling; preparing and using specialized flake tools such as burins, gravers, and bladelets; and the expedient use of whole and broken flakes for a variety of cutting tasks (Marquardt 2005:351).

Most Green River flaked stone artifacts were made from locally available Vienna, St. Louis, or Ste. Genevieve cherts (Gatus 2005). Some of the initial reduction was done close to the source areas, but the many small pieces of shatter indicate that Late Archaic hunter-gatherers also carried out flaked stone tool manufacturing/maintenance activities at river-edge settlements and upland rockshelters (Marquardt 2005:423).

Analysis of temporally diagnostic projectile points from selected Green River middens indicates that their occupational history is quite variable (Jefferies et al. 2005:Table 2). Many middens yield primarily Late Archaic artifacts, but late Middle or even Early Archaic artifacts are common in others. In his detailed analysis of the Carlston Annis lithic materials, Marquardt found that the vast majority of bifaces were stemmed or side notched types (Matanzas, Apple Blossom, Brannon, Karnak, Saratoga, Brewerton), indicating that these locations were important parts of the Green River cultural landscape during the late Middle and/or Late Archaic. The presence of Merom/Trimble variants demonstrates that some of these sites were occupied through the end of the Archaic and possibly later (Marquardt 2005:356).

Most middle Green River Late Archaic bifaces are similar to those made in other parts of the Midsouth and Midwest, but they tend to be shorter, narrower, and thinner than those of similar style from nearby areas. This was not just a matter of more intensive use and rejuvenation, but an actual size difference. In addition, the Green River points tend to be less angular than those from adjacent areas; pronounced barbs or dramatically expanding stems are rare. Overall, the Green River bifaces more closely resemble specimens from Indiana and Illinois than those from the Southeast (Marquardt 2005:405).

In addition to projectile points, Green River Late Archaic hunter-gatherers used bifacial technology to make blanks and preforms, choppers, drills, and endscrapers. Many of the endscrapers appear to be recycled projectile points that had been transversely snapped, then reworked to create a scraping edge on the tool's distal end (Marquardt 2005:353).

As for the rest of the flaked stone technology, middle Green River flakes and cores largely reflect an expedient and informal technology. Most flakes were struck from amorphous cores with little preparation. The resulting small flakes were used for an assortment of scraping and cutting tasks. Thirteen percent of all flakes exhibited evidence of use; and an additional 5 percent had been intentionally retouched. Approximately 3 percent of all flakes were bladelets, bladelike flakes, gravers, or burins, demonstrating that these Late Holocene hunter-gatherers could make such implements when needed (Marquardt 2005:408–410). No significant diachronic differences in flake/core technologies were noted in the Carlston Annis midden; however, river-edge shell middens had more utilized flakes and fewer retouched flakes than most upland rockshelters (Marquardt 2005:410).

Ground Stone

Like many late Middle Archaic hunter-gatherers, Late Archaic people continued to manufacture a diverse inventory of ground and polished stone items (Figure 7.2). The forms and styles of most of these objects were very similar to earlier examples.

Green River Late Archaic hunter-gatherers used ground stone technology to make stylistically distinct atlatl weights, hammerstones, full- and three-quarter grooved axes, grooved mauls, bell-shaped and cylindrical pestles, hoes, and abraders, along with ornamental items such as beads and pendants (Webb 1946:237). Ground stone tools are much less common than flaked stone implements (Marquardt 2005:354), suggesting ground stone technology produced tools with much longer use-lives.

The relative abundance of atlatl weights, one of the components of an atlatl, or spearthrower, signifies the increasing importance of this hunting weapon during the latter part of the Archaic (Figure 7.5). William Webb (1946:237) reported more than 100 drilled prismoidal atlatl weights from his investigations at the Indian Knoll site. C. B. Moore (1916) excavated more than 20 weights

Figure 7.5. Probable Late Archaic bannerstones/atlatl weights from the Indian Knoll (d, e, g, h) and Read (a–c, f) sites (William S. Webb Museum of Anthropology, University of Kentucky).

during his work there 30 years earlier. Based on the presence of unfinished atlatl weights, at least some were manufactured at the site (Webb 1946:267). Other atlatl components, such as antler and bone hooks and handles, provide additional support for the importance of the atlatl "weapons system" to Late Archaic hunters (Figure 7.6, o and p).

Late Archaic atlatl weights were made in many different shapes (butterfly, diamond, bottle, triangular) and from various materials including stone (slate, siltstone, quartz/quartzite, claystone, gneiss, granite, limestone) (Burdin 2004:Figures 4.20, 4.21) and marine shell (Webb 1946). Interestingly, Rick Burdin (2004) found that nearly 80 percent of the Late Archaic atlatl weights he examined were made from quartz, quartzite, gneiss, or granite, marking a dramatic increase in preference for these harder materials compared to earlier times. Polished stone atlatl weights have been documented at numerous Late Archaic sites across the Midcontinent (Burdin 2004; Fowler 1957; Lutz 2000), reflecting the atlatl's widespread use.

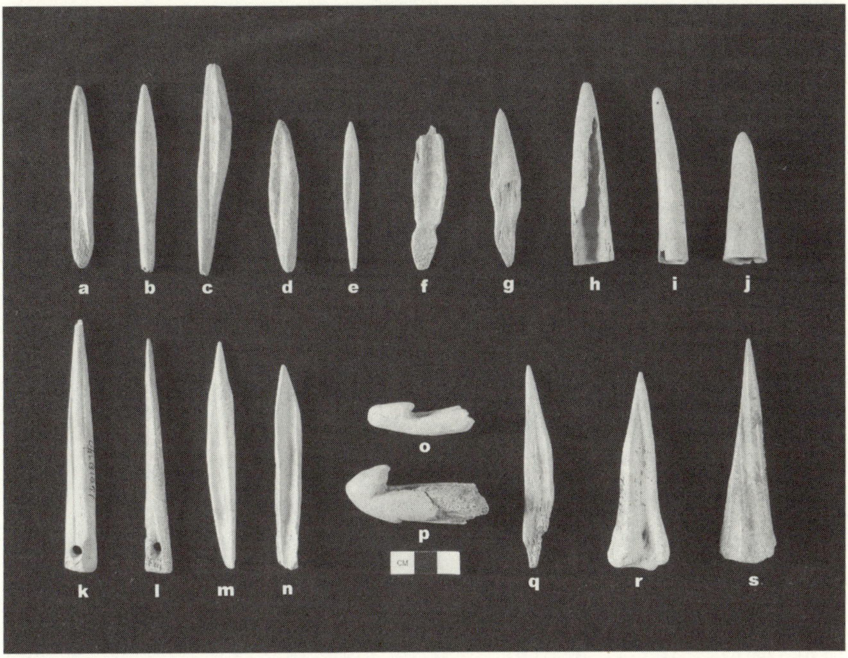

Figure 7.6. Bone artifacts from the Chiggerville site, middle Green River region—possible bone projectile points (a–g), antler tip projectile points (h–j), probable weaving/matting tools (k–1), awls (m–n, q), atlatl hooks (o–p), and probable boring implements (r–s) (William S. Webb Museum of Anthropology, University of Kentucky).

Atlatl weights occur at many Late Archaic sites in the lower Ohio Valley, but some sites have yielded particularly large numbers of them. Melvin Fowler's investigations at the Ferry site (1957; Muller 1986:68–70), an eroded lithic scatter near the Saline River in southeastern Illinois, yielded more than 20 stone atlatl weight fragments representing an assortment of exotic styles and raw materials (Fowler 1957:15–21). Many of the associated projectile points resemble Late Archaic Saratoga cluster points (Muller 1986:69). However, recent reexamination of the Ferry site points suggest that some were Middle Archaic (Butler 2009), as were some of the atlatl weights (Lutz 2000:235). Clearly, more research is needed to clarify the age of the Ferry site artifacts.

Atlatl forms represented at the Ferry site include the Pick, Crescent, Ovoid, Double-Bitted Axe, Ball, and Knobbed Lunate types (Fowler 1957:Figures 8 and 9). Most were made from gray banded slate, but a few were made from gabbro, diorite, or unbanded slate (1957:Table II). Although much of this stone outcrops in southern Canada, it was probably collected from glacial drift deposits located in northern Indiana or southern Wisconsin (1957:21).

It is uncertain exactly why so many atlatl weights were present at the Ferry site. In general, the site resembles many other Archaic lithic scatters found throughout the Lower Ohio Valley. However, local farmers told Fowler about finding plowed-out bone, including human remains, at the site prior to its excavation (Fowler 1957; Muller 1986:69), so it is possible that the atlatl weights represent grave goods that were once associated with the reported Middle to Late Archaic human remains.

Fowler (1957:32) interpreted the numerous bannerstones as meaning that the Ferry site's inhabitants had the leisure time needed to manufacture the ground and highly polished stone objects. He proposed that the different bannerstone shapes had social or ceremonial meaning, possibly as family or clan symbols, and that the Ferry site may have been a gathering place where Late Archaic families held tribalwide rituals and ceremonies (1957:32).

Unusually large quantities of atlatl weights also have been reported from other Midcontinent sites. Of particular note, excavations at the Bullseye site, a late Middle Archaic mortuary site in the lower Illinois River valley, yielded numerous burials and nearly 30 atlatl weights (Hassen and Farnsworth 1987). As Fowler did for Ferry, Jane Buikstra and Douglas Charles suggested that the many atlatl weights found at Bullseye marked a place where scattered hunter-gatherer groups gathered to discuss issues of common concern, to exchange mates, and to bury their dead (1999:208). Several Lower Ohio Valley sites also have yielded large numbers of atlatl weights, like the Crib Mound (Kellar 1956; Lutz 2000:396–394) and some of the middle Green River middens (Webb 1946).

Apparently, the popularity of the atlatl, or at least the ground stone atlatl weight, decreased by the end of the Late Archaic. Winters noted that while atlatl weights were associated with earlier Late Archaic manifestations in the Lower Wabash region, he did not find any in good stratigraphic context in the Riverton middens (1969:26).

Terminal Archaic Riverton culture hunter-gatherers used local sandstone to make a variety of ground stone tools that included pipes, grooved abraders, files, grooved sinkers, "cupstones," hammerstones, manos, and anvils (Winters 1969:26). They frequently used unmodified glacial gravels as manos and hammerstones. Grooved axes and pendants also were made from glacial gravels, but Winters was uncertain if these items were actually part of the Riverton assemblage (1969:26).

Bone

The alkaline sediments of the Green River middens were highly conducive for bone and antler preservation. Excavation of these sites yielded massive quantities of bone and antler artifacts, along with the residues of their manufacture. The WPA excavations at Indian Knoll yielded at least 8,449 bone and 4,325 antler artifacts compared to 8,735 flaked stone and more than 3,000 ground stone items (Webb 1946:237–239). These figures suggest a nearly equal number of bone/antler and flaked/ground stone artifacts at sites where there is good preservation.

Late Archaic inhabitants of Indian Knoll relied heavily on deer bones and antler to make many of their tools, but they also made implements and ornaments from the bones and teeth of raccoon, beaver, turtle, and a variety of birds and fishes. Items manufactured from bone and antler included awls, hairpins, fishhooks, gorges, projectile points, chisels, gravers, scrapers, rattles, rings, beads, flakers, and atlatl handles and hooks (Webb 1946:238–239).

More than 75 percent of the Indian Knoll bone artifacts consisted of awls, or at least objects that Webb called "awls," made from small splinters of unidentified bone (Figure 7.6). While their actual size and shape varied considerably, all were characterized by the sharp points found on one or both ends of the implement (Webb 1946:282). The few awls that still exhibited diagnostic attributes indicated that Indian Knoll residents used deer ulnae, radii, tibiae, scapulae, and mandibles; bird wing and leg bones; raccoon bacula; and fish spines to make these implements. Recent reanalysis of some of the Green River bone "awls" suggests that at least some of these objects are projectile points (Moore 2007). A similar conclusion was drawn by Anne Bader's (1992) analysis of bone objects from KYANG. Some of the bipointed objects are probably fishing gorges (Marquardt, personal communications, 2008).

Karli White's (2005:339) analysis of the Carlston Annis bone and antler artifacts revealed a variety of artifacts including beads, fishhooks and fishhook blanks, incised bird bones, pointed tools (awls, pins, needles, etc.), and modified antlers. Fishhooks were usually made from mammal bone. One of the complete fishhooks had a small groove encircling the end of the shank.

Christopher Moore's (2006) investigation of fishhook production at the Chiggerville and Read sites examined more than 250 fishhooks and pieces of fishhook manufacturing debitage. He identified four distinct manufacturing trajectories, three of which Webb (1950a) recognized during his analysis of the Carlston Annis bone artifacts. Uniformity in the Read and Chiggerville fishhook collections reflects their contemporaneity; however, differences in proportions between stages of fishhook manufacture and fishhook-related tools suggest that the two sites may have been functionally distinct, perhaps signifying different seasons of occupation (Moore 2006).

The Carlston Annis bone artifact inventory also included two incised bone tubes made from the humeri of a large bird. The tubes, both of which were found with burials, were decorated by several narrow bands of cross-hatching that encircled both ends of the tubes (White 2005:Figure 17.3). Previous excavations at Carlston Annis and Indian Knoll yielded similar kinds of engraved bird bone shafts (White 2005:341).

As at Indian Knoll, the most common bone artifacts at Carlston Annis were single- and bipointed implements, many of which were identified as awls, gorges, or points made from mammal longbones (White 2005:Table 17.1, 346). This tool category includes implements with varying degrees of modification. Some simply consist of bone splinters with a sharpened end or ends; others had been more extensively modified by scraping or abrading beyond the tip. Most of these implements were probably used to perforate soft materials such as hide. Other specimens had a hole drilled near their bases, suggesting that they were used for weaving or sewing.

The inhabitants of Indian Knoll and many other middle Green River sites also made hairpins from the bones of deer and other large mammals (Figure 7.7). Unlike the elaborately carved and engraved bone hairpins made by late Middle Archaic hunter-gatherers in southern Indiana and southern Illinois (Jefferies 1997), most of the pins made by people who lived along the middle Green River were relatively plain. Figure 7.7 shows some of the more elaborate examples from Indian Knoll.

My analysis of bone pins from ten of the Green River middens, including Indian Knoll, revealed that while pins were plentiful, most consisted of minimally modified sections of mammal or bird bone (Jefferies 2004). Many specimens retained the morphological characteristics of the bones from which they were made (Jefferies 2004:78).

A few pins with heads resembling those found north of the Ohio River occurred in the Green River sample, but they represented a very small percentage of the hundreds of specimens that I examined in the study. Of these, most had "crutch-top" heads, the least technologically complex and most widely distributed of the late Middle Archaic types (Jefferies 2004:78).

Likewise, engraving of the head and shaft, a common decorative technique

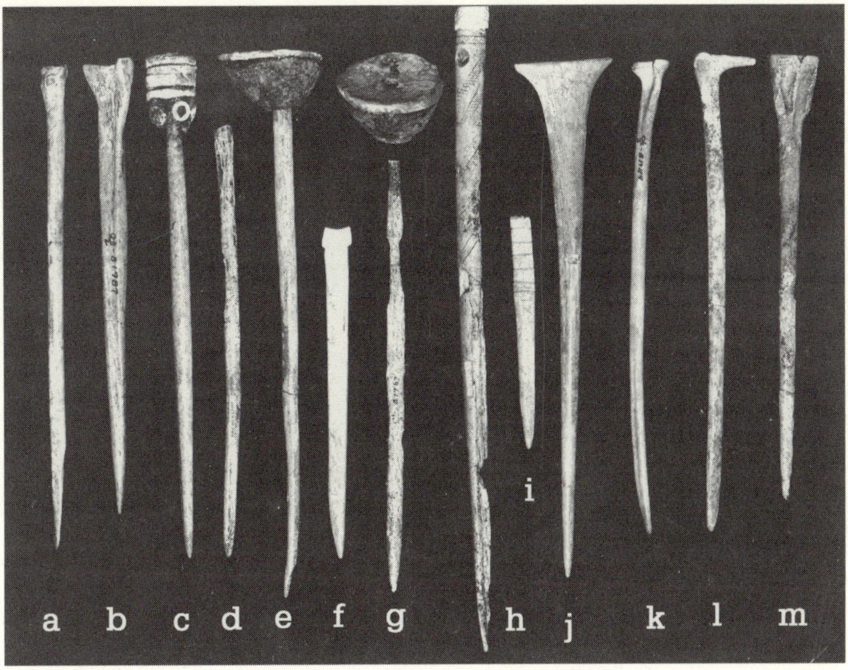

Figure 7.7. Middle to Late Archaic bone pins from Indian Knoll (adapted from Webb 1946:Figure 51A, Photograph 5810H2. William S. Webb Museum of Anthropology, University of Kentucky).

on pins from southern Illinois and southern Indiana, was rare on the Green River pins. A notable exception is an Indian Knoll pin with an engraved spiral design that encircles the pin's shaft several times (Figure 7.7, h). This pin also had a shell bead attached to its head, and the shaft had been transversely drilled below the shell bead (Webb 1946:292, Figure 51A). This pin style has not yet been documented in southern Indiana or Illinois.

When engraving did occur on the Green River pins, it differed in both design and execution from that documented on pins from north of the Ohio River. For example, Webb (1946:Figure 51C) illustrates a section of a bird bone from the Carlston Annis site that had been modified by carving a rectangular hole in its midsection and extensively engraving both ends of the tube.

While rare, some Green River pins were relatively elaborate. Many were associated with burials, often in the head region, supporting their assumed pin function. Some were long (ca. 15 cm) pins with expanded, flat heads formed by extensive grinding and polishing (Figure 7.7, j). Sometimes, the sharp, pointed ends of the pins were highly polished (Webb 1946:291). Webb (1946:238) reported more than 60 of this type from just the Indian Knoll site.

Other kinds of Green River pins were manufactured using technologies not yet documented north of the Ohio River. C. B. Moore (1916) and Webb (1946:Figure 51A) illustrate large bone pins with heads made from various combinations of shell beads and asphaltum (Figure 7.7, c, e, g). In some cases, the shell beads were stacked on top of the pin shaft, and then held in place by the asphaltum. The heads of other pins were formed by inserting the pin shaft into a glob of asphaltum, then placing shell disk beads on the top and sides of the asphaltum.

Occasionally, Green River hunter-gatherers painted the shafts of their hairpins (Webb 1946:292–294, Figure 48C). So far, painted pin shafts have not been documented in the southern Midwest; however, they have been found on Archaic sites in parts of the Cumberland, Tennessee, and Savannah river valleys (Jefferies 2004:83).

While the Green River shell middens, like Carlston Annis (White 2005) and Indian Knoll (Webb 1946), have yielded some of the largest and best preserved Late Archaic bone and antler tool collections, archaeologists have documented similar kinds of artifacts and production strategies at contemporary sites in other parts of the Lower Ohio Valley (Anslinger et al. 1994:88–90; Granger et al. 1992:193–205; Winters 1969:Plates 16–20, 24, 26–28, 33–34, 36–38). These artifacts represent an important part of the Late Archaic tool kit. In addition, Late Archaic people used bone-working technology to make many items of personal adornment, as well as probable ritual objects such as turtle shell rattles and flutes.

Settlement and Mobility

If the number of Late Archaic components can be equated with population, then site file data from the Lower Ohio Valley region of Kentucky, Illinois, and Indiana reflect a substantial population increase during this time. For example, in Kentucky more than twice as many Late Archaic components have been recorded than for the Middle Archaic (Figure 7.8). This trend conforms to an increase in regional populations seen in other parts of eastern North America at about this same time (Milner 2004:Figure 13).

Throughout much of the North American Midcontinent, Late Holocene hunter-gatherer settlement patterns were characterized by groups that were

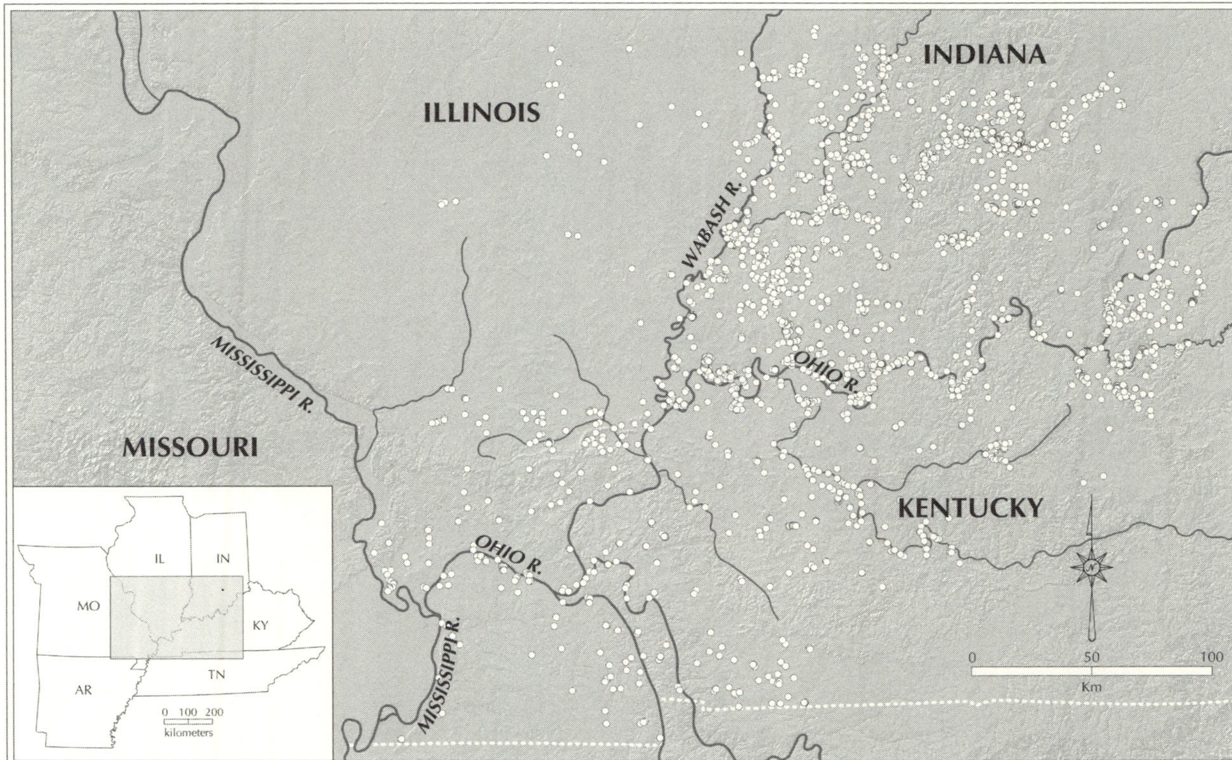

Figure 7.8. Distribution of Late Archaic sites.

"firmly tethered to river or stream valley, semipermanent, dry season residential base camps" (Smith 1986:31). These base camps, many identified by their deep middens, were located in optimum settings where their inhabitants could exploit both shoal-area aquatic species and floodplain plants and animals. Inhabitants of these camps employed a variety of smaller, seasonally occupied sites to exploit more distant interriverine upland resources, as well as those found adjacent to or within river valleys (Smith 1986:31–32).

Bruce Smith's (1986) model is supported by our settlement data from the Cypress Creek watershed in west central Kentucky (Jefferies et al. 2005). Here, midden sites along the edge of the middle Green River continued to be intensively occupied, while at the same time small, more specialized Late Archaic sites appear in parts of the watershed that had not been intensively utilized during earlier times (Jefferies et al. 2005).

Intersite Patterning

The middle Green River landscape was dominated by the river-edge shell middens that have been the subjects of archaeological investigation for nearly 100 years (Marquardt and Watson 2005; Moore 1916; Webb 1946). Many of these sites were adjacent to river shoals inhabited by large populations of mussels. The abundance and reliability of the bivalves promoted intensive use of these river-edge settings for thousands of years (Crothers 1999:249). In contrast, several large middens located away from the Green River, such as Ward and Kirkland (Webb and Haag 1940), indicate that Late Archaic hunter-gatherers also intensively exploited resources found in and along the region's interior wetland and upland settings (Jefferies et al. 2005) (Figure 7.9).

Late Archaic hunter-gatherers aggregated at these river-edge middens, which probably served as base camps, for several months during the warm seasons of the year. While there, they carried out a variety of activities, including the intensive exploitation of river mussels (Marquardt and Watson 2005a:632–635). These sites were also the scenes of many social and economic activities, including mortuary rituals. Over time, these repeatedly occupied base camps probably acquired special significance in the minds of their Late Archaic inhabitants, becoming as much of a part of their cultural landscape as their natural landscape (Crothers 1999:250).

The Green River middens have often been called "shell mounds," but, for the most part, this is a misnomer (Marquardt and Watson 2005a:632). Unlike mounds constructed during the Woodland and Mississippian periods, most of the Green River "shell mounds" had little vertical relief. Instead, they appear

Figure 7.9. Aerial photograph of the Ward site midden (from 1936 Aerial Survey of Paducah Reservoir Area, Engineering Services Division, Tennessee Valley Authority, Knoxville, Tennessee. Sheet No. 2-121-9).

as gentle rises on the local topography. While they undoubtedly were very significant spots on the Late Archaic cultural landscape, most were not highly visible landmarks. In contrast, a few midden sites, such as Ward and Read (Figure 7.10), were strategically situated in elevated topographic settings, providing their inhabitants with good vistas of the surrounding landscape (Milner and Jefferies 1998).

The intensity of Late Archaic activities at the Green River middens is indicated by their size and contents. Middens varied in size, ranging from 50 to 130 m long, 40 to 80 m wide, and up to 2.5 m deep. Although these sites are often referred to as "shell" middens, their actual shell content varied considerably. For example, shell comprised just 17 percent of the Carlston Annis midden, the remaining volume coming from sandstone (15 percent), inorganic matrix (23 percent), and air pores (45 percent). Some of the other Green River shell middens contain more shell, while others have much less. Within the midden matrix are varying densities of ash, charcoal, fire-cracked rock, bone, artifacts, and other cultural or natural materials (Marquardt and Watson 2005a:632).

The relatively few features preserved in the river-edge middens suggest that steaming and/or roasting of mussels was an important activity (Figure 7.11).

Figure 7.10. Depression-era excavations at the Ward site (Photograph 218McL11, William S. Webb Museum of Anthropology, University of Kentucky).

Although researchers working at shell midden sites such as Carlston Annis have identified occupation surfaces, they found very little evidence for actual habitation structures (Marquardt and Watson 2005a:632).

During the cooler, wetter winter and spring seasons, the people who lived at the base camps probably dispersed to smaller camps and rockshelters situated in nearby or remote habitats. These sites are usually characterized by little or no midden, and by a much lower diversity and number of artifacts than found at the midden sites (Marquardt and Watson 2005a:631–633).

The investigation of Archaic hunter-gatherer settlement/subsistence strategies conducted as part of the Cypress Creek Archaeological Project (Jefferies et al. 2005, 2008) helped to place activities represented at the large Green River middens in a broader regional context. Survey of more than 600 ha along Cypress Creek, a major Green River tributary, revealed that Late Archaic hunter-gatherer activities substantially increased over that of the Middle Archaic. This trend is indicated by both higher component and projectile point frequencies when adjusted for the different lengths of the periods (Jefferies et al. 2005:17).

The intensive occupation of several nearby Green River shell middens,

Figure 7.11. Three Late Archaic stone hearths exposed during WPA excavations at the Read site (Photograph 28BT10, William S. Webb Museum of Anthropology, University of Kentucky).

such as Butterfield, Wilson-Seymour, and Hollis, represents a continuation of a settlement strategy initiated by late Middle Archaic hunter-gatherers. Similar intensively occupied sites (Ward) were also located in upland settings where carbonized chenopodium, amaranth, and polygonum seeds (Bonzani 2002) indicate the procurement of certain food resources that were not abundant along the river edge. Hunter-gatherer groups presumably lived at river-edge middens such as Carlston Annis for several months during the year (Marquardt and

Watson 2005a:632). Smaller task groups made short forays from these sites to exploit more remote and scattered resources not found close to the river (Jefferies et al. 2005:17).

In the Cypress Creek area, Late Archaic components appear to cluster along the edges of the uplands that overlook the Cypress Creek wetlands, amplifying a trend that started during the Middle Archaic. Forty-nine percent of all Late Archaic components occur at sites having Middle Archaic components, suggesting the exploitation of a similar variety of resources.

In contrast, approximately one-third of the Late Archaic components occurred at sites first occupied during the Late Archaic. Many of these were situated near the upland-wetland interface, suggesting that the economic significance of interior wetland habitats increased during the Late Archaic. The light artifact scatters found at most newly occupied sites suggest that small groups of hunter-gatherers intermittently occupied them for short periods. The increased emphasis on resource-rich wetlands is consistent with a growing regional population as suggested by the greater frequency of Late Archaic projectile points. Late Archaic use of the uplands also increased relative to that of the Middle Archaic (Jefferies et al. 2005:17–18).

Overall, the settlement strategy employed by Cypress Creek's Late Archaic hunter-gatherers reflects a growing regional population experiencing increasing pressure to more efficiently use all parts of their surrounding environment. Site clusters suggest the disproportionate use of certain sections of the riverbank or swamp edge. Sites were no longer widely distributed across the landscape as they had been earlier (Jefferies et al. 2005:19).

George Crothers and Reinhard Bernbeck (2004:416) have proposed that Green River midden sites are a consequence of what they call "immediate-return hunters and gatherers in a foraging mode of production" (2004:416). They maintain that a midden's large size does not necessarily mean that sedentary groups occupied it or that an occupation represented a group structured along kinship lines. Instead, midden deposition could be attributed to continuous occupation by people whose group composition changed constantly. Frequent short occupations by small forager groups could produce the same kind of midden as created by the activities of one large group.

The fluid social composition of site occupants meant that the midden was an incidental consequence of activities, not an intentionally constructed symbolic marker because the composition of the group using that spot constantly changed. Likewise, the unsystematic placement of graves and the diversity of associated grave goods suggest that these small groups were not bound to a common set of rules about how and where to dispose of their dead. Collectively, these factors may indicate that the Green River middens formed because of successive occupation by small forager groups that did not share identical forms of ritual behavior, not by the continuous occupation by a group of closely related people (Crothers and Bernbeck 2004:414–415).

Like south of the Ohio River, southern Indiana Late Archaic groups continued a settlement strategy initially established by late Middle Archaic hunter-gatherers (Munson and Cook 1980b:704; Stafford 1994). This strategy, which emphasized the establishment of riverine base camps and the exploitation of resource patches located in the lower portions of stream drainage systems, is reflected by an increasing frequency of hafted bifaces found along major rivers.

As in other parts of the Lower Ohio Valley, Late Archaic base camp inhabitants organized logistical forays to procure both upland and near-river resources, which they then brought back to the base camp for storage. Due to increased time and energy considerations, more emphasis was placed on exploiting resources closer to the base camp, although more distant resource areas used by earlier Archaic hunter-gatherers were never completely ignored (Stafford 1994:231–232).

Late Archaic landscape use in southern Indiana's Patoka Lake region, characterized by several distinctive site "types" distinguished by their size, location, tool kits, and season(s) of occupation, included large shell middens along major drainages (summer), fishing camps on small rivers and tributary streams (summer), base camps along small rivers and their tributaries (fall), hunting camps (fall-winter), rockshelter camps (fall-winter), winter habitations, and numerous smaller sites of unknown function (Munson 1980b:678, Figure 25.2). As in many parts of the Lower Ohio Valley (Marquardt and Watson 2005; Winter 1969), this pattern of landscape use represented a continuation of that practiced by late Middle Archaic hunter-gatherers (Munson 1980b:678).

In southern Illinois' Saline River valley, the start of the Late Archaic was marked by the end of intensively occupied sites characterized by deep middens (Jefferies and Butler 1982). Although Late Archaic components are often the most numerous in regional surveys, researchers have not systematically investigated this important period of hunter-gatherer activity. There are no indications of any major subsistence changes during this time, but it appears as though the regional population was more dispersed and more evenly distributed across the landscape than during the late Middle Archaic. For example,

Late Archaic occupations at Carrier Mills, at the nearby West Harrisburg sites, and at the Rose Hotel site near the Ohio River were still quite heavy, but occupation intensity, as measured by artifact frequency and midden deposition, was considerably less than during the late Middle Archaic. Late Archaic projectile point deposition rates at Carrier Mills approach only 50 percent of the Middle Archaic rate, suggesting that, while Late Archaic hunter-gatherers still intensively used the district, the frequency and way the landscape was used appears to have significantly changed (Jefferies 1982a:1486).

While relatively large Late Archaic sites reflecting population aggregation are well documented along the interior portions of the Saline River drainage in southeastern Illinois, archaeologists have failed to locate many similar sites in southwestern Illinois' much more extensive Little Muddy River watershed. Why there was such variation in Late Archaic population distributions in these nearby watersheds remains unclear (Hargrave and Butler 1994b:531).

Southern Illinois' Late Archaic hunter-gatherers occupied all types of topographic settings. In particular, they made extensive use of the many rockshelters found along the region's creek valleys and bluffs. It appears that these groups continued to practice a predominantly logistically organized settlement/subsistence strategy, although the duration and intensity of occupations were quite variable (Butler 2009). In most ways, Late Archaic settlement strategies are similar to those practiced by contemporary hunter-gatherers who lived in adjacent parts of the Lower Ohio Valley.

Terminal Archaic components are identified by the presence of Merom, Trimble, and Robeson points, as well as roughly contemporaneous Buck Creek Barbed points (Seeman 1975; Winters 1967, 1969). Recent radiocarbon dates from the Knob Creek site in southern Indiana suggest that this phase may have persisted from ca. 3500 to as late as 2700 B.P. (Stafford and Cantin 2009).

C. Russell Stafford's (1994) study of Terminal Archaic point distributions in southern Indiana revealed a strong diachronic trend of increasing hafted biface frequencies near major rivers—over 46 percent occurred near a major river and 72 percent were in proximity to streams that flow into a major river (Stafford 1994:231–232). In southern Indiana's Patoka Lake region, situated within Stafford's study area, Terminal Archaic sites tended to be more common in floodplain (24 percent) and ridge spur (12 percent) topographic settings than during the late Middle and early Late Archaic. As in earlier times, however, most Terminal Archaic sites (59 percent) were situated on river and stream terraces (Munson and Cook 1980b:Table 27.3).

Intrasite Patterning

Some of the best-documented examples of intensively occupied Late Archaic sites occur along the middle portion of Kentucky's Green River (Crothers 1999; Marquardt and Watson 2005; Webb 1946; Winters 1974). Nearly 100 years of archaeological investigations have clearly demonstrated that Late Holocene hunter-gatherers carried out a wide range of domestic and ritual activities at these large, complex river-edge and upland sites.

Much of our knowledge about the Green River middens and the people who lived there is due to the exhaustive efforts of Patty Jo Watson and William H. Marquardt, along with their many colleagues who participated in the Shell Mound Archaeological Project (SMAP) (Marquardt and Watson 2005d). SMAP investigations, beginning in 1972, focused on several large shell middens in the river's Big Bend region. Foremost among these was the Carlston Annis site (15Bt5). Webb's (1950a) earlier investigations at Carlston Annis had provided an initial glimpse of the site's complex structure and numerous artifacts, features, and burials. In keeping with the reporting standards of the time, he used a lengthy trait list to characterize Carlston Annis, and then compared it to other nearby and distant Archaic sites. Webb's exhaustive efforts helped mid-twentieth-century archaeologists to better define the still evolving "Archaic" concept in eastern North American prehistory.

Marquardt and Watson initiated SMAP to collect information on Late Archaic subsistence practices at Carlston Annis that they could then compare with data from nearby Mammoth (Watson 1974) and Salts (Watson 1969) caves, where they had found evidence for Early Woodland horticulture (Marquardt and Watson 1983:323–324). Although they found little evidence of gardening at Carlston Annis, this initial effort spawned years of study by dozens of researchers representing many different scientific disciplines. Over the next 30 years, they gradually expanded the SMAP research program by conducting excavations at the Carlston Annis, DeWeese (15Bt6), Haynes (15Bt11), and Bowles (15Oh13) sites, and reanalyzing existing collections from Read (15Bt10) and other previously investigated sites. These studies revealed that the middens formed because of very complex natural and cultural processes, resulting in the accumulation of deposits of shell, rock, soil, and cultural debris of variable extent and thickness.

To learn more about the complex processes responsible for midden development, SMAP pursued a multidisciplinary research program focusing on a broad range of issues that included geomorphology; lithic, faunal, and botanical materials; human skeletal biology; and settlement/subsistence organization (Carstens and Watson 1996). SMAP's final report (Marquardt and Watson 2005) is the most comprehensive overview of middle Green River hunter-gatherers yet produced.

The size of shell middens found along the banks of the middle Green River varies quite a bit, but most range from about 50 to 130 m long, 40 to 80 m wide, and .5 to 2.5 m thick. The majority of the Green River Archaic radiocarbon dates fall between 5800 and 2700 B.P., but a few earlier ones exist (Marquardt and Watson 2005a:632). Examination of temporally diagnostic projectile points indicates that their history of occupation is quiet variable. Initial occupation of some, such as Ward (15McL11), Jimtown Hill (15Oh19), and Baker (15Mu12), clearly date to the late Middle Archaic (Jefferies et al. 2005:Table 2), while intensive use of other river-edge locations, like Indian Knoll, began somewhat later (Hensley 1994:Table 34).

It appears that Late Archaic hunter-gatherers occupied river-side midden sites mainly during the summer and fall. Based on the intensity of midden deposition, artifact frequencies, and the number of burials, it seems likely that many generations of hunter-gatherers occupied these spots. The selection of some site locations appears to have been related to the proximity of economically important river mussel beds that Late Archaic people intensively collected (Morey et al. 2002:526). Many inhabitants probably dispersed to smaller ephemeral camps and rockshelters situated in both wetland and upland settings during the winter and spring (Figure 7.12). The Late Archaic people who lived at the nearby DeWeese, Haynes, Bowles, Chiggerville, and Indian Knoll sites probably pursued a similar settlement strategy (Hensley 1994:222–230; Marquardt and Watson 2005a:635).

Although archaeologists commonly refer to the Green River middens as "shell" middens, the amount of shell found in the midden matrix at different sites is quite variable. Some middens, such as Barrett (15McL4) and Indian Knoll, contain relatively low shell densities; however, Carlston Annis, Bowles, and Jimtown Hill have higher densities (Hensley 1994:Table 36).

Carlston Annis (15Bt5)

The Carlston Annis midden can be attributed to a complex set of natural and cultural processes that took place over several thousand years (Stein 2005). River and lake sediments constitute the midden's noncultural components, while cultural deposits consist of the shell midden, colluvium, and plow zone.

Figure 7.12. Late Holocene hunter-gatherer activities at a Kentucky rockshelter (original artwork by Rex Robinson, courtesy of Rex Robinson and the Kentucky Heritage Council, Frankfort).

Midden stratification is differentiated largely on greater or lesser quantities of shell and sandstone (Stein 2005:121). Shell density, which is highly variable throughout the midden, may be related to proximity to living areas, the effects of alluvial erosion, or other disturbances. The average percentage of shell in different parts of the midden ranged from 1 to 88 percent, with a mean of 17 percent. Shell density was significantly greater in the western part of the deposit, the side facing the river, than on top (47 vs. 8 percent). The shell came from the beds of mussels that lived on the gravel and sandbars or riffle-runs near Carlston Annis (Stein 2005:127).

A distinct cultural deposit, known as the "shell-free midden," was first recognized in 1977. This dark shell-less zone, containing a high density of sandstone, blankets the underlying shell-rich zone. The shell-free midden contains a higher level of organic matter and a lower carbonate content than the underlying shell midden, but similar levels of phosphorus in the two midden zones suggest that they reflect similar activities (Stein 2005:133). Recent radiocarbon dates, combined with detailed chemical and physical analyses, suggest that the shell-free midden is associated with Late Archaic site use, although pottery attributable to later occupations is also present (Marquardt and Watson 2005a:632).

River-edge middens like Carlston Annis appear to have been dynamic places during their occupation. Cultural and natural transformations have continued to impact these sites since their hunter-gatherer inhabitants vacated them. Digging by earthworms, crayfish, and groundhogs have continued to mix the midden, as have modern land use practices, including plowing, building construction, excavating mussel shell to add to chicken feed, and burying trash. Authorized and unauthorized archaeological excavations have added to this disturbance (Marquardt and Watson 1983:329; Stein 1980:170–199).

Archaeological sites marking places where groups stayed for extended periods usually have evidence for substantial structures (Sassaman and Ledbetter 1996), but such is not the case at Carlston Annis. The shelters that were built there must have been relatively flimsy, probably consisting of brush- or hide-covered lean-tos or wattle and daub structures. Even post molds were rare at Carlston Annis (Marquardt and Watson 2005a:632), although Webb did report 19 posts arranged in a parallelogram pattern (Rolingson 1967:358, cited in Stein 2005:131). In addition, he (1950a:271–272) described four "burned clay floors" that were uncovered during the WPA excavations. Webb suggested that inhabitants sometimes brought clean clay to the midden heap where they spread it out on the surface in a thin layer. They then built a fire on the clay "floor" where food was cooked and eaten. The midden surrounding the clay "floor" contained bones, nut hulls, antler, broken artifacts, and shell (1950a:272).

During the late 1970s, SMAP archaeologists exposed six features that resembled Webb's burned clay "floors." One feature, consisting of a red layer overlying an unconsolidated yellow-red matrix, measured 60 x 30 cm and 8 cm thick. The "floor" appeared to have been made from sandy mud probably collected along the banks of the nearby Green River. Once spread on the midden, it served as a hearth for a fire that eventually baked the clay (Stein 2005: 128–129).

WPA excavators exposed other kinds of probable heating and cooking facilities that Webb (1950a:272, Figures 3A and 3B) described as "fire hearths marked by scattered burned stones" and "rock lined fire basins." These "heat-related" features would have served a variety of purposes, including mussel steaming and/or roasting or baking (Marquardt and Watson 2005a:632).

Late Archaic groups responsible for Green River middens were fishers, hunters, gatherers, and gardeners. They fished the waters of the Green River, its tributaries, and the many nearby shallow lakes and swamps. These wetlands also offered many aquatic and terrestrial turtles whose meat was consumed and shells made into a variety of utilitarian and ritual objects. Of course, shellfish were very important food resources at some of the river-edge sites, particularly those located close to shoals or rapids (Crothers 1999; Morey and Crothers 1998). Shellfish collectors exploited at least 36 species of unionids and five species of aquatic gastropods in their pursuit of food.

Shell midden inhabitants focused their hunting efforts on the white-tailed deer, wild turkey, and smaller mammals such as squirrel and rabbit. They also ate many different kinds of seasonally available wild plants, of which acorns and hickory nuts were the most important, along with seeds (knotweed, little barley, foxtail grass, wild rice, chenopod) and fruits (grape, blackberry, strawberry, honey locust, and persimmon). Some people grew a gourdlike squash, probably *Cucurbita pepo* var. *ovifera,* but cultigens grown by other midwestern Late Archaic groups, such as sumpweed, apparently were not grown by Green River people (Marquardt and Watson 2005a:633).

The technological organization of Carlston Annis hunter-gatherers was similar to that practiced by contemporary groups that lived in other parts of the Lower Ohio Valley. They hunted with the atlatl; captured smaller mammals by snaring or trapping; and fished using a hook and line, gorge, or net. They would have used plant fibers to make clothing, footwear, cordage, mats, and nets. Cups, bowls, atlatl weights, beads, and various items of personal adornment were made from marine shell obtained from the coast. Residents would have used animal hides and furs to manufacture clothes, blankets, and bags, while they used animal bones to produce atlatl handles and hooks, awls, projectile points, beads, and pins (Marquardt and Watson 2005a:633).

The Carlston Annis hunter-gatherers collected locally available Vienna, St. Louis, and Ste. Genevieve cherts to make their flaked stone knives, scrapers, projectile points, drills, burins, gravers, and blades. Based on the kinds of debitage found at Carlston Annis, Late Archaic people did most of the initial production activities close to the chert sources, with the final manufacturing stages and tool maintenance being carried out at the site. They used expediently produced flakes struck from amorphous cores for cutting and scraping tasks, and made a few specialized flake implements, like bladelets and bladelike flakes. They employed ground stone technology to make atlatl weights, grooved axes, pestles, and grinding stones (Marquardt and Watson 2005a:633).

Green River hunter-gatherers were well connected to other Lower Ohio Valley hunter-gatherer groups, as well as to those living in more distant parts of the

Midcontinent and beyond. These ties are indicated by objects made from materials not found in the Green River region, such as marine shell, copper, and the bones of certain animals. Marine shell items are the most common, with more than 23,000 shell disk beads documented from five of the Green River shell middens. The number and distribution of these beads suggest that Green River hunter-gatherers maintained relatively strong and sustained socioeconomic ties with groups that lived to the south throughout much of the Late Archaic (Marquardt and Watson 2005a:634).

The much heralded, but rare (n = 13) copper artifacts found at several of the WPA-excavated shell middens, including a bead and a double-pointed awl from Carlston Annis (Webb 1950a:300), indicate that Green River hunter-gatherers also maintained social connections with groups that lived to the north (Watson 2005:623). In addition, a few Carlston Annis burials contained the bones of animals not native to the Green River region. Like copper, these items reflect socioeconomic connections with groups that lived around the Great Lakes (Marquardt and Watson 2005a:634).

Carlston Annis, like many of the Green River shell middens, serves as the final resting place for at least some of its former inhabitants. At least 350 individuals, and probably many more, were buried there (Mensforth 2005:471), indicating that, in addition to being a residence, the site also had a significant mortuary function. Some researchers have proposed that the Green River shell middens functioned as intentionally constructed burial mounds or territorial markers (Claassen 1991a, 1991b, 1992), while others view their mortuary component as reflecting the burial of individuals who died there or nearby, ultimately being buried in the place where they spent at least part of their lives (Milner and Jefferies 1998).

Late Holocene mortuary practices at Carlston Annis resemble those seen at other nearby shell middens. Males and females of all ages were buried there, some singly, others in groups. Most were buried in a flexed or semiflexed position, and about one-third were accompanied by burial objects consisting of items of personal adornment, utilitarian objects, and possible shamanistic paraphernalia. Dogs were also purposely buried there (Marquardt and Watson 2005a:635).

Read (15Bt10)

Like Carlston Annis, the Read site (Milner and Jefferies 1998; Webb 1950b) is another large middle Green River shell midden excavated by WPA crews and is representative of what has become known as the "Green River Shell Mound Archaic." Situated several kilometers downriver and on the opposite bank from Carlston Annis, Read consisted of a large, deep midden on the edge of a sandstone bluff overlooking the Green River. Shell deposits reportedly covered a roughly 61 m by 128 m area of the bluff top, with an estimated volume of 3,700 m^3. WPA investigators noted two discrete shell deposits, one on either side of the clay ridgetop, representing about 53 percent of the estimated midden volume (Milner and Jefferies 1998:120–122; Webb 1950b).

The history of site occupation roughly parallels that at Carlston Annis, with Late Archaic hafted bifaces representing more than 60 percent of the temporally diagnostic flaked stone implements (Hensley 1994:Table 34). However, the bluff top location was also utilized to some extent by Early (8 percent of bifaces) and Middle Archaic (23 percent of bifaces) hunter-gatherers, reflecting ongoing preference for this topographic setting (Milner and Jefferies 1998:125).

Like most other Green River middens excavated during the 1930s, the Read artifact collection is biased by the common practice of Depression-era archaeologists keeping only easily recognizable tools (Milner and Smith 1986). Nevertheless, the diversity of flaked and ground stone, bone, antler, and shell artifacts is consistent with the wide range of domestic tasks commonly performed at habitation sites (Milner and Jefferies 1998:125).

The number and diversity of the Read features also suggest that inhabitants conducted many different tasks (Webb 1950b). Burned shell lenses, shallow circular basins lined with closely laid stones, burned clay areas, and clay-lined basins are indicative of processing activities that involved the use of heat (Webb 1950b:362). Some of these may have been used for cooking.

Areas adjacent to domestic activity areas often contained caches of heavy artifacts, like hammerstones, pestles, and grooved axes, that could not be easily transported. Webb (1950b:363) observed that these caches "were found thus together where they were last used, always in the vicinity of a fire hearth." He (1950b:363) also noted several larger circular pits dug into the hardpan underlying the midden that initially may have functioned as storage pits, but were later used for refuse disposal. Webb documented two areas, each about 10 ft in diameter and covered by chert debitage and broken bifaces and projectile points, that he interpreted as flaked stone tool workshops (1950b:365).

Like most of the Green River shell middens, Read also had a significant mortuary component, represented by the skeletal remains of more than 250 individuals (Webb 1950b:365–377). Graves were unevenly distributed across the site, with most interred on the highest part of the midden along the slope facing the Green River (Herrmann 1996; Milner and Jefferies 1998:123).

Ward (15McL11)

Clearly, Late Archaic hunter-gatherers who lived along and near the middle Green River intensively occupied certain spots along the river's edge, particularly places where mussel beds were easily accessible (Crothers 1999:139–151; Morey and Crothers 1998). People who lived at these river-edge spots were well positioned to take advantage of both aquatic and terrestrial resources, but these were not the only places they intensively occupied. Hunter-gatherers also established major settlements, such as Ward (15McL11), that were situated along the interface between the wetlands and the higher and drier uplands (Jefferies et al. 2008; Webb and Haag 1940).

Hunter-gatherers occupied Ward over the course of at least three millennia during the late Middle and Late Archaic. The large midden is situated along the edge of a high ridge that overlooks the former Cypress Creek swamp (Figure 7.10). The Green River is located approximately 5 km to the northeast.

Excavations demonstrated that Ward's midden closely resembled those of the nearby river-edge sites—lots of artifacts, sandstone, animal bones, burials, and features—reflecting a wide range of tasks typically associated with a base camp (Webb and Haag 1940). Unlike the others, however, Ward only had a light scattering of mussel shells, possibly collected from the riffle-run that once existed in nearby Cypress Creek. If not mussels, then what attracted hunter-gatherers to "inland" settings like Ward and nearby Kirkland?

Our recent investigations at Ward yielded evidence that inhabitants focused their attention on resources found in the woodlands, fields, and swamps near the site (Jefferies et al. 2005, 2008). Botanical samples contained carbonized seeds of wild *Chenopodium berlandieri* var. *bushianum,* as well as knotweed and pigweed (Bonzani 2002). Perhaps, in lieu of reliable, easily collected shellfish, Ward's inhabitants collected the many nutritious starchy and oily seeds that grew on and around the site clearing. Seeds could have been eaten immediately to complement other foods found near the site, or they could have been stored and used when other foods were unavailable. The Ward people also hunted a variety of terrestrial (deer, raccoon, possible black bear) and wetland (fish, aquatic turtle, muskrat) species, and collected a few shellfish from the nearby creek (Jefferies et al. 2005:17–18).

Diversification of subsistence activities, like those documented at Ward, would have minimized the effects of a scarcity of any single resource, thereby buffering the inhabitants from shortages in their usual foods. Establishing complementary base camps at both upland and river-edge settings, each with their own distinctive sets of resources, allowed hunter-gatherers to more easily deal with seasonal and yearly variation in food availability (Jefferies et al. 2005: 17–18).

North of the Ohio

Middle to Late Holocene shell middens also occur north of the Ohio River in southern Indiana. Many of these sites, like Old Clarksville (12CL1), Breeden (12Hr11), Paddy's West (12F146), and Reid (12F11), are located directly on the Ohio River. The Breeden site (Bellis 1968, 1982), a buried shell midden exposed in the eroding riverbank, contained a variety of side notched and stemmed bifaces. Recent examination of the Breeden bifaces indicates that Late Archaic stemmed points were more common in the midden's upper zones, while side notched points, such as Matanzas, Big Sandy II, and Brewerton Eared, were more prevalent in the deeper midden zones. A single radiocarbon date of 4200 +/– 200 B.P. was associated with the lowest excavation level. The date is considerably younger than those usually associated with the side notched point types; however, Breeden's projectile point sequence does indicate a transition to later stemmed types during the occupation (Stafford and Cantin 2009).

The Reid site, a 1.8-m-thick stratified shell midden at the north end of the Knob Creek bottom, yielded radiocarbon dates of 4555 +/– 70 and 5480 +/– 90 B.P. Like many other Lower Ohio Valley shell middens, the majority (62 percent) of the diagnostic bifaces are Late Archaic stemmed forms, such as the Karnak/McWhinney and Benton/Elk River types, but side notched Matanzas points were also common (Stafford and Cantin 2009).

Despite the passage of nearly 40 years since Winters's (1969) investigation of the Riverton culture sites, his work still provides some of the best information on regional Terminal Archaic settlement patterns and site structure. Excavations at Riverton and Swan Island, located on a low terrace of the Wabash River, revealed major midden concentrations covering from .8 to 1.2 ha. The Robeson Hills site, situated on a rise overlooking the Wabash, was about the same size as the other sites, but lacked midden. Based on a variety of indicators, Winters (1969:131) characterized Riverton as a "base camp," Swan Island as a "transient" camp, and Robeson Hills as a "settlement." Although Winters's analysis of site function is not without its problems (Fitting 1973; Thieme

1991), his excavation data provide some useful insights on site structure and organization.

Occupants of the three Riverton sites used both prepared and unprepared hearths for many different processing tasks, as well as for heating and lighting. Shallow clay basins were also common at all three sites. Winters thought the basins were used for cooking mussels and other food items. Terminal Archaic Riverton people frequently buried their dead at Robeson Hills, but they also buried a few individuals at Riverton and Swan Island (Winters 1969:123).

The discovery of numerous post molds and clay and stone floors at several of the Riverton sites provides a glimpse of the kinds of structures built by Lower Ohio Valley Terminal Archaic hunter-gatherers (Winters 1969:Figure 16). At Robeson Hills, arcs of post molds suggest that inhabitants built circular or oval structures ranging from 2 to 10 m in diameter. Small pieces of sandstone or other debris sometimes were placed in the hole around the post, probably to hold it in place while the hole was filled. Other post mold alignments suggest that inhabitants built small rectangular structures or other kinds of facilities (wind breaks, screens, drying racks). Riverton occupants of the Wint site (12B95) in southeastern Indiana also built oval structures using single-post construction techniques (Anslinger 1986, 2001).

In addition to post molds, excavations exposed clay and stone "floors" thought to be associated with structures or other kinds of activity areas. For example, in one part of the Riverton site, at least 10 clay floors were uncovered, all consisting of yellow clay or clay/sand platforms placed directly on the midden surface. Some of the clay layers were up to six inches thick. Several floors showed evidence of having been repaired by patching eroded areas. The floors were generally rectangular, however, several were oval (Winters 1969:97).

Five completely exposed floors had a mean length of 4.6 m, a width of 3.1 m, and a floor area of 15.0 m^2 (Winters 1969:Table 53). Although Winters referred to the clay features as "house platforms," he conceded that despite an intensive search, he found few associated post molds. Similar features were found at the nearby Swan Island site (Winters 1969:93) and at the Read and Carlston Annis sites in Kentucky (Webb 1950a:271–272, 1950b:363) where they were also interpreted as occupation surfaces or floors.

The size of the Riverton floors generally conformed to that of Archaic structures at the Mulberry Creek site in Alabama (Webb and DeJarnette 1942) and the Lamoka Lake site in New York (Ritchie 1932). Interestingly, the Riverton site platforms appear to have been constructed in rows, as opposed to having been randomly scattered about the site (Winters 1969:97, Figure 18). It is difficult to determine whether these clay/stone platforms were actually parts of houses or just more formally designated activity areas. If they were house platforms, the absence of post molds suggests that they were not very substantial and were constructed for short-term use. However, their alignment and regular spacing suggest that their builders had some kind of preconceived community plan in mind.

At the Confluence

Intensive Late Archaic hunter-gatherer occupations also occurred in many other parts of the Lower Ohio Valley area, such as southeast Missouri's Cairo Lowlands near the Ohio-Mississippi River confluence. Like many other parts of the region, the Cairo Lowlands contained a highly diverse biotic community consisting of year-round wetlands, permanent and seasonal swamps, drier uplands, sand dunes, and river levees. Such conditions would have supported highly diverse plant and animal communities, much like those found near the Black Earth site in southern Illinois. Intensively occupied Late Archaic sites were commonly situated along wetland-upland interfaces where site inhabitants could have easily collected seasonally available food resources (Buchner 2003c).

The Clifford LaPlant site is one such location. The site's O'Bryan Ridge phase component appears to represent a seasonal base camp and mortuary site occupied by logistically organized hunter-gatherers. The LaPlant site's Late Archaic component consisted of artifacts and features distributed over a ca. 150 m diameter area along a prominent sand dune and an adjacent level area that overlooked the surrounding wetlands. Artifact and feature diversity indicate that O'Bryan Ridge phase hunter-gatherers conducted a wide range of activities at the LaPlant site. Of particular note, chenopodium seeds found in an earth oven indicate that Late Archaic people were cultivating plants in the Cairo Lowlands as early as ca. 4000 B.P. (Buchner 2003b:468).

The discovery of a human cremation placed on top of the sand dune suggests that this prominent landscape feature not only served as a physical landmark, but may have also functioned as a sacred memorial similar to a mound or earthwork. The distribution of diagnostic O'Bryan Ridge phase artifacts and nonmortuary features suggests that the broad level area north of the sand dune was the scene of many domestic activities (Buchner 2003b:469).

Subsistence

Plant Resources

By the late Middle Holocene (ca. 5000 B.P.), changes in stream flow characteristics throughout much of the Midcontinent, including the Lower Ohio Valley, resulted in the expansion of both backwater and shoal habitats. Simultaneously, pioneer floodplain plants, like goosefoot (*Chenopodium berlandieri*), sumpweed (*Iva annua*), and wild gourd (*Cucurbita pepo* var. *ovifera*), flourished in disturbed portions of the river and stream channels (Smith 1992b:102; Smith and Cowan 2003:110).

The establishment of these rich wetland and floodplain habitats, with their abundant and diverse plant and animal communities, provided Late Holocene hunter-gatherers with important food resources throughout much of the year. Places with particularly abundant food supplies, such as along the middle Green River, were repeatedly occupied for increasingly longer periods of time, resulting in the formation of deep, expansive earth and/or shell middens (Jefferies 1982b; Marquardt and Watson 2005).

Bruce Smith (1992b:102) suggests that, "This annual reoccupation and reuse of specific floodplain locations through the growing season represented the initial emergence of continually disturbed 'anthropogenic' habitats in the interior river valleys of the eastern United States." Repeated human modification of the soil, along with the addition of floodwater sediments, improved the fertility of the already rich midden soils. These anthropogenic habitats provided excellent spots for the colonization of early successional floodplain plant species, with seeds being introduced through both human and natural agents (Smith 1992b:102).

The period from roughly 7000 to 4000 B.P. witnessed the increasing human intervention into the life cycles of some of these wild plants, leading to their eventual cultivation and domestication. This long, slow process is reflected by distinctive morphological changes (increase in achene size, reduction in seed coat thickness) in the seeds of gourd, goosefoot, sumpweed, and sunflower (*Helianthus annuus*), as well as the presence of species beyond their natural range (maygrass—*Phalaris caroliniana*). Collectively, these findings suggest that Late Holocene people were domesticating these plants by no later than 5000 to 4000 B.P. (Smith and Cowan 2003:107).

The accidental human dispersal of seeds near residential sites would have produced dense stands of plants that could have been further manipulated and encouraged by interested hunter-gatherers. Experimentation with these species does not appear to have stemmed from any kind of environmental stress, such as a food shortage. In fact, domestication seems to have taken place during a time of relative economic security (Smith and Cowan 2003:110–111).

Despite the domestication of selected native plants, it is unlikely that they became significant food sources until about 3000 B.P., many centuries after their domestication (Smith 1992b:107–108). Late Middle and Late Holocene hunter-gatherers continued to get the vast majority of their food from the same wild plants (hickory nuts, acorns, fruits, roots, and tubers) and animals (white-tailed deer, turkey, fish, etc.) as did their ancestors, with cultivated native plants having a variable, but small, dietary significance.

C. Margaret Scarry's (2003) general characterization of Ohio Valley Late Archaic plant use, largely based on data from the Falls of the Ohio (Lannie 1979) and northern Kentucky (Lopinot 1988), indicates that hickory nuts, being high in fat and protein, continued to be a major dietary component. Walnuts were also consumed, but their dietary significance is much lower than that of hickory nuts. Acorns and wild and domesticated starchy seeds could have been important sources of carbohydrates but were not widely used by Lower Ohio Valley Late Archaic hunter-gatherers (Scarry 2003:82–85). Late Archaic people would have also eaten fruits, such as grape, blackberry, raspberry, elderberry, pawpaw, and sumac; legumes (wild bean, vetch, and peavine); roots and tubers (groundnut, green brier, cattail, and wild onion/garlic); and greens (purslane, amaranth, chenopod, and knotweed) (Scarry 2003:68–73).

Some of the best data on Lower Ohio Valley Late Archaic plant use come from the more than 30 years of research conducted as part of the Shell Mound Archaeological Project (Marquardt and Watson 2005c). SMAP researchers were first drawn to the Middle Green River region in the hopes of finding a record of plant use that predated the well-documented examples found at nearby Salts (Yarnell 1969) and Mammoth (Yarnell 1974) (Wagner 2005:215) caves. In spite of processing hundreds of Carlston Annis midden samples, the SMAP efforts did not produce any evidence for early horticulture. Instead, hickory nuts, along with a small number of seeds from wild species, dominated the Late Archaic plant inventory. They did, however, find a few pieces of *Cucurbita pepo*, once but no longer thought to be a tropical cultigen. So, while the Carlston Annis botanical material did yield important data on Late Archaic subsistence (Crawford 2005), it did not yield much evidence for early horticulture nor did it conform to ideas concerning the floodplain origin of agriculture (Smith 1987, 1989, 1992a, 1992b) in the North American midcontinent (Marquardt and Watson 2005a:631).

Analyses of Green River plant remains collected from both river-edge (Carlston Annis, Bowles) and upland (Peter Cave) settings (Crawford 2005; Wagner 2005) indicate that Late Archaic people, like their contemporaries throughout the Lower Ohio Valley, gathered wild plant foods on a seasonal basis. Their efforts focused on hickory nuts and acorns, but they also collected an assortment of seeds and fruits like strawberry, blackberry, grape, honey locust, persimmon, knotweed, little barley, foxtail grass, wild rice, and chenopod. They also grew gourd/squash, but did not cultivate some plants grown by their contemporaries to the north and west, most notably sumpweed (Marquardt and Watson 2005a:633).

Carlston Annis's inhabitants collected at least four kinds of edible nuts—hickory, acorn, black walnut, and hazelnut. Hickory, oak, and hazelnut trees often occur in stands, while black walnut trees are scattered across patches of rich, open forest (Wagner 2005:232–233). As at most other Lower Ohio Valley Late Archaic sites, hickory (*Carya* sp.) was the most abundant nut throughout the Carlston Annis midden. Acorn (*Quercus* sp.) shells and meats also occurred, sometimes in large quantities, along with smaller quantities of black walnut (*Juglans nigra* L.) and hazelnut (*Corylus* sp.) (Wagner 2005:221).

The possibility exists that some Late Archaic groups altered the forest composition by selective burning and tree girdling (Wagner 2005:236–237). The selective use of these and other forest management strategies could have sufficiently altered the surrounding natural vegetation, thereby increasing the availability of desirable plants, as well as attracting certain game animals. The opening up of the forests may have also created conditions beneficial to the spread of weedy species that eventually became important parts of the diet (Wagner 2005:238).

While SMAP researchers did not find much evidence for early plant cultivation at Carlston Annis and the other middle Green River middens, the sites did yield the remains of plants that were domesticated a short time later. For example, accelerator radiocarbon dating of cucurbit rind fragments from Carlston Annis and Bowles indicates that regional hunter-gatherers were collecting these small squashes as early as 5730 +/– 640 B.P. (Marquardt and Watson 2005c:Table 6.1). The Green River cucurbits are most likely *Cucurbita pepo,* like those found at contemporary sites in eastern Kentucky and Missouri, since this part of Kentucky is not part of the natural range of other cucurbit varieties (Crawford 2005).

In some parts of the Midcontinent, like west central Illinois, harvesting wild chenopodium was well underway by 4000 B.P. Eastern Kentucky rockshelters have yielded domesticated chenopodium dating to as early as 3400 B.P. (Smith and Cowan 1987). In contrast, the relatively few chenopodium seeds found at Carlston Annis and Peter Cave reflect this plant's minor role in the diets of Middle Green River Late Archaic people.

Based on seed diameter, the Peter Cave chenopodium is probably *C. berlandieri,* an indigenous Midwest species. Some of the seeds appear to be wild varieties, while others may be attributable to a transitional type in the domestication of chenopodium. The Carlston Annis seed sample, which came from upper midden levels, was too small for confident species identification (Crawford 2005:196–197).

The relative abundance of seeds at Peter Cave suggests that Late Archaic people focused their seed collection efforts in the uplands and that the eventual cultivation of these plants took place away from the river's edge. This possibility is supported by the recent recovery of more than 80 carbonized chenopod seeds from the Ward site (15McL11), located in the uplands overlooking the expansive Cypress Creek wetlands (Bonzani 2002). The Ward chenopodium seeds exhibit a relatively narrow range of morphology that generally conforms to the wild morphotype. Based on the seeds' shape, size, and testa appearance, they are probably examples of *Chenopodium berlandieri.* Like some of the Peter Cave chenopod seeds, several of the Ward seeds have testae that are thin enough to be classified as domesticates, but this can occur even within wild populations (Smith 2002).

The relative abundance of the Ward chenopodium suggests that the site's Late Archaic inhabitants were minimally collecting large numbers of these starchy seeds. The presence of a few thin testae examples may mean that Ward inhabitants were involved in the early stages of plant domestication when wild species were tolerated and/or encouraged to grow in the rich, constantly disturbed middens. Ward's few knotweed (*Polygonum* sp.) and pigweed (*Amaranthus* sp.) seeds support the view that Late Archaic inhabitants collected wild seeds at this upland site (Bonzani 2002:167).

Late Archaic plant remains from the Railway Museum site (15Jf630), a large river-edge midden situated near the Falls of the Ohio (Anslinger et al. 1994), suggest a similar pattern as seen along the Green River. Although researchers identified several fragments of nondomesticated *Cucurbita,* they found no evidence for the exploitation of weed seeds, like goosefoot, maygrass, or sumpweed (Crites 1994:337), supporting the idea that Late Archaic seed collecting/cultivating strategies focused on upland settings.

Recent work at the Hedden (15McN81) and Highland Creek sites (15Un127) provides important new information on Late Archaic wetland resource use in the western part of the Lower Ohio Valley (Rossen 2000, 2006). The Hedden site was most intensively occupied from about 5100 to 3600 B.P. (Rossen 2000:Table 1). Because of the site's topographic setting and age, it provided investigators with an opportunity to study the relative importance of wetland plant use versus intensive nut exploitation during a time of decreased hunter-gatherer mobility (Brown and Vierra 1983; Stafford 1991, 1994).

Nuts were an important part of the Hedden diet. Thick-shelled hickory (*Carya* sp.) was by far the most important species, accounting for more than 98 percent of all nutshells. In addition to being an important subsistence resource, the abundance of carbonized hickory nutshells may mean that inhabitants used them for fuel (Rossen 2000:9). Acorns (*Quercus* sp.) were also important, despite containing only one-half the protein and one-third the fat of hickory nuts and requiring special processing to make them edible. The relative ease of collecting acorns and their abundance appear to have compensated for these nutritional disadvantages. Hedden inhabitants also collected other kinds of nuts when available, like black walnut (*Juglans nigra*), pecan (*Carya illinoinensis*), hazelnut (*Corylus* sp.), and butternut (*Juglans cinerea*) (Rossen 2000:9–10).

Small quantities of carbonized persimmon (*Diospyros virginiana*) and grape (*Vitis* sp.) seed reflect the Late Archaic use of these fruits. Hedden inhabitants also occasionally collected small quantities of wild sunflower (*Helianthus* sp.), chenopod (*Chenopodium* sp.), and amaranth (*Amaranthus* sp.) seeds. Although squash (*Cucurbita* sp.) has been found at some western Kentucky Late Archaic sites, its absence from Hedden suggests that its use was sporadic (Rossen 2000:13–14).

The Hedden inhabitants' use of wetland plants was indicated by carbonized bulrush (*Scirpus* sp.), chufa sedge (*Cyperus* sp.), flatsedge (*Cyperus* sp.), pondweed (*Potamogeton* sp.), horned pondweed (*Zannichellia palustris*), sedge (*Carex* sp.), and spikerush (*Eleocharis* sp.) seeds. All of these plants have edible seeds, stems, and/or roots. Although seed frequency was low, species diversity suggests the exploitation of a broad spectrum of edible wetland plants. Many of these plants retain their seeds for much of the year, which made them available as a substitute food during lean periods when more preferred species were unavailable (Rossen 2000:13).

By occupying the Hedden site, its Late Archaic inhabitants positioned themselves to take advantage of several habitats using a highly diversified plant exploitation strategy. This strategy combined the specialized collecting and processing of nuts with a more generalized collecting of wetland plants, fleshy fruits, and starchy and oily seeds (Rossen 2000:17).

Highland Creek (15Un127), situated some 100 km up the Ohio River from Hedden, was also surrounded by densely forested backwater swamps and sloughs. Radiocarbon dates indicate that much of the midden was deposited over a ca. 400-year period between 4500 and 4100 B.P. (Maggard and Pollack 2006:Table 4.2).

Highland Creek's Late Archaic inhabitants pursued a similar plant exploitation strategy as those who lived at Hedden (Rossen 2006:73–84). Nutshell, dominated by hickory, black walnut, and acorn, comprised most of the plant remains. Other plant materials included a few fruit pits (grape, persimmon, chokeberry), starchy and oily seeds (erect knotweed [*Polygonum erectum*], marshelder [*Iva annua*], and seeds from wetland-oriented plants (pondweed, hawthorn [*Crataegus* sp.]).

Although some Late Archaic people cultivated marshelder and erect knotweed, the Highland Creek specimens appear to have come from wild species. Unlike at Hedden, Highland Creek inhabitants grew small quantities of two possible cultigens, gourd (*Lagenaria* sp.) and squash (*Cucurbita* sp.). Overall, the Highland Creek plant remains suggest that the site's Late Archaic inhabitants had "a stable plant economy based on intensive nut collection and extensive, broad spectrum plant collecting from various ecozones" (Rossen 2006:84).

The Hedden and Highland Creek botanical inventories collectively contain four of the six recognized plants comprising the "Eastern Agricultural Complex." Seed size, morphology, and low frequency of occurrence clearly indicate that they are from wild plants; however, their presence indicates that the potential for domestication existed (Rossen 2006:83).

Deeply stratified deposits at the Little Muddy Rock Shelter in southwestern Illinois have provided relatively fine-grained data on Late Archaic plant use to the north of the Ohio River (Cremin 1992). Like south of the Ohio, hickory nuts were an important food throughout the shelter's Late Archaic occupation. Thick-shelled hickory, including shagbark hickory, dominated the Late Archaic I (ca. 4500 B.P.) botanical samples, and all of the identified nutshell from the Late Archaic II component (ca. 4000 B.P.) came from thick-shelled varieties of hickory. The Late Archaic II zone also yielded five squash rind fragments, demonstrating the widespread, but discontinuous significance of this species (Moffat 1992:502–504).

Thick-shelled varieties of hickory nut continued to be important during the Late Archaic III (ca. 3700 B.P.) occupation, but trace amounts of black walnut, butternut, and acorn suggest a growing importance of other nut-bearing trees (Moffatt 1992:502). The significance of black walnuts continued to increase during the Late Archaic IV (ca. 3700 B.P.) occupation (Moffatt 1992: 501–502).

By the Terminal Archaic Au Vase phase (3000–2600 B.P.), thick-shelled hickory nuts (shagbark and pignut) were supplemented by small amounts of thin-shelled hickories, like bitternut hickory (*C. cordiformis*), along with some black walnut. Carbonized squash fragments (*Cucurbita pepo*) were also common (Moffat 1992:500–501).

Like at the Little Muddy Rock Shelter, Late Archaic plant remains from the Hills Branch Rock Shelter in southeastern Illinois indicate that plant collecting focused on nuts. More than 80 percent of nutshell came from thick-shelled hickory (*Carya* sp.), along with lesser quantities of black walnut (*Juglans nigra*), hazelnut (*Corylus americana*), acorn (*Quercus* sp.), and pecan (*Carya illinoinensis*). The high density and wide distribution of carbonized nutshells, a by-product of burning them as fuel, suggests the consistent local harvesting and processing of hickory nuts, along with the occasional collection of other mast resources. Several large stone mortars found in the shelter's Late Archaic midden support the importance of nut collecting and processing at the shelter, as does the low frequency of seeds and the absence of starchy and oily seeds that were cultivated by some Late Archaic groups (Parker 2000:150–153).

Although many regional Late Archaic groups chose not to cultivate starchy seeds, others did. Excavations at the Dirk site (23Mi605) in southeastern Missouri yielded nearly 2,000 small starchy seeds, primarily chenopodium. Many specimens had thin seed coats and equatorial banded to rounded margins, indicating that they were on the road to domestication. The number and ubiquity of seeds suggest that chenopodium was an important cultivated crop for some Late Archaic groups that lived near the Ohio-Mississippi confluence. Despite the relative abundance of chenopodium, however, nuts were still the predominant food source (Lopinot 2000, cited in Powell and Lopinot 2003:434).

Animal Exploitation

Like for plant use, animal procurement strategies practiced by Late Archaic hunter-gatherers were similar throughout the Lower Ohio Valley. Late Archaic groups that lived along the middle Green River focused their efforts on an assortment of mammals, birds, fish, shellfish, and reptiles that inhabited the extensive riverine, wetland, and upland habitats found in west central Kentucky (Crothers 1999, 2005; Glore 2005). Much of what we know about the dietary significance of specific animals, the strategies used to capture them, and the time of the year in which it was done is based on the efforts of SMAP archaeologists at the Carlston Annis site, supplemented by additional data from the nearby DeWeese and Haynes sites (Claassen 2005; Crothers 1999). While direct comparisons between the multiple collections of faunal materials from these sites, or even within one site, is difficult because of the different recovery techniques used, general trends in the relative importance of different species is largely consistent (Glore 2005:315). George Crothers's (2005:295) analysis, which included both small specimens from flotation samples and screened fractions, found that white-tailed deer, freshwater drum, squirrel, box turtle, catfish, musk turtle, rabbit, raccoon, wild turkey, and softshell turtle were the most common food species represented at Carlston Annis. The abundance of deer, squirrel, rabbit, and raccoon was expected, but the poor representation or absence of turkey and other large fowl was surprising (2005:313).

The hunting strategies of Carlston Annis's Late Archaic inhabitants focused on the seasonal procurement of the many white-tailed deer that lived along the forest edges near the site. They probably hunted deer using an atlatl and spear (Glore 2005:327–329). Based on the bones present at Carlston Annis, it appears that they rarely brought the entire deer carcass back to the site. Instead, they removed the limbs, desirable cuts of meat, and the internal organs at the kill site, then left the remaining parts where they killed the animal (Crothers 2005:312–313).

Many smaller animals, such as rabbit, fox, raccoon, opossum, squirrel, and groundhog, were captured using traps, snares, and perhaps dogs. In some cases, animal furs may have been as important as their limited meat (Glore 2005:327–329). The foot bones, rostra, mandibles, and teeth of small animals accompanying some of the Carlston Annis and Indian Knoll burials (Webb 1950:336–343) may have been part of the contents of "medicine bags" (Crothers 2005:309).

Late Archaic people collected many kinds of aquatic and terrestrial turtles for their meat. They also used their shells to make rattles, cups, and other containers. Box turtles and other terrestrial turtles could have been easily captured by hand. In contrast, aquatic species were probably netted, speared, or trapped. They probably used a hook and line to catch some turtles and fish, such as freshwater drum and bullhead/catfish. However, while they did eat fish, the relatively small percentage of fish bone found at some Green River middens, including small species only represented in flotation samples, suggests that fish were not as important as shellfish and deer. Turkey would have been available in the forests that surrounded Carlston Annis (Glore 2005:329).

The most obvious food source for middle Green River Late Archaic people was, of course, river mussels (Morey et al. 2002; Patch 2005). Their shells account for 17 percent of the Carlston Annis midden volume, reflecting the significant and relatively stable role of this important food resource (Glore 2005:329). While shellfish collectors took mussels from a variety of river habitats, most came from shallow, fast-moving riffle-runs where they could be easily collected by hand while wading. Apparently, shellfish collectors rarely overexploited the riffle-run near Carlston Annis since the same species occurred in roughly the same frequencies throughout the midden (Patch 2005:269).

Mussels were probably steamed, roasted, or simmered to open the shell and expose the meat, then eaten or dried for storage and future consumption. Not surprisingly, analysis of growth rates indicates that mussel collecting was usually done during the warm part of the year. Similar patterns of shellfish procurement were detected at both Carlston Annis and DeWeese (Claassen 2005:290). Small gastropods that were collected could have been stewed to make a nutritious broth (Glore 2005:329).

SMAP researchers examined variation in the bone frequency and species diversity at Carlston Annis to determine whether any significant changes occurred in animal exploitation strategies (Crothers 2005; Glore 2005). Overall, the data suggested relatively uniform deposition across the site and that inhabitants maintained relatively stable animal exploitation strategies during the Middle to Late Holocene (Crothers 2005:313–314). Smaller animal species, like catfish, rabbit, and softshell turtle, may have been somewhat more important during the earliest period of occupation. However, the three most common species, white-tailed deer, drum, and squirrel, occurred in roughly consistent amounts throughout the midden.

A decrease in mussel shell in Carlston Annis's upper midden levels, referred to as the Shell Free Midden (SFM) (Wagner 2005), suggests that by about 3000 B.P., many of the prominent shell middens were still occupied, but mussels were no longer a significant part of the diet (Marquardt and Watson 2005a:639). In contrast, a few locations, such as the Bowles site at the confluence of the Mud and Green rivers, still had ample mussel and fish resources.

Late Archaic people who lived along the main course of the Ohio River pursued similar animal exploitation strategies as did the Green River groups. The inhabitants of the Highland Creek site exploited a broad spectrum of terrestrial and aquatic species (Maggard and Pollack 2006:99). Deer and turkey provided a substantial quantity of meat for food and bone for tools. Unlike at many other Late Archaic sites, however, small mammals such as weasel, gray squirrel, woodchuck, swamp rabbit, and opossum, while present, were less common than expected. The bones of mallard ducks and geese are indicative of hunting seasonally available migratory waterfowl, while fish bones and mussel shells reflect the importance of riverine resources. Collectively, these remains suggest that Highland Creek was occupied throughout the year, although repeated short-term occupations seem more likely than an extended, multiseason occupation (Breitburg 2006:87; Maggard and Pollack 2006:100).

Howard Winters (1969) found that Terminal Archaic Riverton hunter-gatherers also exploited a wide range of animals including deer, turtle, migratory birds, fish, and river mussels. A comparison of faunal materials from the three sites that formed the nucleus of his Riverton settlement system (Robeson Hills, Riverton, and Swan Island) revealed considerable variability that Winters attributed to different site functions. Deer were well represented at all three sites, underscoring their central role in Riverton subsistence. Nevertheless, there was still considerable variation, with the percentage of deer bone ranging from 27 percent at Swan Island to 68 percent at Robeson Hills. In contrast, fish and turtle accounted for nearly one-half of the Swan Island bone, where these species were abundant in the nearby river and wetlands, and only 13 percent at Robeson Hills. Small mammal remains were also more common at Swan Island than at the other two sites. Winters suggested that this reflected a foraging pattern emphasizing locally available fish, turtles, raccoons, squirrels, and rabbits at Swan Island (1969:121–123).

An average of nine species was represented in the Robeson Hill excavation levels, 16 at Riverton, and 26 at Swan Island. Winters maintained that Swan Island was occupied for relatively short periods during which its inhabitants conducted a variety of foraging activities. In contrast, Robeson Hills and Riverton were occupied for longer periods, and there was an emphasis on larger species at the expense of fish, turtles, and small mammals. River mussels were an important part of the diet at all three sites and were collected from a wide range of aquatic habitats. In the broader context of the Riverton settlement system, Winters characterized the Swan Island subsistence pattern as one expected for a transient group, Riverton with a more sedentary group, and Robeson Hills with a rather settled group that practiced more specialized subsistence activities that may have employed communal strategies (1969:123).

Social Organization

In some parts of eastern North America, Late Archaic hunter-gatherers developed more complex forms of social organization, as indicated by the construction of large earth or shell mounds and rings, the exchange of "exotic" materials and artifacts, and the more elaborate treatment of the dead (Sassaman 2004). A prime example of this increased complexity is the Poverty Point "culture" in the lower Mississippi River valley (Gibson 1996, 2001). The Poverty Point site, located in northeastern Louisiana, consists of large earthen mounds and a series of concentric earthworks built between ca. 3200 and 2700 B.P. Once considered an anomaly in Southeastern prehistory, more recent investigations in the surrounding region have determined that the Poverty Point site is a more elaborate Late Holocene version of earlier mound complexes constructed during the Middle Holocene (Russo 1996; Sassaman and Heckenberger 2004). The quantity and diversity of nonlocal materials found at Poverty Point indicate that residents participated in far-reaching and intensive trade relationships with nearby and distant groups (Gibson 2001). The characterization of the organization of the societies responsible for building these early mounds has varied from mobile, generalized foragers with no obvious economic indicators for social differentiation or inequality to chiefdom-level societies (Gibson 1974; Sassaman 2004:254).

In other parts of the Southeast, particularly along the South Atlantic and Gulf coasts, large circular or arc-shaped shell middens are interpreted as evidence for more complex forms of social organization (Sassaman 2004:261–262). While some archaeologists think that these impressive shell features are simply domestic middens attributable to circular villages occupied by egalitarian groups (Thompson et al. 2005; Trinkley 1985), others (Russo 2002; Saunders 2002) maintain that the large, asymmetrical shell deposits seen at many of the rings are the remains of competitive feasting practiced by more complex societies (Sassaman 2004:261).

As for most of the Southeast and Midwest, little archaeological evidence exists as to the level or levels of social complexity exhibited by Lower Ohio Valley Late Archaic hunter-gatherers. The closest definite Archaic mounds are small, bluff-top burial mounds located to the north along the lower Illinois River valley (Charles and Buikstra 1983). As we have seen, large shell middens occur along the middle Green River, the lower Wabash River, and even parts of the Ohio River, but most archaeologists consider them to have been formed by the deposition of midden resulting from the activities of daily life rather than intentional constructions (Milner and Jefferies 1998). Other than the possible structures found in the Riverton shell middens in southern Illinois (Winters 1969), we know little about the size and distribution of Late Archaic domestic or ceremonial buildings. As elsewhere, most reconstructions of Late Archaic social organization are based on the analysis of mortuary practices and the differential treatment of the dead.

Despite the absence of Late Archaic mounds in the Lower Ohio Valley, research at the Burkett site, located in southeastern Missouri near the Ohio-Mississippi confluence, uncovered a possible Late Archaic O'Bryan Ridge phase mound. Although now destroyed, a 1931 map clearly depicts a ca. 1.5-m-high, roughly circular mound at Burkett. Excavations conducted in 1999 exposed a stratigraphic profile that conformed to that expected for a mound (Thomas et al. 2004).

Efforts to determine the possible mound's age using the Oxidized Carbon Ratio (OCR) dating technique have produced perplexing results, but suggest that O'Bryan Ridge phase groups constructed at least part of it. If it is a Late Archaic mound, its builders placed it on top of an earthquake fissure. Site investigators consider placement of the mound on top of the earthquake feature as deliberate, suggesting that seismic activity had a prominent role in hunter-gatherer ritual, political, and social life. Such a supposition is not unreasonable considering the frequency and magnitude of seismic activity along the New Madrid fault (Thomas et al. 2004).

One of the first attempts to examine Late Archaic hunter-gatherer social issues, such as status, interaction, and gender distinctions, was Winters's (1968) study of grave goods associated with burials at some of the Green River middens. He (1968:209) noted that special artifacts, like marine shell beads and copper items that did not originate locally, were associated with only a very small proportion of the burial population (Figure 7.13). Winters suggested that the limited distribution of these exotic items reflected the differential distribution of wealth among living members of these societies. He (1968:203) interpreted the presence of exotic items with infants, children, and adults as reflecting ascribed rather than achieved status. Based on the restricted distribution of shell items and presence of shell-working kits in some burials, Winters (1968:209) proposed that Indian Knoll culture expressed the initial elements of craft specialization.

Nan Rothschild's (1979) analysis of Indian Knoll mortuary practices suggested to her that at least limited social distinctions existed among middle Green River hunter-gatherers. She identified two status dimensions—one applying to people buried with multiple types of artifacts and the other to those buried with only one artifact. Social positions indicated by burials having mul-

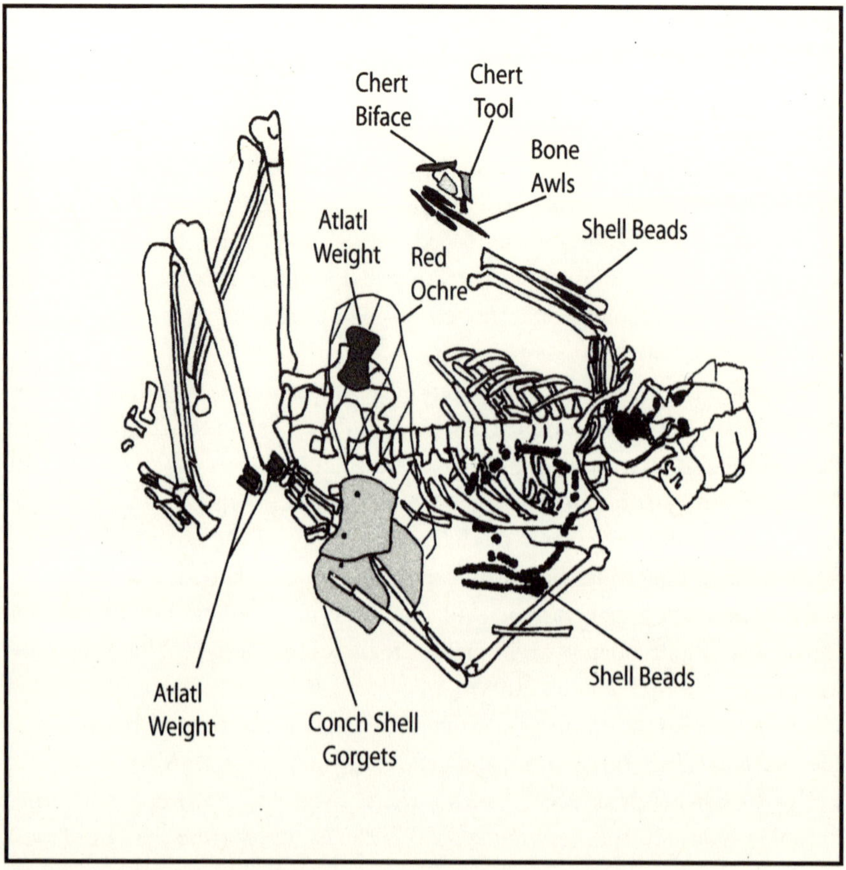

Figure 7.13. Artist's rendering of Read site Burial 66 and associated artifacts (adapted from Jefferies and Fenton 2001:Slide 16).

tiple artifact types were held by adult males and children. Individuals having just one artifact type were all adults, which Rothschild interpreted as reflecting the activities in which people participated while living (1979:671).

Our reanalysis of the Read site (15Bt10) skeletal series provides some insights into the social aspects of Late Archaic middle Green River societies (Milner and Jefferies 1998). Dating of the skeletal series, consisting of more than 250 individuals, is somewhat problematic since the site's temporally diagnostic artifacts span the Archaic period. However, nearly 70 percent of the hafted bifaces are Late Archaic, indicating that much of the site's occupation dates from ca. 5000 to 3000 B.P. Also, radiocarbon dates from all three dated burials fall in this temporal range (Herrmann 2002:Table 3.1), suggesting that Late Archaic individuals and mortuary practices are well represented in the site's burial population.

The shell and midden deposits found along the middle Green River have traditionally been interpreted as debris heaps formed by the repeated occupation of these locations by small hunter-gatherer groups. While living at these spots, group members carried out a wide range of activities, including the burial of their dead (Marquardt and Watson 2005a:635–637). Recently, some archaeologists have suggested that instead of being debris piles, some shell heaps represent intentionally constructed mortuary facilities or territorial markers (Claassen 1992, 1996a; Sassaman 1995).

Recent reevaluation of the Read site midden contents, along with the distribution of features and graves, are more consistent with refuse heaps than with an intentionally constructed burial mound. The burial distribution does not seem to represent a formal cemetery, since there was no fixed and easily recognizable plan for the placement of the deceased. Most likely, when a person died he or she was buried in a culturally acceptable location near where they lived. Over time, the number of people buried at Read steadily increased as hunter-gatherers continued to return to this site, reflecting the cultural significance of this part of the Green River landscape. Clearly, Read's prominent location on a high sandstone bluff overlooking the Green River valley would have been visible from a distance and afforded an excellent view of the surrounding landscape (Milner and Jefferies 1998:125).

Read site inhabitants buried most of their dead in a flexed position; the remaining ones were either extended or too disturbed to determine positioning. Males were more common than females (61 vs. 39 percent), but problems of identifying female skeletons may have created a bias toward males. With the exception of an underrepresentation of infants, age distribution is generally consistent with that expected for prehistoric North American hunter-gatherers (Milner and Jefferies 1998:129).

At Read, 26 percent of the burials contained some kind of grave good, roughly the same proportion as found at the late Middle Archaic Black Earth site (Lynch 1982). The percentage of juveniles (< 15 years) with artifacts was similar to that for adults (30 vs. 25 percent), as was the percentage of adult males and females with grave goods (30 vs. 28 percent). With the exception of infants, examples of weapons (atlatl weights and hooks, chert and antler projectile points), utilitarian items (bone pins/awls; chert blades, scrapers, drills, flakes; stone pestles; antler or bone fishhooks, chisels, and flakers), and ornaments (shell, bone, stone, and cannel coal beads; shell and cannel coal pendants; shell gorgets; drilled teeth) were found with individuals of all ages and both men and women.

Nevertheless, a person's age and sex did have some influence on their burial ritual. Infants (0–1 year) were accompanied only by ornaments, particularly shell beads. Adults who died during the prime of life more often received burial goods than the very young or the very old. The few individuals having an unusually large number of artifacts were either juveniles or young adults. Most of the atlatl weights and hooks were associated with adult males. Some of these items were positioned as if they were attached to the atlatl's wooden shaft at the time of burial. Undoubtedly, the atlatls represented by these preserved parts were prized possessions of their now deceased owners (Milner and Jefferies 1998:129).

The kinds and distribution of grave goods seen at Read generally conform to patterns seen at other late Middle and Late Archaic sites throughout the Midcontinent (Charles and Buikstra 1983; Rothschild 1979; Winters 1968). At least as suggested by the preserved grave goods, the social and economic significance of both men and women in their family and local community peaked between the ages of 15 and 45 years. This would have been the time in their lives when they made the greatest contributions toward group well-being, security, and survival. They also would have been the ones most likely involved in intergroup exchange and interaction, establishing and/or maintaining social, economic, and political ties with nearby and distant groups. In the middle Green River region, evidence for such interaction is indicated by copper and marine shell artifacts and by artifact styles shared by other regional groups (Goad 1980; Jefferies 2004; Marquardt 1985; Winters 1968). As people aged, their status appears to have declined, at least as indicated by the preserved burial goods. Overall, mortuary practices at Read reflect a person's ability to make a significant material contribution to group survival (Milner and Jefferies 1998:129).

New insights on the biological relationships, and presumably the social connections, among the hunter-gatherers who once inhabited the middle Green River region comes from the analysis of human skeletal remains (Herrmann 2002). Nicholas Herrmann's biological distance study examined the nonmetric cranial traits of over 1,300 individuals from six Green River middens (Carlston Annis, Read, Barrett, Ward, Chiggerville, and Indian Knoll), as well as a sample from the Eva site (40Bn12) in Tennessee. Based on his study, inhabitants of the Eva site are biologically distinct from those who lived along the middle Green River. Samples from Indian Knoll, Carlston Annis, and Read, all situated near the Big Bend of the Green River, are similar, as are samples from

Ward and Barrett, located to the west in the Cypress Creek/Pond River watershed. Interestingly, the Chiggerville samples are quite different from the other Green River samples.

The biological distance structure of the Green River data is consistent with a pattern of geographic isolation where people usually selected their mates from nearby groups. Interaction was less likely between groups that lived in the Big Bend region and the Pond River area to the west. This pattern is consistent with a down-the-line mode of exchange that characterized many Late Archaic economies. Mate selection may have taken place when small, scattered local groups periodically gathered at the large midden sites to conduct various socially and ritually integrative activities (Herrmann 2002:170, 173). There is some evidence suggesting that female mobility was greater than that of males (Herrmann 2002:172).

Late Archaic mortuary practices at the Rosenberger site were similar to those seen along the middle Green River, but there were some important distinctions (Driskell 1979:755). Like at the Green River sites, the more than 200 Rosenberger graves were distributed across the entire site, suggesting that there was no designated burial area or formal cemetery. In addition, the Rosenberger graves span a considerable time, reflecting the repeated occupation of this spot by Late Archaic hunter-gatherers.

Most of the Rosenberger skeletons were placed in a flexed or semiflexed position; however, there were a few extended burials. Flexed burials tended to be placed in circular to slightly oval pits that were about one meter in diameter. Sixty percent were interred with their head toward the east or south (Wolf and Brooks 1979:Table 9.6). Unlike some of the Green River sites, very few graves contained more than one individual. Associated projectile points included stemmed (primarily McWhinney), Wade, Flint Creek, and Merom-Trimble types, indicating that people were buried at Rosenberger throughout the Late/Terminal Archaic (Driskell 1979:Table 7.12).

Fifty-two percent of the Rosenberger burials had associated grave goods, a considerably higher figure than for many other Late Archaic mortuary programs. In addition to hafted bifaces, associated objects included stone and bone atlatl weights, bone atlatl handles, grooved axes, bone fishhooks, net weights, and a drilled canine tooth. Clearly, the kinds of items placed in the Rosenberger graves are quite distinct from those associated with some of the middle Green River burials. For example, many of the Green River burials contained shell artifacts, but none were found at Rosenberger. The presence of bone artifacts suggests that this was not a matter of preservation. In addition, Green River people frequently placed ornamental/ceremonial items, such as shell beads, cups, and pendants, with their dead. In contrast, archaeologists classified 98 percent of the Rosenberger grave goods as technological items, although some of the associated turtle shell may have come from rattles (Driskell 1979:774).

Conforming to trends seen elsewhere in the region (Lynch 1982; Milner and Jefferies 1998), age and gender appear to have been the primary factors distinguished in the Rosenberger mortuary program. Few subadults had grave goods, so a person's status was probably based more on their personal abilities and accomplishments than on who they were or to whom they were related. Like at Read, young and middle-aged individuals, the major contributors to group survival, often had grave goods buried with them. Grave goods, particularly shell beads, were commonly placed with Green River children, but not at Rosenberger. These distinctions in burial treatment led project archaeologists to suggest that mechanisms for status differentiation and social conditions at Rosenberger differed from those in the middle Green River area (Driskell 1979:774).

At the Habich site (15Jf550), situated on a terrace ridge near the Falls of the Ohio, excavations revealed 12 mortuary features scattered throughout the habitation area. Most contained one or two individuals, but one large pit held the remains of at least seven people (Granger et al. 1992:124–126). The lowest zone contained the disarticulated remains of five individuals—three adults and two subadults. There were no associated artifacts, but red ocher accompanied all the burials. The skeletal material appears to represent a secondary inhumation, possibly an ossuary. Later, a layer of rock was placed on top of the five individuals, separating the initial interments from subsequent ones. Some time later, a single subadult burial was placed on top of the rock layer. Like the underlying burials, a rock layer sealed it too. The final burial was that of a subadult positioned on top of the second rock layer. Other than a small white stone with possible incised marks found with the second interment, there were no definitive grave goods. Based on the appearance of the feature fill, all of the individuals seem to have been placed in the pit at roughly the same time (Granger et al. 1992:135).

Clearly, the study of human skeletal remains provides important insights on the organization and social dynamics of Lower Ohio Valley hunter-gatherers. In addition, the well-preserved human skeletons found in Middle to Late Holocene middens like Indian Knoll provide important insights on intergroup social relations (Mensforth 2001; Milner 1999; Smith 1996). Much of this information comes from evidence for deliberate trauma, like scalping and decapitation, resulting from raiding or other forms of interpersonal violence (Mensforth 2001; Smith 1996).

Anthropologists have shown that eastern North American hunter-gatherers participated in raiding and warfare since the Early Holocene; however, intergroup conflict became more common during the late Middle and Late Holocene periods. This was certainly true in the Lower Ohio Valley where human skeletons with projectile points lodged in their bones have been reported for nearly 100 years (Moore 1916:Figure 20; Webb 1946). In fact, there is more evidence for conflict from the region's shell heaps and middens than most other parts of the Midcontinent (Milner 1999:121). Spatial variability in the level of intergroup violence has been attributed to differences in group mobility; the reliability, diversity, and distribution of food; and population density (Milner 1999:121; Smith 1996:144).

Raiding probably caused very few deaths, but a few graves containing multiple victims suggest that raids occasionally resulted in small-scale massacres (Milner 1999:121). Intergroup trade and exchange, reflected by the distribution of exotic materials and artifacts, may have helped reduce the likelihood for violence, but these negotiations did not always settle disagreements.

Late Archaic hunter-gatherers who once lived along the southwestern margin of the Lower Ohio Valley in what is now Kentucky Lake practiced scalping and dismemberment as part of their raiding strategy. Only males showed evidence of postmortem violence, except for a mass grave at the Cherry site in western Tennessee, where several juveniles and a young adult female were buried with a young adult male having an inflicted point. Individual longbones and crania found with some burials in west Tennessee have been interpreted as "trophy" items (Smith 1996:143).

War-related traumata in the Kentucky Lake reservoir were not uniformly distributed, with a few sites like Cherry having an exceptionally high number of incidences. Over 10 percent of the site's skeletal sample died from violence, as opposed to only 1 percent at the six other sites that were studied. Maria Smith (1996:144) suggests that the Cherry site's location, on an upland tributary stream as opposed to the main river channel, put it in a more remote location that made residents more vulnerable to raiding. The site may have also been situated on or near a territorial frontier that was occasionally subject to dispute (Smith 1996:144).

Skeletal materials from three Green River midden sites (Ward, Carlston Annis, and Indian Knoll) exhibit similar evidence for Late Archaic hunter-gatherer conflict and trauma (Mensforth 2001). Green River hunter-gatherers having perimortem violent injuries were often found in graves with multiple burials. Adult males were the primary victims of that conflict.

The kinds of trauma inflicted on the Green River individuals varied quite a bit. Five, all adult males, had been scalped. Two of the three Ward scalping victims had been dismembered, variously missing legs, arms, and hands. Cut marks on the cervical or upper thoracic vertebrae indicate that people were sometimes decapitated. Many of these burials were also missing some limb bones, suggesting that they had been removed as trophies (Mensforth 2001:117–129).

Forty-four individuals from the three sites displayed inflicted injuries consisting of projectile point penetration wounds and stabbing/puncture wounds. Many of these victims died from their wounds, but bone growth associated with others shows that a few survived. Stone or antler projectile points caused most of these injuries, but other individuals apparently had their throats slit, were hit in the head with a spiked club, or were stabbed with a stone knife or spear point (Mensforth 2001:120–121). At the Carlston Annis site, all projectile point–induced traumata were found in males (Belovich 2005:511).

These studies of Late Archaic hunter-gatherer conflict demonstrate a marked increase in intergroup violence during the Late Holocene, probably associated with decreased group mobility, more formalized territories, and concern over access to critical resources. They also suggest that patterns and practices of warfare used by Native Americans during the late prehistoric and historic eras (scalping, trophy-taking) were well established by the start of the Late Holocene. Victims of violent death found in graves with multiple burials are probably attributable to raids conducted by multimale war parties, since it is unlikely that a single person could kill three or four members of the targeted group (Mensforth 2001:124–125).

Paleodemography and Bioarchaeology

Much of what we know about the biological and demographical characteristics of Lower Ohio Valley Late Holocene hunter-gatherers comes from nearly a century of research on the Green River skeletal collections. Most of this material was collected during the mid-twentieth century (1930–1950) when many of the sites were excavated as part of Depression-era work-relief projects (Haskins and Herrmann 1996; Milner and Smith 1986).

The excellent bone preservation found at these sites contributed to the recovery of very large, highly representative samples of all those buried there, including the fragile remains of infants and children not usually preserved at Ohio Valley Archaic sites. Indian Knoll, with its many (*n* = 880) well-preserved individuals, is the best known of these sites, but WPA archaeologists also excavated large, scientifically significant collections at Ward, Kirkland, Barrett, Carlston Annis, Read, Butterfield, and Chiggerville. The application of increasingly more sophisticated analytical techniques to the study of these collections has provided a much clearer understanding of the demography, pathology, nutrition, growth and development, and social organization of the Late Holocene people who once inhabited the middle Green River Valley (Haskins and Herrmann 1996:Table 10.2).

Mary Lucas Powell's (1996) overview of Green River Archaic health and disease is an excellent starting point for examining Late Holocene hunter-gatherer paleodemography. Like among many other hunting and gathering societies, childhood death was a relatively common event. On average, one out of every five infants died before reaching the age of one year. Children who survived to celebrate their first birthday had a somewhat better chance of living, most of them surviving into puberty. An increase in mortality around this time probably was associated with a child entering the world of adult activities, like pregnancy and childbirth for girls and hunting and warfare for boys. Adult mortality increased as people reached their thirties, but some individuals lived well into their sixties. Compared to some modern western European societies, the life expectancy of the Green River hunter-gatherers was short; however, on average, they lived longer than did the members of many late prehistoric Mississippian or Fort Ancient agricultural groups (Powell 1996:120).

Osteological analysis of more than 300 sets of human remains from the Carlston Annis site demonstrated that the age of the individuals ranged from 7 months in utero (unborn) to more than 70 years old. Life expectancy at birth was slightly over 22 years. Infant mortality was very high with about 22 percent of all individuals dying before the age of 1 year. For those who reached the age of 15 years, they could expect to live an average of about 20 more years, with approximately 62 percent of the population surviving to adulthood. A mean adult age of 35 years old is approximately one-half that of people living in the twenty-first-century United States. Peak levels of adult mortality occurred between the ages of 25 and 35 years. Like in much of the world today, Carlston Annis women tended to outlive men, although the difference between males and females was not statistically significant (Mensforth 2005).

Estimated age distributions at the nearby Ward site were very similar to those at Carlston Annis—21 percent of the people died within their first year, nearly 32 percent were dead by the age of five, and only 61 percent survived to the age of 15 years. Ward men generally did not live as long as women did, probably because of high levels of violence. Based on the skeletal sample, the Ward population had an annual growth rate of about 2.5 percent (Meindl et al. 2001:93–97).

Comparison of Carlston Annis demographics with those from nearby and contemporary Indian Knoll revealed generally similar trends. However, while nearly 20 percent of the Carlston Annis people lived more than 40 years, only 3 percent of the Indian Knoll people lived that long. Variation in the mean age at death between the two skeletal collections may reflect actual differences in mortality experiences, or it may be attributable to methodological differences in the study of the two collections (Mensforth 2005).

Green River people experienced various degrees of skeletal trauma and osteoarthritis, reflecting their physically strenuous lifestyle. Gender differences in the anatomical location of severe arthritis may be due to the different kinds of activities performed by men and women. At Chiggerville, females exhibited higher rates of spinal arthritis in the thoracic and lumbar regions, probably attributable to the mechanical stress of pregnancy and childbirth (Sullivan 1977, cited in Powell 1996:124–125). The long-term effects of atlatl use may have caused a high frequency of arthritic wrist and elbow joints for Indian Knoll inhabitants (Powell 1996:124).

Green River hunter-gatherers were not strangers when it came to broken bones; in fact, many had at least one bone fracture during their lifetimes (Powell 1996:125). At Carlston Annis, nearly one out of every two individuals had at least one broken bone, with men having slightly more fractures than did women. About 22 percent of adults had one or more longbone fractures and nearly 33 percent broke other bones. Five Carlston Annis people had three fractures, and six individuals had four or more. All but one of these accident-prone individuals were males. In contrast, only about 2 percent of the subadults had fractures (Belovich 2005).

Many of the longbone and multiple fractures experienced by men were probably attributable to environmental encounters (stumbling over rocks, slipping in mud, falling off cliffs, etc.) while hunting and fishing or during forays to distant lands. In contrast, fractures represented in older women are consistent with those associated with osteoporosis fracture syndrome (Belovich 2005).

For most Green River hunter-gatherers, there was a strong relationship between age and the kinds of fractures a person experienced. Most longbone fractures occurred before a person was 40 years old, but many fractures to other bones happened later in life. Clearly, the longer a person lived, the higher the probability that they would break a bone (Belovich 2005).

While the Carlston Annis people commonly broke bones, those with fractures did not seem to be at any greater risk of dying than those with no broken bones. Many fractures showed indications of remodeling and healing, suggesting that, in most cases, they received adequate treatment for their injuries. The many misaligned healed fractures indicate that Carlston Annis caregivers usually did not try to set fresh fractures. A few fractures may have been the results of interpersonal conflict, but most appear to have been the consequence of accidents (Belovich 2005).

Carlston Annis inhabitants experienced a very high frequency of auditory exostoses (AE)—smooth, rounded bony nodules that form on the walls of the ear's external auditory canal. Sometimes contributing to hearing loss, these bony nodules are most commonly caused by brief (2–3 minutes) exposures of the ear to cold water (below 65 degrees F). They are much more common for men than women (Mensforth 2005).

Along the Green River, Late Archaic males had very high frequencies of AE. The highest rates were at Indian Knoll (35 percent) and Carlston Annis (22 percent), both situated directly on the Green River. Frequencies at upland sites, like Ward, Kirkland, and Barrett, were lower (13–15 percent), but still high compared to other groups (Mensforth 2005).

The economic significance of shellfishing at Indian Knoll and Carlston Annis is reflected by the middens' high densities of mussel shell. It seems likely that shellfish collecting would have been an important subsistence activity at these sites during warm weather when the water level was low and mussels were easily accessible. In addition, however, the high incidence of AE among site inhabitants suggests that these people also collected shellfish during the late fall, winter, and/or early spring when the water was deeper (Mensforth 2005: 465–467).

Like many other Lower Ohio Valley hunter-gatherers (Bassett 1982), the middle Green River people experienced heavy tooth wear. In some cases, teeth were worn down to the pulp cavity, opening them to the possibility of infection. Grit-filled food, particularly the freshwater mussels, was the probable cause of much of this abrasion (Powell 1996:125–127). Particles from stone mortars, pestles, and grinding stones contributed additional grit. At Carlston Annis, most individuals who reached their mid-forties were commonly left with only their canines and incisors for chewing. A few people were toothless for much of the later part of their lives (Ward 2005).

Abscessing associated with excessive wear often extended into the sinuses, causing additional problems. The high rate of abrasion, infection, and abscessing meant that many of these people experienced a high rate of tooth loss. At Carlston Annis, inhabitants (that is burials) over the age of 35 years who still had their teeth had at least one active abscess at the time of death (Ward 2005). In contrast, the high grit content removed food traces trapped on or around the teeth, resulting in relatively few dental caries (Powell 1996:125–127). Caries were present in fewer than 5 percent of all teeth at Carlston Annis (Ward 2005).

Green River hunter-gatherers exhibited few skeletal indications of metabolic disorders. No cases of scurvy, caused by vitamin C deficiency, have been observed, and only a few signs of rickets, attributable to a shortage of vitamin D/calcium, are known. Severe iron deficiency anemia was also rare, at least as indicated by preserved skeletal indicators (Powell 1996:127–128).

Late Archaic hunter-gatherers generally experienced less osteoporotic bone loss than did later prehistoric horticultural groups, but, in both cases, losses were greater for women than men (Powell 1996:127–128). At Carlston Annis, the extremely low occurrence of enamel hypoplasias, macroscopic dental lesions caused by interruptions in growth and development, suggests that these hunter-gatherers did not suffer from long-term dietary stress like many late prehistoric groups that relied on horticulture for their subsistence (Ward 2005).

The demographic and health characteristics of Late Archaic people who lived near the Falls of the Ohio were similar to those of many other Lower Ohio Valley hunter-gatherers. Much of this consistency is due to the similar environments in which these groups lived and the similar mortality risks they experienced (Wolf and Brooks 1979:938).

At the Rosenberger site, 78 percent of the burials were adults, 7 percent adolescents, 4 percent children, and 10 percent infants and neonates. Unfavorable bone preservation conditions probably contributed to the underenumeration of infants and juveniles. Individuals for which sex could be determined were evenly divided between males and females (Wolf and Brooks 1979:933).

Like at some of the middle Green River sites, the first peak in death took place among the very young, with approximately 10 percent of the population dying before the age of 10 years. This is about one-half the frequency at Carlston Annis and Ward, but is probably attributable to the underrepresentation of subadults at Rosenberger. The highest proportion of people died between the ages of 26 and 40 years old, closely conforming to the Carlston Annis population (25 and 35 years). Ninety-six percent of the people died before reaching 40 (Wolf and Brooks 1979:933–935).

Women experienced higher mortality than did men during adolescence and young adulthood, attributable to death during pregnancy and childbirth. In contrast, more men died during middle age than did women, possibly reflecting their participation in risky hunting and raiding activities. Despite the risks of birthing, women had a greater life expectancy than did men—three Rosenberger women lived more than 40 years, but only one man accomplished this feat (Wolf and Brooks 1979:935).

Overall, Rosenberger inhabitants were somewhat shorter in stature than many other Late Archaic groups. Men attained a mean height of only 5′ 4″; women were somewhat shorter, averaging about 5′ 1″. As a group, their overall body size most closely resembled those of individuals who inhabited Indian Knoll (Wolf and Brooks 1979:927).

Pathological conditions experienced by Rosenberger inhabitants resembled those documented at middle Green River sites. Dental abscesses were common, particularly among middle-aged and old adults. Alveolar absorption, associated with tooth loss, was also common, particularly among older people. Many of the Rosenberger inhabitants exhibited excessive tooth wear, primarily caused by eating grit-filled food. Among some middle-aged and old adults, teeth had been worn down to the gum line or lower. The Rosenberger people averaged about two caries per person, most of which were in the lower molars (Wolf and Brooks 1979:914–916).

Auditory exostoses occurred in some Rosenberger residents, primarily males (Wolf and Brooks 1979:916). Unlike many of the middle Green River middens, however, shellfishing does not appear to have been a major activity at this site.

Like their Green River contemporaries, people at Rosenberger suffered from the effects of osteoarthritis. Approximately 40 percent of the analyzed individuals showed signs of this painful affliction, with symptoms ranging from slight lipping to fusion of several vertebrae. Symptoms increased with an individual's age and were more common in males (67 percent) than females (32 percent). At least 13 people showed signs of osteoporosis, a disease often linked to dietary stress or nutritional deficiencies, and a few may have suffered from osteitis or rickets (Wolf and Brooks 1979:920–921).

Similarities in the ways of life between Rosenberger and Carlston Annis inhabitants led to a similar range of traumatic injuries. Broken bones, reflected by improper settings and misalignments, were common at Rosenberger. Inhabitants most commonly incurred fractures of the hands and feet, but they also occasionally broke femora, ribs, fibulae, and scapulae. An extreme example of an unfortunate environmental encounter is represented by the remains of a middle-aged woman with a poorly set and healed femur, broken ribs on her right side, and altered muscle attachments on her right scapula. Inspection of the right leg and foot bones disclosed that this individual was active after her injury healed. Based on the pattern of her injuries, it seems likely that she fell from a considerable height and survived (Wolf and Brooks 1979:921). Like the late Middle Archaic man at the Black Earth site, the likelihood that this woman required long-term, if not life-long, care suggests that at least part of the Rosenberger group was relatively sedentary, remaining at the site for much of the year.

Unlike their middle Green River contemporaries, Rosenberger inhabitants showed very few signs of interpersonal violence. Whether this absence is indicative of a lower level of conflict or simply the poorer preservation of the Rosenberger skeletal material is unclear (Wolf and Brooks 1979:923).

Over the years, archaeologists working in the Lower Ohio Valley have investigated many other sites containing the skeletal remains of Late Archaic hunter-gatherers (Bader and Granger 1989; Herrmann 2006; Ross-Stallings 1994; Wilson 1993). Some of these collections have never been systematically analyzed, and many of those that have been studied usually consist of a few poorly preserved individuals. Despite these limitations and possible biases, the demographic and biological characteristics of those smaller collections that have been investigated using modern bioarchaeological techniques generally conform to those of the larger, better-preserved ones (see Ross-Stallings 1994).

Interaction and Exchange

The past existence of social networks is sometimes marked by the distribution of certain raw materials (copper, marine shell, chert), or items manufactured from those materials, far from the materials' sources (Goad 1980). In other cases, the geographic extent of these once far-reaching intergroup networks is indicated by the broad spatial distribution of similar stylistic traits, like those seen on late Middle Archaic bone pins (Jefferies 2004), atlatl weights (Burdin 2004), and flaked stone implements (Johnson and Brookes 1989).

Interaction networks, like those discussed for the Middle Holocene (Jefferies 1997; Johnson and Brookes 1989), continued to serve as conduits for social interaction and exchange into the Late Holocene. The kinds of materials and items that flowed through these networks, as well as the intensity and direction of interaction, varied both temporally and spatially throughout the Lower Ohio Valley. Nevertheless, they all reflected the efforts of individuals and groups to establish, maintain, or reinforce social, economic, and political ties with nearby and distant hunter-gatherer groups. Despite the presence of some materials hundreds of miles from their source, most examples of Late Archaic exchange can be best characterized as "geographically restricted down-the-line" trade, although a few raw materials or items reflect occasional "long-distance direct individual exchange" (Brose 1994:219; Lafferty 1994:187).

As with many aspects of Late Archaic hunter-gatherer life, some of the best evidence for intergroup interaction and exchange comes from the middle Green River valley (Goad 1980; Marquardt and Watson 2005a; Moore 1916; Webb 1974; Winters 1968). Excavation of several of the large river-edge or upland middens yielded marine shell and copper objects, along with a few bones of nonnative animals, documenting the movement of goods between Lower Ohio Valley groups and those that lived to the south and north (Marquardt and Watson 2005a). Both copper and marine shell have been described as occurring at these sites in "large numbers" (Goad 1980:11), but while WPA excavations at eight of these sites yielded more than 40,000 shell artifacts, only 13 copper artifacts are known from three Green River sites (Watson 2005:Table 23.12).

Green River hunter-gatherers obtained marine shell, either directly of indirectly, from groups that lived along the south Atlantic and Gulf coasts. They made beads from *Marginella* and *Olivella* shell, and pendants, vessels, gorgets, pins, bars, discs, and atlatl weights from conch and whelk shell. The few copper items include bars, pendants, pins, a bead, and an awl made using a "cold hammering" technology (Goad 1980:3) from native copper that probably originated in the Lake Superior region (Watson 2005:Table 23.12). Like many of the shell items, most of the copper artifacts were nonutilitarian. The bones of animals that lived far to the north, like the marten found at Carlston Annis and Indian Knoll, could have come from the same region as the copper (Marquardt and Watson 2005a).

The quantity and diversity of Green River shell artifacts compared to those made from northern copper suggest that Indian Knoll and Carlston Annis groups intensively interacted with groups that lived to the south, reflecting centuries, or even millennia, of trading forays in that direction. In contrast, all of the copper could have been transported to the Green River region in one trip (Marquardt and Watson 2005a). Analysis of bone pins from several of the middle Green River middens also suggested a southern focus (Jefferies 2004). Jack Nance (1988:147) took a similar stance for groups that lived to the west of the middle Green River valley, maintaining that Late Archaic hunter-gatherers in the lower Tennessee-Cumberland region were most closely affiliated with groups that lived to the south.

Numerous researchers have addressed questions concerning the economic processes responsible for movement of marine shell and copper from their distant sources to the Middle Green River area (Goad 1980; Marquardt 1985; Winters 1968). Howard Winters (1968) proposed three possible scenarios for this to have transpired. First, Green River hunter-gatherers may have regularly left the region and moved to coastal areas as part of their annual rounds. No evidence for this has ever been found.

Second, Green River groups may have maintained trading relationships with neighboring groups who obtained shell from more distant groups as part of an elaborate intergroup exchange system that ultimately linked Green River hunter-gatherers with coastal groups. Winters (1968:216) considered this a possible explanation; however, such a reciprocal network should have resulted in a steady flow of goods, not the cyclical flow that he proposed for Indian Knoll. Other researchers have favored this second scenario, contending that exotic goods were transported to the middle Green River region through what has been described as "linear" or "chain-like" exchange (Wright and Zeder 1977). Based on their computer simulation of such a system, Henry Wright and Melinda Zeder (1977:238–239) proposed that the hypothetical trading cycles Winters described could be attributed to demographic and/or social fluctuations in the coastal groups that supplied the shell. The small amount of copper moving counter to the shell simply reflected the greater distance to copper sources.

Winters's third scenario proposed that Green River groups had direct ties with a redistribution center that provided raw materials and/or finished goods for inland groups. He favored this alternative, even though no such Lower Ohio Valley centers have ever been found. Winters (1968:218) maintained that such trading centers would share many characteristics with Poverty Point, particularly the presence of a variety of imported raw materials (copper, steatite, crystal quartz).

Sharon Goad (1980) expanded on Winters's (1968) trade center alternative, suggesting that middle Green River hunter-gatherers lived at the interface of two exchange networks—a northern copper network and a southern shell network. Individuals who participated in this hypothetical exchange would have accumulated these materials, the majority of which were kept as indicators of his or her own social status. The location of the Green River between the two exchange networks effectively made these individuals "middlemen" who facilitated copper and shell exchange far beyond the expected range of normal linear reciprocity. Apparently, the desire for marine shell accounted for the willingness of northern groups to exchange copper with middle Green River people, even though they lived beyond the area normally included in such a network (Goad 1980:12). The very few copper artifacts found in the Green River region do not support the entrepreneurial role of some Green River hunter-gatherers (Marquardt 1985:80).

Despite the lack of evidence for long journeys to the coast, distant trading partners, or redistributive centers, marine shell and copper artifacts found their way to the Green River because of some kind or kinds of interpersonal relationships. However, just why the hunter-gatherers maintained these ties is still unclear.

Marquardt (1985) maintained that exchange relationships should be viewed at several different levels—the role of exotic goods in defining local leadership structures, the relation between access to information and access to resources at the interregional level, and the multilevel relation between the economic-technological and religious-economic sectors of society. By the Late Archaic, if not earlier, certain group members established socioeconomic relationships with trading partners from other nearby and distant groups. In addition to trading for desired goods, these far-ranging travelers negotiated alliances, arranged marriages, and exchanged information about their respective physical and social worlds. The exchange of goods and information was mutually beneficial and could be used to the advantage of both partners after returning home. Trading partners might use the esoteric knowledge and goods obtained while traveling, like copper or marine shell, to increase their social standing within their groups, and practical information could contribute to group survival and prosperity. In the end, the differential access to goods and information might lead to a person's enhanced leadership role, and, perhaps, a permanent leadership position (Marquardt 1985:81).

In addition to marine shell and copper, Late Holocene people also acquired a variety of lithic materials and artifacts through these networks, including the finely crafted atlatl weights or bannerstones used by some Green River hunter-gatherers. Examination of several hundred of these elaborately shaped and carefully drilled objects from different Lower Ohio Valley sites revealed that more than 40 percent were made from quartz or quartzite, and that many of the others were fabricated from either granite (18 percent) or gneiss (16 percent) (Burdin 2004:Figure 4.21). None of these types of stone is native to the region, suggesting that they were acquired from distant sources through exchange, or perhaps during logistical forays to their sources. Understanding the procurement strategies used to acquire the atlatl weights, or at least the stone to make them, is made more difficult since these apparently "foreign" lithic materials also occur in glacial till found north of the Ohio River, far from their primary geological sources (Fowler 1957:21).

Evidence for exchange also occurs in other parts of the Lower Ohio Valley. In southeast Missouri, interregional exchange appears to have been limited to certain cherts used to make flaked stone implements. At the Clifford LaPlant site, chert source data suggest that the Late Archaic inhabitants maintained the strongest connections with groups that lived to the northwest and north in the Ozark uplift region and St. Francis Mountains of Missouri and Shawnee Hills in southern Illinois (Buchner 2003b:486).

While the geological sources of these cherts suggest the direction of interaction, like with marine shell and copper, they do not provide much information on just how the chert arrived at the site. Was it through exchange with groups living near the chert sources or did the LaPlant inhabitants collect it during their travels through these regions as part of their seasonal rounds?

Late Archaic groups that lived near the mouth of the Ohio River do not appear to have had a strong relationship with Poverty Point–affiliated groups that lived far to the south. Projectile points diagnostic of Poverty Point culture, like Motley, Pontchartrain, and Delhi, are rare, and, while clay "Poverty Point Objects" occur, they differ in shape from most of those found at Poverty Point. Likewise, other hallmarks of Poverty Point culture (steatite and the lamellar microlith and lapidary industries) are poorly represented in this part of the region (Thomas et al. 2004:125–126).

Symbolism and Ideology

Compared to the technological and social realms, we know little about Late Archaic symbolism and ideology. Objects that primarily functioned in the ideological world are sometimes difficult to identify once removed from their symbolic context. It seems reasonable to assume that some middle Green River items made from "exotic" materials functioned in the symbolic or ideological spheres of hunter-gatherer society (Johnson 1994:109). An individual's ability to acquire these exotic goods and any associated esoteric knowledge probably contributed to his or her success in establishing a position in his or her group's political-ideological leadership structure (Marquardt 1985:81). Nevertheless, examples of hunter-gatherer symbolism and ideology in the Lower Ohio Valley are difficult to find.

An exception to this is the recent discovery of mud and charcoal drawings inscribed on the floors and walls of several caves in the west central Kentucky karst region (DiBlasi 1996). Examination of the drawings revealed that the same motifs occurred in different caves, and that the motifs were clearly different from late prehistoric Mississippian motifs found in caves in eastern Tennessee (Faulkner 1986; Faulkner et al. 1984). Charcoal from a cane torch fragment found in Adair Glyph Cave yielded a radiocarbon date of 3560 B.P. +/– 110 years, indicating that the Kentucky glyphs date to the Late Archaic period. Most of the motifs are geometric forms; however, zoomorphic and herptemorphic figures, random lines, and stroke marks occur (DiBlasi 1996:43).

The Adair Glyph Cave motifs, found more than 1 km from the entrance, included zigzags, chevrons, and cross-hatching designs. These motifs were drawn in the damp mud floor using a pointed or sharp-edged implement such as a stick, a piece of stone, or the edge of a mussel shell. The glyphs appear to have been drawn with care since they do not overlap and the individual motifs are generally symmetrical (DiBlasi 1996:42). Interestingly, these motifs are similar to many of those engraved on Archaic bone pins found in parts of the Southeast (Waring 1968:169) and southern Midwest (Jefferies 1997).

One of the most common motifs was the zigzag, consisting of a series of parallel lines spaced two to three centimeters apart. Zigzags occur singly or in pairs. The paired zigzags, appearing as mirrored images, are very symmetrical. A second kind of zigzag, represented by a series of single overlapping lines connected at the top, form a Christmas-tree-like design. Zigzag motifs range from 50 to 100 cm long and from 20 to 30 cm wide, and often occur in parallel rows (DiBlasi 1996:42).

Glyphs found in Salts Cave were drawn using charcoal from cane torches. Three figures placed on a large stone slab consisted of a turtle, a lizard or salamander, and a possible anthropomorphic or zoomorphic figure (DiBlasi 1996: Figure 5.1). The turtle, approximately 10 cm in diameter, has a round carapace, four extended legs, a tail, and a head. The herptemorphic image may represent a "horned" lizard or salamander. Its long, thin body, including the tail, is 10 cm long and 1 cm wide, with four extremities extending from the body. A down-turned crescent form extends from the round head, possibly representing the creature's horns. The third figure, measuring 10 cm tall and 4 cm wide, is the least distinct of the three. The short, stocky body is topped by a "head." Four short, stocky appendages and a tail extend from the body (DiBlasi 1996:43–45).

Cross-hatching is also common in parts of Salts Cave. Many of these motifs occur in areas that experienced extensive Late Archaic/Early Woodland foot traffic. Often, these crosshatched motifs seem to have been placed near the intersection of cave passages (DiBlasi 1996:45). Cross-hatching motifs have also been documented in Fisher Ridge Cave in Hart County, Kentucky. Associated radiocarbon dates place the prehistoric cave use from ca. 3200 to 2800 B.P. (Kennedy, Hensley-Martin, and Watson 1983:22; cited in DiBlasi 1996:46). Several charcoal drawings are also known for Mammoth Cave; however, they do not resemble those described for other Kentucky caves (DiBlasi 1996:46).

At least as early as the Late Archaic, hunter-gatherers who inhabited Kentucky's karst region used its caves for ceremonial or ritual purposes (Watson 1996:161). While caves probably had an important role in hunter-gatherer rituals for millennia, interpreting what the placement of these motifs in caves meant is difficult. At one extreme, they may represent Late Holocene "doodling" with no deep symbolic meaning; at the other they may have a much closer connection to the ideology and belief systems of the hunter-gatherers who put them there. Most of the motifs are simple in design, but the fact that the same designs occur in several caves, and that similar motifs were placed on other items like bone hairpins suggest that their "meaning" or "importance" was understood and communicated by people who inhabited much of the Lower Ohio Valley and beyond. Whatever their meaning(s), caves continued to be visited for a variety of purposes for more than 1,000 years, suggesting that they remained a significant part of the Lower Ohio Valley cultural landscape for later Woodland people.

Summary

Late Archaic hunter-gatherers lived in the Lower Ohio Valley during the early part of the Late Holocene. Environmental conditions resembled those of today, as did the plants that they collected and the animals they hunted. The only exceptions are species, like passenger pigeon and elk, which were quickly eradicated by Euroamerican hunters and sportsmen.

In many places, the hunter-gatherer way of life established during the late Middle Holocene continued unabated. Many groups still focused their social, economic, and ritual activities at places near reliable, diversified, and predictable food resources, such as swamps and rivers. This was particularly true for groups that lived along Kentucky's Green River. In other parts of the valley, groups appear to have adopted strategies that could be more efficiently carried out by dispersed groups.

More numerous examples of intergroup exchange and interaction suggest that local groups were expanding the size of their social networks, and, in some cases, creating large, more complex social groups. Members of these larger regional-scale societies expressed their common identity by sharing artifact styles and decorative techniques.

Late Archaic cave use suggests that the complexity of their ritual lives continued to increase as people sought new places to explore the more esoteric aspects of hunter-gatherer life. Many other places also had major ritual significance, but the archaeological correlates of those activities are more difficult for us to see compared to the symbolism expressed in the dark reaches of caves.

Perhaps the most significant change to hunter-gatherer life during the Late Holocene was the gradual adoption of horticulture as part of the subsistence strategy. While not as important as in some parts of the North American Midcontinent, early indications of the domestication of local cultigens occur at some Lower Ohio Valley Late Archaic sites. Although the dietary importance of local cultigens increased over time, traditional hunter-gatherer subsistence strategies continued to provide the core of the diet for millennia to come.

In the next chapter, I explore the evolution of the physical landscape on which the people lived and the cultural landscape they created and modified for nearly 10,000 years. From this perspective, we can examine how different components of hunter-gatherer lifeways evolved from the Late Pleistocene into the Late Holocene.

8 / Hunter-Gatherer Landscapes in Space and Time

Introduction

More than 12,000 years have passed since the first hunter-gatherers found their way to what we now call the Lower Ohio Valley. Small hunter-gatherer groups that moved into the Lower Ohio Valley at the end of the Pleistocene appear to have had little immediate impact on the valley's landscape. However, as their numbers increased and their mobility decreased during the Holocene, their descendants left increasingly noticeable cultural "footprints" behind. Figure 8.1 illustrates the increasing intensity of hunter-gatherer occupation from the Late Pleistocene through the Late Holocene, based on numbers of recorded sites. Gradually, hunter-gatherers superimposed the technological, organizational, and ritual aspects of their cultures on the land around them, slowly transforming physical spaces into meaningful places (Anschuetz et al. 2001:161).

As discussed in Chapter 1, the spatial and temporal distributions of cultural materials across the Lower Ohio Valley landscape provide ways to view changes in the patterning of hunter-gatherer cultural traditions as realignments of human behavior. This approach to the study of prehistoric hunter-gatherers provides insights into the interactive economic, social, and ideational strategies that these people used to ensure their survival in the face of continually changing environmental conditions (Anschuetz et al. 2001:187–191). Collectively, these strategies shaped the Lower Ohio Valley's cultural landscape.

Kurt Anschuetz and colleagues (2001:176) suggest that three areas of research are particularly relevant to the study of past cultural landscapes—

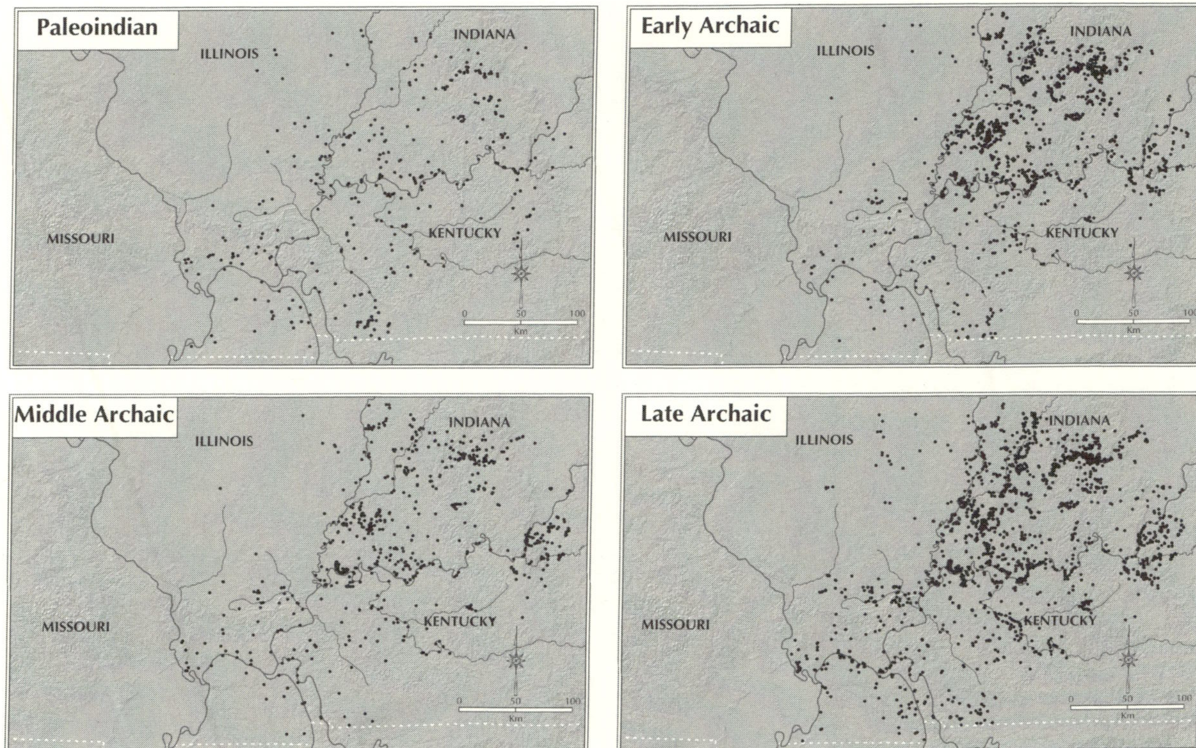

Figure 8.1. Distribution of Lower Ohio Valley Paleoindian and Archaic sites.

settlement ecology, ritual landscapes, and ethnic landscapes. The unification of these three approaches is linked by four major anthropological concepts—culture, tradition, vernacular knowledge, and sociocultural change.

In the course of a group's daily activities, they change their physical surroundings into meaningful places. In the process, their conceptualization of the surrounding landscape becomes an important part of their cultural heritage, providing members with a particular sense of place and time. These cultural traditions unify how people create and occupy their landscapes. This worldview, in turn, is defined and informed by vernacular knowledge consisting of those universal propositions held by the entire group. Changes in a group's vernacular knowledge, traditions, and worldview occur either as a breakdown or revolutionary overthrow of traditional cultural frameworks, marked by major discontinuities in the archaeological record, or, more commonly, as gradual changes that maintain the group's fundamental structure and organization (Anschuetz et al. 2001:181–184).

Settlement ecology, derived from settlement pattern and system approaches, is based on the assumption that landscapes are a product of people's interactions with their environment. This continuum is defined by a matrix of land-extensive and labor-intensive tactics and strategies. Settlement ecology emphasizes natural environmental variables such as subsistence resources, raw materials required for comfort and health, and trade items. Various cultural and natural factors contribute to the unpredictable availability of these resources through time and space. Settlement ecology investigates how a community uses its economic, social, and ideational technologies to manage this risk. For example, a series of small, gradual diachronic changes in archaeological patterns could represent adjustments that mitigated the effects of recurrent environmental perturbation. In contrast, rapid, large-scale change might represent a major realignment of strategies for interacting with the environment (Anschuetz et al. 2001:177).

Ritual landscapes are the product of stereotyped actions representing socially prescribed orders that define, legitimize, and sustain a group's occupation of their traditional territory (Anschuetz et al. 2001:178). Studies of ritual landscape are based on the examination of patterns in the spatial distribution of ritual features including public buildings, monuments, squares or plazas, petroglyphs or pictographs, and vernacular markers. Such patterns conceivably relate to changes in the deep patterning of "ideationally informed guidelines" that help condition the underlying structure of behavior (Anschuetz et al. 2001:179).

Ethnic landscapes are created by the spatial and temporal constructs defined by communities whose members make and manipulate material culture and symbols to signify ethnic or cultural identity and boundaries that are sometimes based on nothing more than tradition (Anschuetz et al. 2001:179). Since ethnic identity involves both inclusionary and exclusionary behaviors, a group's morphological, stylistic, or spatial dimensions could reveal evidence for differential ethnic representation. When a network of interacting people is focused in a particular territory, groups can maintain a coherent sense of affiliation and group membership with place (Anschuetz et al. 2001:180–181).

Anschuetz and colleagues (2001:181) maintain that these three approaches to the study of landscapes—settlement ecology, ritual landscapes, and ethnic landscapes—share the view that people are more than passive recipients of change dictated from outside the cultural system. Collectively, they contribute useful, relevant insights into a society's past behavioral dynamics and patterns of change.

The remainder of this chapter will examine the nature of change in the Lower Ohio Valley's Late Pleistocene and Holocene cultural landscapes. Particular emphasis is placed on patterns of diachronic change and continuity in technology, subsistence, demography, settlement, social organization, interaction, and exchange. Ultimately, these trends are related to how people defined, shaped, and used space at particular times.

The Physical Landscape

The Lower Ohio Valley's physical landscape has experienced considerable change over the past 15,000 years, much of which is attributable to the retreat of the most recent Pleistocene glacier. For example, about 10,000 years ago, as the glacial front retreated northward into what is now Canada, the outflow of glacial meltwater gradually decreased. As it did, the Ohio River along with many of its northern tributaries, like the Wabash River, slowly evolved from highly eroding landscapes created by the rapidly flowing meltwater to aggrading landscapes that promoted the development of food-rich floodplains, sloughs, and swamps. Prior to this transition, these rivers would not have been very hospitable places for people to live (Gray 1971; Stein 1980).

Southern tributaries, like the Green River, slowly began to down-cut through deep lakebed sediments deposited by glacially related flood episodes. In some cases, erosion of the Pleistocene channels exposed deeply buried geological formations, resulting in the development of rapids and shoals. Remnants of some

Pleistocene slack-water lakes continued to exist well into the Holocene in the form of resource-rich swamps and wetlands (Gray 1971; Stein 1980:43–51).

By the end of the Pleistocene, many of the animals that had roamed the boreal landscape for thousands of years, like mammoths and mastodons, were well along the road to extinction. The loss of many Pleistocene species contributed to a decrease in mammal population diversity across the Lower Ohio Valley at the start of the Holocene. However, many other animal species that were well adapted to the postglacial conditions, like the white-tailed deer and turkey, rapidly expanded in both number and range, filling in the landscape vacated by the Pleistocene fauna. Essentially precontact animal populations inhabited the region by about 10,000 years ago. Nevertheless, Holocene climatic change and regional differences in vegetation contributed to minor fluctuations in the number and distribution of some animal species throughout the remainder of the Holocene (Semken 1983).

As the climate warmed and seasonal precipitation patterns shifted during the Holocene, new plant communities emerged on the landscape. By the Early Holocene, the range of many temperate deciduous tree species expanded northward, creating new species-rich forests that gradually replaced the retreating boreal taxa. Later, Middle Holocene drying promoted the spread of deciduous forests and grasslands. The return of more moist conditions during the Late Holocene eventually led to the development of the plant communities that covered the landscape when the first Europeans entered the region.

The evolving Holocene landscapes provided numerous new habitats for many plants and animals that eventually became important parts of the hunter-gatherer diet. Middle Holocene changes in river hydrology promoted the establishment of food-rich habitats that were not available to earlier hunter-gatherers. Of particular importance was the development of shoals or riffles that supported large populations of river mussels. In parts of the Lower Ohio Valley, like along the middle Green and lower Wabash rivers, late Middle through Late Holocene hunter-gatherers heavily exploited these abundant and readily available shellfish (Morey and Crothers 1998).

In other areas, the deposition of sediments adjacent to many rivers contributed to the formation of expansive floodplains containing abundant wetlands. Like the shoals, these newly formed swamps, lakes, and sloughs supported new combinations of plant and animal communities that greatly increased the quantity, diversity, and reliability of foods available to the region's hunter-gatherer populations. Many nut-bearing trees, like oaks and hickories, increased in frequency and distribution, providing important food sources for people, as well as many of the animals they hunted.

Collectively, these environmental changes produced a diverse physical landscape whose plant and animal communities could markedly differ over relatively short distances. Differences in hunter-gatherer archaeological assemblages found throughout the Lower Ohio Valley in part reflect the different adaptive strategies these groups used to get the resources they needed to survive.

Settlement Ecology

As in most other parts of eastern North America, Lower Ohio Valley archaeologists have little information about when the first hunter-gatherers arrived or how they interacted with the late Pleistocene landscape. Part of this poor understanding is due to the very dynamic physical landscape on which they lived—much of it having been buried, eroded, or extensively remodeled by more than 12,000 years of natural and cultural activity.

Whoever the first inhabitants were, they encountered a landscape that was unoccupied by other people, a situation seldom experienced by most humans. The environment was very harsh, affected by the influence of the Late Pleistocene glacial front. Plant and animal communities were quite different from those of today, dominated by cold-adapted species. Undoubtedly, this was a time of exploration and colonization as these earliest hunter-gatherers made their way through the Lower Ohio Valley (Anderson et al. 1996).

Although small groups of people may have arrived earlier, the first good evidence for human presence in the region consists of fluted Clovis projectile points dating to ca. 12,500 B.P. Clovis lithic technology reflected these peoples' highly mobile way of life. They made many tools, particularly projectile points, from high-quality stone that extended a tool's use-life when used in combination with bifacial technology. In addition, tool kits were highly "flexible," meaning that Clovis toolmakers designed many flaked stone tools to perform a variety of tasks. Attributes like these would have been advantageous for highly mobile people who might not have access to desirable lithic resources for extended periods (Tankersley 1996).

Certain parts of the Lower Ohio Valley appear to have higher densities of Clovis points than do others. These artifact concentrations probably mark places where Clovis people focused their activities because of more abundant and reliable food, higher quality lithic materials, or easier access to important communication and transportation routes (Anderson 1990).

Clovis points are often found hundreds of kilometers from the sources of the stone from which they were made. The wide distributions of these exotic materials suggest that Clovis hunting territories covered thousands of square kilometers and that Clovis groups pursued a highly mobile way of life that was at least partially determined by the movement of the game they hunted. Clovis groups based part of their diet on hunting migratory Pleistocene animals that they killed on an "encounter" basis. However, smaller and less mobile animals were also important, as were some plant foods (Tankersley 1996).

In the Lower Ohio Valley, most Clovis sites are small, contain few artifacts, and offer little evidence for structures or other facilities. Many sites are situated near lithic source areas or near places where game could be observed, hunted, and processed. The fact that these locations are archaeologically visible suggests that at least some sites were occasionally revisited (Tankersley 1989).

The similarity of tool assemblages found at widely scattered Clovis sites suggests that late Pleistocene groups shared a common cultural tradition as they dispersed over the rapidly evolving landscape. Interaction among different Clovis groups probably occurred during brief periodic encounters or when several groups gathered for short periods, perhaps near mutually recognized landmarks.

Middle Paleoindian (ca. 11,000 to 10,500 B.P.) hunter-gatherers lived in a world that was experiencing rapid environmental change at the Pleistocene-Holocene interface. Lower Ohio Valley plant communities were in a state of flux, reflecting major climatic transitions associated with rapid glacial retreat. In addition, many of the Late Pleistocene mammals were disappearing because of a variable combination of environmental and cultural factors. Middle Paleoindian groups responded to these dynamic habitat changes by adopting a mixed foraging strategy that targeted both large and small game. The broader distribution of Middle Paleoindian artifacts compared to those of earlier Clovis groups may be attributable to this shift in hunting strategies.

As with earlier Lower Ohio Valley hunter-gatherers, what we know about the settlement and mobility of Middle Paleoindian groups is largely based on the distribution of diagnostic bifaces. Unlike earlier times, however, Middle Paleoindian hunter-gatherers manufactured a variety of stylistically distinct projectile points that included both fluted and unfluted varieties. This stylistic variability may reflect the fissioning of earlier Clovis groups as they gradually increased in size. Once split, the cultural strategies of each group diverged as they differentially responded to the variability of their respective physical and social worlds (Anderson et al. 1996).

In addition to stylistic differences, many Middle Paleoindian lithic assemblages are characterized by the use of lower-quality stone. In some places, bipolar technology replaced the core and blade technology used by Early Paleoindian hunter-gatherers. Both of these strategies may be attributable to the smaller size of hunting territories and the reduced access to suitable lithic resources.

Although greater regional differences are evident by Middle Paleoindian times, these groups continued to interact for purposes of managing risk and uncertainties, obtaining marriage partners, and maintaining the flow of information about their physical and social environments needed to help ensure group survival.

Late Paleoindian (10,500 to 10,000 B.P.) hunter-gatherers did not make fluted points, but instead manufactured a variety of stylistically diverse types that ranged in form from long and narrow to short and wide. Foremost among these was the Dalton point, characterized by its lanceolate shape, steeply beveled lateral margins, and serrated edges. Although Dalton sites occur throughout the Lower Ohio Valley, they are most common in western Kentucky and southern Illinois. The greater number of sites, compared to earlier Paleoindian components, suggests a regionwide population increase.

Dalton hunter-gatherers utilized locally available lithic sources to manufacture many of their flaked stone implements, reflecting a continued reduction in mobility associated with smaller hunting territories. In contrast, some large Dalton bifaces, often occurring as caches, were made from high-quality nonlocal stone, particularly Crescent Hills chert from eastern Missouri. These ritual caches may be associated with strategically situated and favored places where Dalton groups gathered to conduct rituals and reinforce intergroup social, economic, and political connections (Walthall and Koldehoff 1998).

Compared to earlier hunter-gatherers, Dalton groups appear to have significantly reorganized their mobility strategies as a way of adapting to the northward spread of Holocene deciduous forests and the disappearance of migratory Pleistocene animal species. In some parts of the Midcontinent, Dalton groups adopted a more circumscribed range of movement based on an annual round shaped by topographical and seasonal changes in resource distributions (Walthall 1998a). The repeated use of more circumscribed hunting territories contributed to an increased familiarity with the local landscape and the identification and reoccupation of particularly favorable locations. Over time, the repeated use of smaller hunting territories may have led to the increased cultural significance of certain places within a group's home range, as well as their identification with those places, ultimately converting space to place. The Sloan cemetery in northeastern Arkansas is an example of one of these special places.

In many ways, the Early to early Middle Holocene (10,000 to 6500 B.P.) cultural landscape closely resembles those created by earlier Dalton groups. In parts of the Lower Ohio Valley, Early Archaic projectile point distributions reflect a residential mobility strategy that involved the movement of foraging groups from one source area to the next, procuring needed resources on an encounter basis. The occurrence of flaked stone tools made from nonlocal cherts suggests that these residential groups frequently moved and that they still occupied fairly extensive home ranges (Stafford 1994:232).

Early Holocene sites usually consist of short-duration, widely scattered camps located in a variety of environmental settings. Hunter-gatherer groups favored some locations more than others, but they were not tethered to specific places. While most occupations were brief, they repeatedly occupied a few upland and floodplain sites where the diversity and abundance of subsistence resources permitted.

The technological organization of Early to early Middle Holocene hunter-gatherers also was similar to that of Late Paleoindian groups. Most of what we know about hunter-gatherer technology during this time is based on the study of flaked stone artifacts. Occasional ground stone, bone, and antler implements occur, but they are rare. Pitted stones and digging tools occur on some sites, probably marking the increased importance of plant foods, or at least new ways to process them.

Perhaps the greatest distinction from earlier technologies was the use of basal-, corner-, or side notching for attaching bifaces to their wooden or bone shafts or handles. The reason(s) behind the adoption of new hafting techniques is unknown, but may have to do with the use of new hunting equipment, like the atlatl. An atlatl would have significantly extended a hunter's "killing range" and been particularly effective against white-tailed deer in open habitats.

A variety of archaeological indicators suggest that by the late Middle or early Late Holocene, some Lower Ohio Valley hunter-gatherer groups started to significantly alter the way they moved about the landscape, obtained their food, and exploited the uplands and river valleys. These groups appear to have gradually shifted from a residential mobility strategy to one emphasizing logistical mobility (Stafford 1994:221). This transition represents a major change in the organization of hunter-gatherer societies that more closely linked groups to specific parts of the landscape. The repeated occupation of these places resulted in their greater cultural significance, particularly once the group buried their dead there. Researchers have proposed many explanations for this socioeconomic restructuring, including environmental change (Jefferies 1983; Nance

1988:147), population increase (Brown 1985), and the adoption of more specialized settlement-subsistence strategies (Stafford 1994).

This shift in mobility strategies is reflected in some parts of the Lower Ohio Valley by the establishment of multiseasonal base camps characterized by deep earth, rock, and/or shell middens, a high diversity of features and tools, burials, and evidence of interregional exchange (Brown 1985). Many of these impressive sites were situated near wetlands that offered an abundant, diverse, and reliable food supply, especially when used in conjunction with resources found in drier upland settings. Other kinds of sites also occur, but their wide distribution and generally low artifact content suggest that late Middle Archaic hunter-gatherers occupied them only for short periods when performing task-specific activities.

Archaeologists working in some parts of the Midcontinent have noted a decreased emphasis on upland habitats and an increased importance of river and bottomland settings during the late Middle Archaic (Ahler 1984; Cook 1976:118; Jefferies 1983; Nance 1988). This refocusing of settlement-subsistence strategies appears to have been tied, at least in part, to climatic changes during the Hypsithermal interval that lowered upland plant food productivity. These changes in population distribution also may be associated with increased nut production attributable to the transition from a mixed deciduous to a xeric oak-hickory forest and a more heterogeneous distribution of nut-bearing trees (Stafford 1994).

The appearance of thick middens on the late Middle Archaic landscape indicates that certain strategically positioned spots were the scenes of intensive or repeated cultural activities. Midden formation may be attributable to one group that inhabited a place for an extended period or, in contrast, the repeated occupation of the same place by several different groups. Regardless of occupation history, the high level of activities resulted in the deposition of organic and inorganic constituents that significantly altered the landscape's physical and chemical properties. A greater emphasis on local lithic resources is also indicative of decreased mobility and the "entrenchment" of groups into spatially restricted territories.

In addition to altering the area around their settlements, hunter-gatherers likely impacted other parts of the surrounding landscape through the removal of forest cover for firewood and building materials or selectively clearing or burning vegetation to facilitate the growth and productivity of desirable trees (nut- and fruit-bearing species), bushes (fruit-bearing species), and annual plants (starchy and oily seed-bearing species) (Wagner 2005).

Late Middle Holocene hunter-gatherers maintained a highly diversified tool kit, consisting of flaked stone, ground stone, bone, antler, and shell implements that were well suited for performing a variety of extractive and maintenance tasks. As in earlier times, perishable woods and fibers would have played important technological roles as well. A few Lower Ohio Valley groups were aware of copper working, but whether they actually made the items or just acquired them through trade is unclear.

Establishment of multiseasonal base camps meant that it was feasible for hunter-gatherers to invest more time and energy into making larger, less portable tools and facilities not used by more mobile groups. This investment of time and energy may have contributed to the more efficient processing and storage of subsistence resources, like the bulk processing of large quantities of nuts, as well as elevating the greater ritual prominence of these places on the cultural landscape.

The increased abundance and diversity of ground stone tools suggests that these implements had become integral parts of the late Middle Holocene tool kit. Like some kinds of processing and storage facilities, the production of ground stone tools may have initially required more time and effort, but since they could be quickly resharpened with little loss of mass, they had long use-lives. In some cases, formal tools replaced those that were expediently manufactured in earlier times. For example, many conical, cylindrical, and bell-shaped ground stone pestles had similar functions as informal grinding stones made earlier.

The increased use-lives of some ground stone implements like axes and adzes may have made them preferable to their flaked stone counterparts, particularly in the face of decreased mobility and the reduced availability of suitable lithic resources. The development of regionally distinct styles of some ground stone artifacts, such as atlatl weights, probably reflects the emergence of more formal social networks and some peoples' desire to express their shared social connections and identity.

In other parts of the region, some late Middle Archaic hunter-gatherers appear to have maintained a way of life reminiscent of earlier times, continuing an existence based on small group size, high mobility, and a simple, portable technology. Whether this variation in social complexity was by necessity or choice is still open to question (Sassaman 1995:187).

Although not always a good indicator of population, a substantial increase in the number of Late Archaic sites in some parts of the Lower Ohio Valley suggests that regional hunter-gatherer populations were growing during the early Late Holocene (Figure 8.1). For example, 2,583 Late Archaic components are recorded for the Lower Ohio Valley study area compared to the 1,650 and 818 documented for the Early and Middle Archaic subperiods, respectively. This represents a substantial increase in component frequency, paralleling trends seen throughout much of eastern North America during the Late Holocene (Milner 2004).

The distribution of Late Archaic sites generally coincides with that of the late Middle Archaic. However, Late Archaic sites are more numerous and clustered on the landscape, contributing to fewer empty spaces or buffer zones between different hunter-gatherer groups. These subtle changes in site distributions may reflect one or more ongoing processes related to more permeable social boundaries, greater intergroup cooperation and interaction, and/or greater group mobility within larger territories (Anderson 1996b). Increased evidence for weapon-related trauma in some areas may point to greater inter- or intragroup conflict during this time (Smith 1996).

In some parts of the Lower Ohio Valley, settlement strategies initiated in the late Middle Holocene continued well into the Late Holocene (post-5000 B.P.). In the middle Green River area, many midden sites that were first intensively occupied during the late Middle Archaic continued to be inhabited throughout much of the Late Archaic. In addition, the increased number of Late Archaic sites in other habitats, particularly interior wetlands, suggests a steadily increasing population that was exploiting a wider variety of environmental resources. In contrast, some floodplain sites in other parts of the Lower Ohio Valley experienced more intermittent and less intensive activities during the Late Holocene (Jefferies et al. 2005).

The number of Late Archaic components, the richness of the early Late Holocene archaeological record, and the intensity of research on Late Holocene hunter-gatherers has enabled Lower Ohio Valley archaeologists to define a number of subregional archaeological "cultures" or phases (Riverton, Lone Hill, Indian Knoll, O'Bryan's Ridge, Au Vase). Recognition of these archaeological manifestations is based on the sharing of common technologies, settlement-subsistence strategies, and social organization by Late Archaic hunter-gatherers who inhabited a geographically defined portion of the valley.

Groups that shared these cultural traits probably also maintained closer social, economic, and ritual connections with one another than they did with geographically distinct groups that exhibited different cultural traits. Ultimately, members of these subregional groups developed stronger ties to their portion of the regional landscape and a stronger sense of social identity.

Nuts appear to have formed the core of most Late Archaic diets throughout the Lower Ohio Valley, but groups adapted their subsistence strategies to take advantage of plant resources native to their specific territories. For example, groups that lived near extensive wetlands, like those near the Highland Creek and Hedden sites in western Kentucky, made extensive use of nearby aquatic and circum-aquatic plant resources.

In many parts of the North American Midcontinent, Late Archaic groups were experimenting with plant cultivation, and, in some places, cultivating gardens where they grew several domesticated starchy and oily seeded native plants. Despite extensive research at numerous Lower Ohio Valley sites, researchers have found little evidence for the cultivation of domesticated plants. Groups that inhabited the western part of the Lower Ohio Valley, particularly in southwestern Illinois and southeastern Missouri, seemed to have been more active in these efforts than those that inhabited other parts of the region.

Clearly, the settlement ecology of Lower Ohio Valley hunter-gatherers radically changed from the Late Pleistocene to the Late Holocene. Several closely related long-term trends are evident, including smaller home ranges, decreased group mobility, population increase, and the more intensive use of local resources. These trends are revealed by diachronic changes in site frequencies, the duration of site occupations, the kinds of cultural objects represented at those sites, and the sources of the materials from which those objects were made.

Ritual Landscapes

Ritual landscapes are products of peoples' actions that define, legitimize, and sustain occupation of their traditional territories. These landscapes are revealed through the spatial distribution and patterning of ritual features such as public buildings, squares or plazas, monuments, petroglyphs or pictographs, and vernacular markers (Anschuetz et al. 2001:178). Indications of ancient hunter-gatherer ritual landscapes are difficult to detect because most groups were highly mobile and did not make long-lasting modifications to the landscapes in which they lived. Of those cultural modifications that they did make, most are no longer archaeologically visible.

Studies of historically documented hunter-gatherers indicate that many have very complex ritual or sacred landscapes consisting of both cultural and natural features and modifications. Importantly, extensive ethnographic data exist for many Australian groups. These data can provide further clarification of the relationships between landscape perceptions and conceptions (Tacon 1999).

In parts of Australia, analyses of the location and organizational structure of rock art have provided important insights on the relationships between people and places. In addition, certain kinds of natural landscape features invoke feelings of power, awe, and respect. These ritually significant places include: (a) locations where major natural transformations occurred (volcanoes, mountain ranges, gorges); (b) at major transitions in the physical landscape (waterfalls or where there is a sudden elevation change); (c) where there is an unusual landscape feature (a cave, peak, or sinkhole); and (d) places offering panoramic views of interesting landscape features (Tacon 1999:34).

Another important source of information on ritual landscapes comes from the investigation of hunter-gatherer mortuary practices (Buikstra 1981; Buikstra and Charles 1999; Buikstra and Seddon n.d.). In parts of the North American Midcontinent, places where hunter-gatherers buried their dead were used for thousands of years, often dividing and marking the landscape in ways that had religious, social, and political meanings (Buikstra and Charles 1999:203). For example, in the Lower Illinois Valley, archaeologists identified three distinctive patterns for disposing of deceased Middle Archaic people. One of these involved the interment of middle-aged and young adults in bluff-crest knolls near their villages. As groups added more and more of their dead to the knolls, the landforms were culturally transformed into "mounds." In addition to serving as mortuary facilities, they also served as territorial markers where deceased ancestors legitimized a group's access to local resources (Buikstra and Charles 1999:207–208; Charles and Buikstra 1983).

When the first hunter-gatherers entered the Lower Ohio Valley some 12,500 or more years ago, there was no ritual landscape. No one had seen, much less modified, the landscape, and none of the natural features had any cultural significance. This situation would quickly change as more people occupied the region and they interpreted the world in which they lived through their specific cultural perceptions. As the nature of hunter-gatherer societies changed over time, so did the ritual significance of the natural features that surrounded them. In addition, people increasingly altered their world, creating culturally significant places that, along with the natural elements, collectively formed their ritual landscape.

The complexity of ritual landscapes increased as hunter-gatherers became more familiar with the parts of their world that they either directly or indirectly used. Highly visible natural features like prominent ridges probably served as markers or reference points, perhaps indicating places where groups periodically aggregated to exchange information, seek companionship, or maintain social connections that could be used in times of need. Places that offered good visibility for spotting game might also have acquired ritual significance. Many other kinds of natural settings probably had varying degrees of ritual importance, but determining just which places did and what their ritual significance was would be purely speculative.

The investigation of early Lower Ohio Valley hunter-gatherer ritual landscapes is made more difficult by the low frequency of petroglyphs, pictographs, and other forms of rock art that have been positively associated with the region's earliest inhabitants. Rock art that has been documented in Kentucky appears to date from the Late Archaic to the historic periods (Coy et al. 1997:151), with virtually none associated with Paleoindian and Early to Middle Archaic hunter-gatherers. Obviously, there is no ethnohistoric or ethnographic information for this early period of Lower Ohio Valley settlement.

The most obvious components of Lower Ohio Valley hunter-gatherer ritual landscapes are mortuary areas where groups interred the remains of their deceased members. Although a possible Dalton cemetery has been reported in northeastern Arkansas (Morse 1997a), so far archaeologists have not located any in the Lower Ohio Valley. The partially cremated remains of a woman at the Olive Branch site indicates that Dalton people buried at least some of their dead in places where they lived (Gramly 2002:91).

Some Early Holocene hunter-gatherers buried their dead at what appear to have been specialized mortuary sites like Jerger and McCullough's Run in southern Indiana (Cochran et al. 1998; Tomak 1979). Sand dunes or ridges situated along the margins of wetlands seem to have been particularly favored locations for interment. These groups cremated some of their deceased prior to placement in the burial facility; others are represented by inhumation burials. Objects of nonlocal origin accompanied some of the dead, suggesting that even during the Early Holocene, highly mobile hunter-gatherers maintained social connections with other groups that probably served both social and economic purposes.

The placement of burials at sites like Jerger and McCollough's Run meant that these locations held some significance on the Early Holocene ritual landscape. However, these sites generally contain only a few individuals, suggesting that they were used for a short time, reflecting the mobile lifestyle of their survivors.

Some Early Holocene groups that lived in north central Kentucky buried their dead in caves and rockshelters (DiBlasi 1981), demonstrating that these kinds of places also acquired ritual significance. The fact that the woman buried at Ashworth Cave met a violent death may have influenced where she was buried.

By the late Middle Holocene and early Late Holocene, the Lower Ohio Valley's ritual landscape(s) was experiencing significant change. The appearance of sites having thick middens and hundreds of burials indicates hunter-gatherers were becoming "entrenched" in their movements and spending more time at fewer places. Although it is hard to determine if the burials are attributable to one or several groups, the many people interred there suggest that hunter-gatherers frequently returned to these places. Such strategically located sites initially may have been important because of nearby food sources, like the river mussels found near many of the Green River middens. However, over time these sites acquired ritual significance in the worldviews of regional hunter-gatherers, particularly with respect to issues of ancestor veneration and cultural identity (Crothers 1999:107, 250).

Some groups placed their mortuary areas so that they offered an expansive view of the surrounding landscape. For example, the Middle to Late Archaic Ward site is situated on a high bluff overlooking the Cypress Creek wetlands and the more distant Green River (Jefferies et al. 2005, 2008).

While most elements of the Middle Holocene ritual landscape consisted of surface features, by the early Late Holocene, subterranean regions of the Lower Ohio Valley were also being incorporated (DiBlasi 1996). The best evidence for these activities comes from the karst region of west central Kentucky where archaeologists have found geometric and zoomorphic images incised or drawn on cave floors and walls. The presence of these images far from cave entrances indicates that Late Archaic hunter-gatherers were making long, dangerous trips into the dark regions of these caves. At least some of these forays were made to collect various minerals, like gypsum and mirabilite, used for a variety of domestic and ritual purposes. Hunter-gatherers may have made other caving trips as parts of vision quests or rites of passage that were components of various ritual activities. Similar use of the region's cave systems continued for at least 1,000 years.

The possible construction of a mound over an earthquake fissure at the Burkett site in southeastern Missouri suggests that Late Holocene ritual activities in this region incorporated both cultural and natural landscape elements. The direct association of the mound and the earthquake fissure suggests that seismic activity had a major role in shaping the beliefs and rituals of the region's Late Archaic hunter-gatherers (Thomas et al. 2004:125).

The ritual landscape as perceived by Lower Ohio Valley hunter-gatherers continued to evolve throughout the Holocene. Prior to the arrival of the first groups at the end of the Pleistocene, there was no ritual landscape simply because the land was uninhabited by humans. As the first hunter-gatherers entered the region, they gradually assigned meanings to the land and to the natural features that surrounded them. As time progressed, more restricted group mobility and smaller home ranges resulted in a greater familiarity with the local environment and the increased ritual significance of local features and landmarks. These hunter-gatherers culturally modified their surroundings by clearing the landscape, spending more time in one place, and burying their dead in central locations, increasing the ritual significance of the places where they lived, worked, and died.

Ethnic Landscapes

Like ritual landscapes, past hunter-gatherer ethnic landscapes are difficult to discern in lieu of ethnographic or ethnohistoric data. Ethnic landscapes are marked by the temporal and spatial distributions of a community's material culture and symbols, resulting in the definition of ethnic or cultural identities and boundaries that reflect customs or shared ways of thinking. When a network of interacting people focuses these activities in a particular territory, the group can maintain a coherent sense of affiliation and membership with place (Anschuetz et al. 2001:179–181).

Clearly, it is a big leap going from the spatial/temporal distribution of stylistic attributes and artifacts to ethnicity and cultural identity. However, the fact that groups shared cultural characteristics suggests that they maintained social connections and shared some kind of social identity.

Robert Whallon (2006:261) suggests that these social connections represented a safety net that enabled groups to survive in times of local resource scarcity. Intergroup social ties provided the means for people to move from their own areas of scarcity to other places where there were sufficient resources to support them through the rough periods. Over time, groups that maintained regular social connections would develop a shared social identity that distinguished them from other groups with which they had fewer or no social contacts (Wobst 1977:323). The existence of these regional-scale social networks should be marked in the archaeological record by a greater similarity in artifact manufacturing technologies and decorative styles (Wobst 1977).

Archaeological evidence for Late Pleistocene and Early Holocene intergroup social connections is rare (Ahler 1991; Cochran et al. 1998; Tomak 1979). Those transactions that did exist probably took place during brief encounters of isolated, mobile groups or during occasional multigroup gatherings (Anderson and Hanson 1988:266). Group members could exchange goods, swap information, seek mates, or simply enjoy the company of different people during these brief encounters (Anderson 1990, 1996a:44; Anderson and Hanson 1988; Sassaman 1995:179). Once these groups parted, the time and place of future meetings was problematic.

By the late Middle Holocene, some Lower Ohio Valley hunter-gatherer societies had undergone significant social and economic changes that contributed to decreased seasonal mobility, smaller home ranges, and the establishment of multiseasonal base camps (Anderson 1996b:52; Brown 1985). As mobility and territory size decreased, traditional ways of handling environmental and social risks became less effective. Of critical importance was maintaining social connections with nearby and distant groups needed for group survival in uncertain environments (Whallon 2006:260).

I think that one possible response to decreased hunter-gatherer mobility and associated changes in group interaction was the amplification of more formal social connections through a network of trading partners (Kelly 1991:143–145; Marquardt 1985). Partners maintained ties with different groups, perhaps through marriage, creating a web of social ties across the landscape that linked the "home" group with any number of nearby or distant groups. The resulting network provided the social "pipeline" through which needed materials, information, and mates moved from one group to another (Jefferies 2004:72).

My analyses of bone hairpins from the Lower Ohio Valley and adjacent parts of the Midcontinent have provided important insights into the emergence of hunter-gatherer ethnic landscapes (Jefferies 1996a, 1997, 2004). The bone pin study revealed that late Middle to early Late Archaic hunter-gatherers who lived north of the Ohio River employed generally similar manufacturing technologies and decorative styles. This area extends approximately 500 km east to west, from the Falls of the Ohio to eastern Missouri, and 300 km south to north, from the Ohio River to west central Illinois. In contrast, most pins from south of the Ohio River had different head shapes and decorations, suggesting a relatively low level of interaction between groups that lived on opposite sides of the river.

The similarity of bone pins manufactured by groups north of the Ohio River suggests that the efforts of more sedentary, spatially circumscribed hunter-gatherers to establish and/or reaffirm their social connections with other regional groups were successful and that the social networks formed by those ties covered a significant part of the North American Midcontinent (Jefferies 1997:447).

The distribution of pin styles seen north of the Ohio River suggests several possibilities. On the one hand, if members of each local group made a specific pin style, then the mixture of styles documented at many sites may reflect exchange among members of different groups or the movement of people who wore the pins from one group to another—perhaps indicating the movement of marriage partners. On the other hand, different styles at one site may simply indicate that interacting groups shared a common set of ideas about how to make and decorate bone pins (Jefferies 1997).

Whatever the mechanism, long-term interaction among these hunter-gatherers resulted in the strengthening of social ties and the broadening of cultural identities, leading to the emergence of larger regional groups comprised of interdependent local groups. Radiocarbon dates associated with some of the bone pins indicate that these groups maintained some level of interaction for approximately 1,000 years, spanning the period from ca. 6000 to 5000 B.P. (Jefferies 2004:76–84).

Although similarities in artifact styles and manufacturing technologies suggest that groups shared some kind of social identity, they do not indicate the closeness of those ties or whether or not interacting groups had a common origin. In the case of Lower Ohio Valley hunter-gatherers, what it appears to reflect is the coalescence of local groups into a more highly integrated regional group in which participating members shared numerous traits through their increased levels of interaction. At this point, it is difficult to say whether or not these similarities in material culture reflect true "ethnic" groups (shared common language, origin, religion, etc.) or just an intermediate stage in the ethnogenesis of more complex intermediate level societies (tribes and chiefdoms). Regardless, they chose to identify themselves with others living in the surrounding region and to distinguish themselves from those who lived to the south across the Ohio River.

Differences in bone pin styles and technologies used by groups that lived north and south of the Ohio River suggest that distinct social networks existed in these two areas. Not surprisingly, late Middle and early Late Holocene hunter-gatherers who lived south of the Ohio River appear to have maintained closer social connections with contemporary groups that lived to the south and west than with those to the north. In contrast, those who lived north of the Falls of the Ohio maintained long-term social ties with groups that live to the north and west.

I think these trends indicate the emergence of larger, more complex societies consisting of more closely integrated local groups. These groups maintained their social ties by participating in common social networks that created a safety net critical to group survival in times of local resource scarcity or failure (Whallon 2006:261). The social connections maintained with other groups allowed people to move from their own areas of scarcity to other groups situated in places where there were enough resources available to support them through hard times. Periodic reaffirmation of these social connections was accomplished by the exchange of ritually significant objects through the network (Whallon 2006:261; Wobst 1977). Over time, groups that maintained these regular social connections likely developed a shared social identity that distinguished them from other groups with which they had less or no social contact (Wobst 1977:323).

Postscript

The cultural landscape constructed by hundreds of generations of Lower Ohio Valley hunter-gatherers reflects many dynamic changes in the technological, social, economic, political, and ideological aspects of their societies. Perhaps foremost among these developments was the gradual transition from a highly mobile way of life to one characterized by an increasingly sedentary lifestyle. As hunter-gatherers became more stationary for longer periods, the diversity and breadth of their subsistence base increased, along with the increased dietary importance of smaller and more prolific food sources.

Coeval with the greater resource intensification was plant husbandry, food storage, and the emergence of the institutional means for regulating a more settled way of life (Brown 1985:202). The shared cultural identities and perceived social distinctions apparent among Late Holocene hunter-gatherer groups were based on a long history of decision-making about how to successfully cope with their changing natural and cultural worlds. Over time, these people extended the spatial extent of their social and economic networks, with some individuals maintaining important intergroup and interpersonal connections that spanned much of eastern North America. Although these people may have lived in what many twenty-first-century Americans would call the "wilderness," they were "well-connected" to their local neighbors and distant acquaintances. The world constructed by Late Holocene hunter-gatherers formed the cultural and natural foundations that would support the Lower Ohio Valley's future Woodland and Mississippian societies.

References

Adams, Jenny L.

2002 *Ground Stone Analysis: A Technological Approach.* University of Utah Press, Salt Lake City.

Adovasio, J. M., J. Donahue, and R. Stuckenrath

1990 The Meadowcroft Rockshelter Radiocarbon Chronology 1975–1990. *American Antiquity* 55:348–354.

Adovasio, J. M., D. R. Pedler, J. Donahue, and R. Stuckenrath

1998 Two Decades of Debate on Meadowcroft Rockshelter. *North American Archaeologist* 19(4):317–341.

Adovasio, J. M., and D. R. Pedler

2005 A Long View of Deep Time at Meadowcroft Rockshelter. In *Paleoamerican Origins: Beyond Clovis,* edited by Robson Bonnichsen, Bradley T. Lepper, Dennis Stanford, and Michael R. Waters, pp. 23–28. Center for the Study of the First Americans, Texas A&M University, College Station.

Ahler, Steven R.

1984 *Archaic Settlement Strategies in the Modoc Locality, Southwest Illinois.* Unpublished Ph.D. dissertation, Department of Anthropology, University of Wisconsin, Milwaukee.

1991 Modoc Matting and Beads. *The Living Museum* 53:3–6.

1993 Stratigraphy and Radiocarbon Chronology of Modoc Rock Shelter, Illinois. *American Antiquity* 58:462–489.

Albertson, Donald G., and Douglas K. Charles

1988 Archaic Mortuary Components. In *The Archaic and Woodland Cemeteries at the Elizabeth Site in the Lower Illinois Valley,* edited by D. K. Charles, S. R. Leigh, and J. E. Buikstra, pp. 29–40. Research Series 7. Center for American Archaeology, Kampsville, Illinois.

Allen, Roger C.

1971 Archaic of the Mississippian Plateau of Kentucky. Manuscript on file, Office of State Archaeology, Lexington, Kentucky.

1976 *Archaeological Investigations at Two Sites in the U.S. Interstate Highway 24 Right of Way in Marshall County, Kentucky.* Unpublished Master's thesis, Department of Anthropology, University of Kentucky, Lexington.

Amick, Daniel S., and Philip J. Carr

1996 Changing Strategies of Lithic Technological Organization. In *Archaeology of the Mid-Holocene Southeast,* edited by Kenneth E. Sassaman and David G. Anderson, pp. 41–56. University Press of Florida. Gainesville.

Anderson, David G.

1990 The Paleoindian Colonization of Eastern North America: A View from the Southeastern United States. In *Early Paleoindian Economies of Eastern North America,* edited by Kenneth B. Tankersley and Barry L. Isaac, pp. 163–216. Research in Economic Anthropology Supplement 5. JAI Press, Greenwich, Connecticut.

1996a Approaches to Modeling Regional Settlement in the Archaic Period Southeast. In *Archaeology of the Mid-Holocene Southeast,* edited by K. Sassaman and D. Anderson, pp. 157–176. University Press of Florida, Gainesville.

1996b Models of Paleoindian and Early Archaic Settlement in the Lower Southeast. In *The Paleoindian and Early Archaic Southeast,* edited by David G. Anderson and Kenneth E. Sassaman, pp. 29–57. University of Alabama Press, Tuscaloosa.

2001 Climate and Cultural Change in Prehistoric and Early Historic Eastern North America. *Archaeology of Eastern North America* 29:143–186.

Anderson, David G., and Glen T. Hanson

1988 Early Archaic Settlement in the Southeastern United States: A Case Study from the Savannah River Valley. *American Antiquity* 53:262–286.

Anderson, David G., and Kenneth Sassaman

1996 Modeling Paleoindian and Early Archaic Settlement in the Southeast: A Historical Perspective. In *The Paleoindian and Early Archaic Southeast,* edited by D. Anderson and K. Sassaman, pp. 16–28. University of Alabama Press, Tuscaloosa.

Anderson, David G., and Kenneth E. Sassaman (editors)

1996 *The Paleoindian and Early Archaic Southeast.* University of Alabama Press, Tuscaloosa.

Anderson, David G., Lisa D. O'Steen, and Kenneth E. Sassaman

1996 Environmental and Chronological Considerations. In *The Paleoindian and Early Archaic Southeast,* edited by David G. Anderson and Kenneth E. Sassaman, pp. 3–15. University of Alabama Press, Tuscaloosa.

Anderson, Eve

1991 New Perspectives on Archaic Mortuary Behavior at Modoc Rockshelter. Paper presented at the 56th Annual Meeting of the Society for American Archaeology, New Orleans.

Andrefsky, William, Jr.

1994 Raw-Material Availability in the Organization of Technology. *American Antiquity* 59:21–34.

Anschuetz, Kurt F., Richard H. Wilshusen, and Cherie L. Scheick

2001 An Archaeology of Landscapes: Perspectives and Directions. *Journal of Archaeological Research* 9:157–211.

Anslinger, Michael

1986 *The Riverton Culture: Lithic Systems and Settlement Parameters.* Unpublished Master's thesis, Department of Anthropology, Washington State University, Pullman.

2001 The Archaic Period in the Falls of the Ohio Region of Kentucky. Electronic document, http//www.crai-ky.com/falls.html#introduction, accessed September 2007. Cultural Resource Analysts.

Anslinger, Michael, Albert M. Pecora, Charles M. Niquette, and Jonathon P. Kerr

1994 *Salvage Excavations at the Railway Museum Site (15Jf630), Jefferson County, Kentucky.* Contract Publication 94–15. Cultural Resource Analysts Inc., Lexington.

Armelagos, George J.

1969 Disease in Ancient Nubia. *Science* 163:255–259.

Armelagos, George J., David S. Carlson, and Dennis P. van Gerven

1982 The Theoretical Foundations and Development of Skeletal Biology. In *A History of American Physical Anthropology, 1930–1980,* edited by F. Spenser, pp. 305–328. Academic Press, New York.

Asch, David L., and Nancy E. Asch

1985 Prehistoric Plant Cultivation in West-Central Illinois. In *Prehistoric Food Production in North America,* edited by Richard I. Ford, pp. 149–203. Anthropological Papers No. 75, Museum of Anthropology, University of Michigan, Ann Arbor.

Asch, Nancy, Richard I. Ford, and David L. Asch

1972 *Paleoethnobotany of the Koster Site: The Archaic Horizons.* Reports of Investigations No. 24, Illinois State Museum, Springfield.

Ashmore, Wendy, and A. Bernard Knapp

1999 *Archaeologies of Landscape: Contemporary Perspectives.* Blackwell Publishers, Malden, Massachusetts.

Atwater, Caleb

1820 *Description of the Antiquities Discovered in the State of Ohio and Other Western States: Communicated to the President of the American Antiquarian Society.* American Antiquarian Society, Worcester, Massachusetts.

Bader, Anne T.

1992 *An Analysis of Bone and Antler Tool Use Patterns from the Kentucky Air National Guard Site.* Unpublished Master's thesis, Department of Anthropology, University of Kentucky, Lexington.

2004 The Meyer Site: An Inadvertent Discovery in Spencer County, Indiana. *Currents of Change* 2:17–29.

2005 Why Dig Yet Another Late Archaic Site in the Falls of the Ohio River Region? Insights into Corporate Identity through an Examination of Mortuary Behavior at the KYANG Site, Jefferson County, Kentucky, and the Meyer Site, Spencer County, Indiana. Paper presented at the 22nd Annual Kentucky Heritage Council Archaeological Conference, Lexington.

Bader, Anne T., and Joseph E. Granger

1989 *Recent Archaeological Investigations on the Kentucky Air National Guard Site (15Jf267), Jefferson County, Kentucky.* Granger Consultants, Louisville.

Bassett, Everett J.

1982 Osteological Analysis of Carrier Mills Burials. In *The Carrier Mills Archaeological Project: Human Adaptation in the Saline Valley, Illinois,* edited by Richard W. Jefferies and Brian M. Butler, pp. 1029–1114. Center for Archaeological Investigations Research Paper No. 33, Southern Illinois University Carbondale, Carbondale.

Bassett, John L.

1980 Chert Resources of the Upper Patoka Drainage Basin. In *Archaeological Salvage Excavations at Patoka Lake, Indiana: Prehistoric Occupations of the Upper Patoka River Valley,* edited by Cheryl A. Munson, pp. 72–95. Glenn A. Black Laboratory of Archaeology, Indiana University, Bloomington.

Bellis, James O.

1968 *An Appraisal of the Archaeological Resources of the Lafayette and Patoka Reservoirs and a Report of the Test Excavation at Site Hr11 in Harrison County, Indiana.* Glenn Black Laboratory of Archaeology, Indiana University, Bloomington.

1982 Test Excavations Conducted at the Breeden Site, 12Hr11, in Harrison County. *Proceedings of the Indiana Academy of Science* 91:78–92.

Belovich, Stephanie J.

2005 Fractures in the Carlston Annis Shell Mound Skeletal Population. In *Archaeology of the Middle Green River Region, Kentucky,* edited by William H. Marquardt and Patty Jo Watson, pp. 505–514. Monograph 5, Institute of Archaeology and Paleoenvironmental Studies, University of Florida, Gainesville.

Binford, Lewis R.

1979 Organization and Formation Processes: Looking at Curated Technologies. *Journal of Anthropological Research* 35:255–273.

1980 Willow Smoke and Dogs' Tails: Hunter-Gatherer Settlement Systems and Archaeological Site Formation. *American Antiquity* 45:4–20.

Binford, Lewis R., and Sally R. Binford

1966 A Preliminary Analysis of Functional Variability in the Mousterian of Levallois Facies. In *Recent Studies in Paleoanthropology,* edited by J. D. Clark and F. C. Howell, pp. 238–295. *American Anthropologist* 68:Part 2.

Boisvert, Richard A.

1979 Excavations at the Spadie Site (15Jf14). In *Excavations at Four Archaic Sites in the Lower Ohio Valley, Jefferson County, Kentucky,* Vol. 2, edited by

Michael B. Collins, pp. 804–882. Occasional Papers in Anthropology No. 1, Department of Anthropology, University of Kentucky, Lexington.

1984 *Kentucky Salt Licks: A Preservation Planning Perspective.* Report on file, Office of State Archaeology, University of Kentucky, Lexington.

Boisvert, Richard A., Boyce N. Driskell, Kenneth W. Robinson, Steven D. Smith, and L. F. Duffield

1979 Materials Recovered. In *Excavations at Four Archaic Sites in the Lower Ohio Valley, Jefferson County, Kentucky,* Vol. 1, edited by Michael B. Collins, pp. 60–470. Occasional Papers in Anthropology No. 1, Department of Anthropology, University of Kentucky, Lexington.

Bonzani, Renee M.

2002 Plant Remains from Test Excavations at the Ward site. In *Cypress Creek Archaeological Project: Archaic Adaptive Strategies in West Central Kentucky,* by Richard W. Jefferies, Victor D. Thompson, George R. Milner, Renee M. Bonzani, and Tanya M. Peres, pp. 158–174. Department of Anthropology, University of Kentucky, Lexington.

Braun, E. Lucy

1950 *Deciduous Forests of Eastern North America.* Blakiston, Philadelphia.

Breitburg, Emanuel

1982 Analysis of Area A Fauna. In *The Carrier Mills Archaeological Project: Human Adaptation in the Saline Valley, Illinois,* edited by Richard W. Jefferies and Brian M. Butler, pp. 863–957. Center for Archaeological Investigations Research Paper No. 33, Southern Illinois University Carbondale, Carbondale.

2006 Faunal Remains. In *The Highland Creek Site: Middle to Late Archaic Wetland Utilization in Western Kentucky,* edited by Greg Maggard and David Pollack, pp. 85–87. Research Report No. 5, Kentucky Archaeological Survey, Lexington.

Brose, David S.

1994 Trade and Exchange in the Midwestern United States. In *Prehistoric Exchange Systems in North America,* edited by Timothy G. Baugh and Jonathon E. Ericson, pp. 215–240. Plenum Press, New York.

Brown, James A.

1985 Long-Term Trends to Sedentism and the Emergence of Complexity in the American Midwest. In *Prehistoric Hunter-Gatherers: The Emergence of Cultural Complexity,* edited by T. Douglas Price and James A. Brown, pp. 201–231. Academic Press, Orlando.

1986 Food for Thought: Where Has Subsistence Analysis Gotten Us? In *Foraging, Collecting, and Harvesting: Archaic Period Subsistence and Settlement in the Eastern Woodlands,* edited by Sarah W. Neusius, pp. 315–330. Occasional Paper No. 6, Center for Archaeological Investigations, Southern Illinois University Carbondale, Carbondale.

Brown, James A., and Robert K. Vierra

1983 What Happened in the Middle Archaic? Introduction to an Ecological

Approach to Koster Site Archaeology. In *Archaic Hunters and Gatherers in the American Midwest,* edited by James L. Phillips and James A. Brown, pp. 165–195. Academic Press, New York.

Broyles, Bettye J.

1971 *Second Preliminary Report: The St. Albans Site, Kanawah County, West Virginia, 1964–1968.* Report of Archaeological Investigations No. 3, West Virginia Geological and Economic Survey, Morgantown.

Buchner, C. Andrew

2003a Cultural Background. In *Data Recovery Excavations at the Clifford LaPlant Site (23Nm561) on Barnes Ridge, New Madrid County, Missouri,* by C. Andrew Buchner, Eric S. Albertson, Emmanuel Breitburg, Gina S. Powell, and Neal H. Lopinot, pp. 31–90. Panamerican Consultants, Inc., Memphis.

2003b Discussion. In *Data Recovery Excavations at the Clifford LaPlant Site (23Nm561) on Barnes Ridge, New Madrid County, Missouri,* by C. Andrew Buchner, Eric S. Albertson, Emmanuel Breitburg, Gina S. Powell, and Neal H. Lopinot, pp. 461–492. Panamerican Consultants, Inc., Memphis.

2003c Environmental Setting. In *Data Recovery Excavations at the Clifford LaPlant Site (23Nm561) on Barnes Ridge, New Madrid County, Missouri,* by C. Andrew Buchner, Eric S. Albertson, Emmanuel Breitburg, Gina S. Powell, and Neal H. Lopinot, pp. 7–29. Panamerican Consultants, Inc., Memphis.

Buchner, C. Andrew, Eric S. Albertson, Emmanuel Breitburg, Gina S. Powell, and Neal H. Lopinot

2003 *Data Recovery Excavations at the Clifford LaPlant Site (23Nm561) on Barnes Ridge, New Madrid County, Missouri.* Panamerican Consultants, Inc., Memphis.

Buikstra, Jane E.

1981 Mortuary Practices, Paleodemography and Paleoanthropology: A Case Study from the Koster Site (Illinois). In *The Archaeology of Death,* edited by Robert Chapman, Ian Kinnes, and Klavs Randsborg, pp. 123–132. Cambridge University Press, Cambridge.

Buikstra, Jane E., and Douglas K. Charles

1999 Centering the Ancestors: Cemeteries, Mounds, and Sacred Landscapes of the Ancient North American Midcontinent. In *Archaeologies of Landscapes: Contemporary Perspectives,* edited by Wendy Ashmore and A. Bernard Knapp, pp. 201–228. Blackwell, Malden, Massachusetts.

Buikstra, Jane E., and M. T. Seddon

n.d. *The Bullseye Site: a Middle Archaic Cemetery Complex in the Lower Illinois River Floodplain.* Kampsville Studies in Archeology and History No. 3, Center for American Archeology, Kampsville, Illinois.

Burdin, Sheldon R.

2004 *Interaction, Exchange, and Social Organization among Hunter-Gatherers in the Midcontinent—Evidence from the Lower Ohio River Valley: Bannerstone*

Use from 6500 to 3000 B.P. Unpublished Master's thesis, Department of Anthropology, University of Kentucky, Lexington.

Burnett, Richard

1963 Lone Hill. *Central States Archaeological Journal* 10:84–90.

Butler, Brian M.

1994 Biophysical Setting. In *Archaeological Investigations on the Bankston Fork of the Saline River: The AMAX West Harrisburg Project,* edited by Michael L. Hargrave and Brian M. Butler, pp. 10–31. Technical Report 1994-3, Center for Archaeological Investigations, Southern Illinois University Carbondale, Carbondale.

2000 Archaeological Setting. In *Archaeological Investigations at Dixon Springs State Park: The Hills Branch Rock Shelter, Pope County, Illinois,* by Mark J. Wagner and Brian M. Butler, pp. 11–15. Technical Report 00-2, Center for Archaeological Investigations, Southern Illinois University Carbondale, Carbondale.

2009 Land between the Rivers: The Archaic Period of Southern-Most Illinois. In *Archaic Societies: Diversity and Complexity across the Midcontinent,* edited by Thomas Emerson, Dale McElrath, and Andrew Fortier, pp. 607–634. State University of New York Press, Albany.

Butler, Brian M., and Ernest E. May (editors)

1984 *Prehistoric Chert Exploitation: Studies from the Midcontinent.* Occasional Paper No. 2, Center for Archaeological Investigations, Southern Illinois University Carbondale, Carbondale.

Butler, Brian M., JoAnne M. Penney, and Cathy A. Robison

1981 *Archaeological Survey and Evaluation for the Shawnee 200 M.W.A.F.B.C. Plant, McCracken County, Kentucky.* Research Paper 21, Center for Archaeological Investigations, Southern Illinois University Carbondale, Carbondale.

Butler, Brian M., and Mark J. Wagner

1999 Summary and Synthesis. In *Archaeological Investigations at the Rose Hotel (11-Hn-116), Hardin County, Illinois,* by Mark J. Wagner and Brian M. Butler, pp. 443–461. Technical Report 99-3, Center for Archaeological Investigations, Southern Illinois University Carbondale, Carbondale.

2000 Summary and Synthesis. In *Archaeological Investigations at Dixon Springs State Park: The Hills Branch Rock Shelter, Pope County, Illinois,* by Mark J. Wagner and Brian M. Butler, pp. 168–176. Technical Report 00-2, Center for Archaeological Investigations, Southern Illinois University Carbondale, Carbondale.

Cable, John S.

1982 Organizational Variability in Piedmont Hunter-Gatherer Lithic Assemblages. In *The Haw River Sites: Archaeological Investigation at Two Stratified Sites in the North Carolina Piedmont,* assembled by Stephen R. Claggett and John S. Cable, pp. 637–688. Report R-2386. Commonwealth Associates, Jackson, Michigan.

Caldwell, Joseph R.
1958 *Trend and Tradition in the Prehistory of the Eastern United States.* American Anthropologist Memoir 88.

Cambron, James W.
1957 Some Early Projectile Point Types from the Tennessee Valley. *Journal of Alabama Archaeology* 3:17–19.

Cambron, James W., and David C. Hulse
1969 *Handbook of Alabama Archaeology. Part I Point Types.* Alabama Archaeological Society, Huntsville.
1975 *Handbook of Alabama Archaeology. Part I Point Types* (revised edition), edited by David L. DeJarnette. Alabama Archaeological Society, Huntsville.

Canouts, Veletta, Ernest E. May, Neal H. Lopinot, and Jon D. Muller
1984 *Cultural Frontiers in the Upper Cache Valley, Illinois.* Research Paper 16, Center for Archaeological Investigations, Southern Illinois University Carbondale, Carbondale.

Cantin, Mark
1989 Diachronic Model of Projectile Point Chert Usage in Southwestern Indiana. In *Research Trends in Midwest Archaeology,* edited by K. McGowan and P. Kreisa, pp. 37–71. *Journal of the Steward Anthropological Society* 18, Steward Anthropological Society, University of Illinois, Champaign-Urbana.
1994 Provenience, Description, and Archaeological Use of Selected Cherts of Indiana. Manuscript on file. Indiana State University Anthropology, Terre Haute.
2000 Comparative Analysis of Thebes and Kirk Lithic Technology and Home Range Implications in Southwestern Indiana. Unpublished master's thesis, Department of Geography, Geology, and Anthropology, Indiana State University, Terre Haute.

Carr, Lucien, and Nathanial S. Shaler
1876 On the Prehistoric Remains of Kentucky. *Memoirs of the Kentucky Geological Survey* (4). Frankfort.

Carstens, Kenneth C.
1980 *Archeological Investigations in the Central Kentucky Karst.* Unpublished Ph.D. dissertation, Department of Anthropology, Washington University, St. Louis.

Carstens, Kenneth C., and Patty Jo Watson (editors)
1996 *Of Caves and Shell Mounds.* University of Alabama Press, Tuscaloosa.

Catlin, George
1841 *Letters and Notes on the Manners, Customs, and Conditions of the North American Indians.* Published by the author at Egyptian Hall, London.

Chapman, Carl
1948 A Preliminary Survey of Missouri Archaeology, Part IV: Ancient Cultures and Sequence. *Missouri Archaeologist* 10:133–164.
1975 *The Archaeology of Missouri I.* University of Missouri Press, Columbia.

Chapman, Jefferson
1973 *The Icehouse Bottom Site: 40Mr23.* Report of Investigations 13, Department of Anthropology, University of Tennessee, Knoxville.

1975 *The Rose Island Site and the Bifurcate Point Tradition.* Report of Investigations 14, Department of Anthropology, University of Tennessee, Knoxville.

1976 The Archaic Period in the Little Tennessee River Valley: The Radiocarbon Dates. *Tennessee Anthropologist* 1:1–12.

1977 *Archaic Period Research in the Lower Little Tennessee River Valley, 1975: Icehouse Bottom, Harrison Branch, Thirty Acre Island, Calloway Island.* Report of Investigations 18, Department of Anthropology, University of Tennessee, Knoxville.

1978 *The Bacon Farm Site and a Buried Site Reconnaissance.* Report of Investigations 23, Department of Anthropology, University of Tennessee, Knoxville.

Charles, Douglas O., and Jane E. Buikstra

1983 Archaic Mortuary Sites in the Central Mississippi Drainage: Distribution, Structure, and Behavioral Implications. In *Archaic Hunters and Gatherers in the American Midwest,* edited by James L. Phillips and James A. Brown, pp. 117–145. Academic Press, Orlando.

Charles, Douglas O., Steven R. Reigh, and Jane E. Buikstra

1988 *The Archaic and Woodland Cemeteries at the Elizabeth Site in the Lower Illinois Valley.* Center for American Archeology Research Series, Vol. 7. Kampsville, Illinois.

Claassen, Cheryl P.

1991a Gender, Shellfishing, and the Shell Mound Archaic. In *Engendering Archaeology: Women and Prehistory,* edited by J. M. Gero and M. W. Conkey, pp. 276–300. Blackwell, Oxford.

1991b New Hypotheses for the Demise of the Shell Mound Archaic. In *The Archaic Period in the Mid-South,* edited by C. H. McNutt, pp. 66–71. Archaeological Report 24, Mississippi Department of Archives and History, Jackson.

1992 Shell Mounds as Burial Mounds: A Revision of the Shell Mound Archaic. In *Current Archaeological Research in Kentucky,* Vol. 2, edited by D. Pollack and A. G. Henderson, pp. 1–11. Kentucky Heritage Council, Frankfort.

1996a A Consideration of the Social Organization of the Shell Mound Archaic. In *Archaeology of the Mid-Holocene Southeast,* edited by K. E. Sassaman and D. G. Anderson, pp. 235–258. University Press of Florida, Gainesville.

1996b Research Problems with Shells from Green River Shell Matrix Sites. In *Of Caves and Shell Mounds,* edited by Kenneth C. Carstens and Patty Jo Watson, pp. 132–139. University of Alabama Press, Tuscaloosa.

2005 An Analytical Study of Shellfish from the DeWeese Mound, Kentucky. In *Archaeology of the Middle Green River Region, Kentucky,* edited by William H. Marquardt and Patty Jo Watson, pp. 279–293. Monograph 5, Institute of Archaeology and Paleoenvironmental Studies, University Press of Florida, Gainesville.

Claggett, Stephen R., and John S. Cable (assemblers)

1982 *The Haw River Sites: Archaeological Investigations at Two Stratified Sites in the North Carolina Piedmont.* Report R-2386, Commonwealth Associates, Jackson, Michigan.

Clapper, Taryn

2006 *The New World and the Natufian: Musculoskeletal Stress Markers of Hunter-Gatherer Lifeways.* Unpublished Master's thesis, Department of Anthropology, Southern Illinois University Carbondale, Carbondale.

Clark, John E.

2004 Surrounding the Sacred: Geometry and Design of Early Mound Groups as Meaning and Function. In *Signs of Power: The Rise of Cultural Complexity in the Southeast,* edited by J. L. Gibson and P. J. Carr, pp. 162–213. University of Alabama Press, Tuscaloosa.

Clay, R. Berle, and Douglas W. Schwartz

1963 The Archaeology of Barkley Basin and Adjacent Regions: A Synthesis. Manuscript on file, Webb Museum of Anthropology, University of Kentucky, Lexington.

Cochran, Donald R., Paige Waldron, Ron Richards, Beth McCord, and Debbie Rotman

1998 *McCullough's Run: A Bifurcate Tradition Cemetery in Central Indiana.* Report of Investigations 44. Archaeological Resources Management Service, Ball State University, Muncie, Indiana.

Coe, Michael D., and F. William Fischer

1959 Barkley Reservoir—Tennessee Portion, Archaeological Excavations 1959. Manuscript on file, National Park Service, Regional Office, Richmond, Virginia.

Coe, Joffre L.

1964 *The Formative Cultures of the Carolina Piedmont.* Transactions of the American Philosophical Society, Philadelphia.

Cole, Faye Cooper, Robert Bell, John Bennett, Joseph Caldwell, Norman Emerson, Richard MacNeish, Kenneth Orr, Roger Willis

1951 *Kincaid: A Prehistoric Illinois Metropolis.* University of Chicago Press, Chicago.

Collett, John

1879 *Geological Report on Harrison and Crawford Counties, Indiana, 1878.* Indiana Geological Report, 1876–1878. Indianapolis Journal Co., Indianapolis.

Collins, Michael B. (editor)

1979a *Excavations at Four Archaic Sites in the Lower Ohio Valley, Jefferson County, Kentucky.* Occasional Papers in Anthropology No. 1, Department of Anthropology, University of Kentucky, Lexington.

Collins, Michael B.

1979b The Longworth-Gick Site (15Jf243). In *Excavations at Four Archaic Sites in the Lower Ohio Valley, Jefferson County, Kentucky,* Vol. 2, edited by Michael B. Collins, pp. 471–589. Occasional Papers in Anthropology No. 1, Department of Anthropology, University of Kentucky, Lexington.

Collins, Michael B., and Boyce N. Driskell

1979 Summary and Conclusions. In *Excavations at Four Archaic Sites in the Lower*

Ohio Valley, Jefferson County, Kentucky, Vol. 2, edited by Michael B. Collins, pp. 1023–1042. Occasional Papers in Anthropology No. 1, Department of Anthropology, University of Kentucky, Lexington.

Conaty, Gerald T.

1985 *Middle and Late Archaic Mobility Strategies in Western Kentucky.* Unpublished Ph.D. dissertation, Department of Archaeology, Simon Fraser University, Burnaby, British Columbia.

1987 Patterns of Chert Use during the Middle and Late Archaic in Western Kentucky. *Southeastern Archaeology* 6:140–147.

Cook, Della Collins, Sandra K. Parker, Robert G. McCullough, Timi L. Barone, Gregory Cook, Douglas S. Blank, Christine Shaneyfelt, Kristin Hedman, and Christine A. Marvin

1996 Little Pigeon Creek Cemetery (12W340): An Archaic Midden-Mound in Warrick County, Indiana. Electronic document, http://www.gbl.indiana.edu/abstracts/86/cook_et al_86.html, accessed March 21, 2006. Glenn A. Black Laboratory of Archaeology.

Cook, S. F., and Robert F. Heizer

1965 *Studies on the Chemical Analysis of Archaeological Sites.* University of California Press, Berkeley.

Cook, Thomas G.

1976 *Koster: An Artifact Analysis of Two Archaic Phases in West-Central Illinois.* Prehistoric Research Records, Archaeology Program, Northwestern University, Evanston.

1980a The Cultural Use of Chert in the Patoka Lake Region. In *Archaeological Salvage Excavations at Patoka Lake, Indiana: Prehistoric Occupations of the Upper Patoka River,* edited by Cheryl A. Munson, pp. 608–644. Research Reports No. 6, Indiana University, Glenn A. Black Laboratory of Archaeology, Bloomington.

1980b Preliminary History of Prehistoric Occupations at Patoka Lake. In *Archaeological Salvage Excavations at Patoka Lake, Indiana: Prehistoric Occupations of the Upper Patoka River,* edited by Cheryl A. Munson, pp. 684–697. Research Reports No. 6, Indiana University, Glenn A. Black Laboratory of Archaeology, Bloomington.

Cotter, John L.

1937 The Occurrence of Flints and Extinct Animals in Pluvial Deposits near Clovis, New Mexico, Part IV, Report on the Excavations at the Gravel Pit in 1936. *Proceedings of the Academy of Natural Sciences of Philadelphia* 89:2–16.

Coy, Fred E., Thomas C. Fuller, Larry G. Meadows, and James L. Swauger

1997 *Rock Art of Kentucky.* University Press of Kentucky, Lexington.

Crawford, Gary W.

1982 Late Archaic Plant Remains from West-Central Kentucky: A Summary. *Midcontinental Journal of Archaeology* 7:205–224.

2005 Plant Remains from Carlston Annis (1972, 1974), Bowles, and Peter

Cave. In *Archaeology of the Middle Green River Region, Kentucky,* edited by William H. Marquardt and Patty Jo Watson, pp. 181–212. Monograph 5, Institute of Archaeology and Paleoenvironmental Studies, University Press of Florida, Gainesville.

Cremin, William H.

1992 Botanical Analysis. In *The Little Muddy Rock Shelter: A Deeply Stratified Prehistoric Site in the Southern Till Plains of Illinois,* by Charles R. Moffat, Brad Koldehoff, William M. Cremin, Terrance J. Martin, Mary Carol Masulis, and Mary R. McCorvie, pp. 375–425. Cultural Resource Management Report 186, American Resources Group, Ltd., Carbondale.

Crites, Gary D.

1994 Appendix G: Plant Remains from 15JF630. In *Salvage Excavations at the Railway Museum Site (15Jf630), Jefferson County, Kentucky,* by Michael Anslinger, Albert M. Pecora, Charles M. Niquette, and Jonathon P. Kerr, pp. 334–348. Contract Publication 94–15, Cultural Resource Analysts, Inc., Lexington, Kentucky.

Crothers, George M.

1999 *Prehistoric Hunters and Gatherers, and the Archaic Period Green River Shell Middens of Western Kentucky.* Unpublished Ph.D. dissertation, Department of Anthropology, Washington University, St. Louis.

2005 Vertebrate Fauna from the Carlston Annis Site. In *Archaeology of the Middle Green River Region, Kentucky,* edited by William H. Marquardt and Patty Jo Watson, pp. 295–314. Monograph 5, Institute of Archaeology and Paleoenvironmental Studies, University Press of Florida, Gainesville.

Crothers, George M., and Reinhard Bernbeck

2004 The Foraging Mode of Production: The Case of the Green River Kentucky Archaic Shell Middens. In *Hunters and Gatherers in Theory and Archaeology,* edited by George M. Crothers, pp. 401–422. Occasional Paper 31, Center for Archaeological Investigations, Southern Illinois University Carbondale, Carbondale.

Daniel, J. Randolph, Jr.

1996 Raw Material Availability and Early Archaic Settlement in the Southeast. In *The Paleoindian and Early Archaic Southeast,* edited by David G. Anderson and Kenneth E. Sassaman, pp. 84–91. University of Alabama Press, Tuscaloosa.

Deevy, Edward S., and Richard F. Flint

1957 Postglacial Hypsithermal Interval. *Science* 125:182–184.

DeJarnette, David L., E. Kurjack, and J. Cambron

1962 Excavations at the Stanfield-Worley Bluff Shelter. *Journal of Alabama Archaeology* 8.

DelCastello, Brian G.

2000 Analysis of Lithic Artifacts. In *Archaeological Investigations at Dixon Springs State Park: The Hills Branch Rock Shelter, Pope County, Illinois,* edited by

Mark J. Wagner and Brian M. Butler, pp. 60–123. Technical Report 00-2, Center for Archaeological Investigations, Southern Illinois University Carbondale, Carbondale.

DelCastello, Brian G., and Brian M. Butler

1999 Lithic Analysis. In *Archaeological Investigations at the Rose Hotel (11-Hn-116), Harding County, Illinois,* by Mark J. Wagner and Brian M. Butler, pp. 167–243. Technical Report 99-3, Center for Archaeological Investigations, Southern Illinois University Carbondale, Carbondale.

2000 *Archaeological Investigations at 11Js321, Cache River Scenic Natural Area Visitor Center, Johnson County, Illinois.* Technical Report 00-1, Center for Archaeological Investigations, Southern Illinois University Carbondale, Carbondale.

Delcourt, Paul A., and Hazel R. Delcourt

1981 Vegetation Maps for Eastern North America: 40,000 Years B.P. to the Present. In *Geobotany,* edited by R. Romans, pp. 123–166. Plenum Press, New York.

1987 *Long-Term Forest Dynamics of the Temperate Zone: A Case Study of Late-Quaternary Forests in Eastern North America.* Springer-Verlag, New York.

Denny, Sidney G.

1972 *The Archaeology of the Big Muddy River Basin of Southern Illinois.* Unpublished Ph.D. dissertation, Department of Anthropology, Southern Illinois University Carbondale, Carbondale.

DiBlasi, Philip J.

1981 *A New Assessment of the Archaeological Significance of the Ashworth Site (15Bu236): A Study of the Dynamics of Archaeological Investigation in Cultural Resource Management.* Unpublished Master's thesis, Interdisciplinary Studies, University of Louisville, Kentucky.

1996 Prehistoric Expressions from the Central Kentucky Karst. In *Of Caves and Shell Mounds,* edited by Kenneth C. Carstens and Patty Jo Watson, pp. 40–47. University of Alabama Press, Tuscaloosa.

Dillehay, Tom D.

1997 *Monte Verde, A Late Pleistocene Settlement in Chile: Paleoenvironment and Site Context,* Vol. 1. Smithsonian Institution Press, Washington, D.C.

2000 *The Settlement of the Americas: A New Prehistory.* Basic Books, New York.

Dorwin, John T.

1966 Fluted Points and Late-Pleistocene Geochronology in Indiana. *Prehistoric Research Series* 4:1141–1188. Indiana Historical Society, Indianapolis.

Driskell, Boyce N.

1979 The Rosenberger Site (15JF18). In *Excavations at Four Archaic Sites in the Lower Ohio Valley, Jefferson County, Kentucky,* Vol. 2, edited by Michael B. Collins, pp. 697–803. Occasional Papers in Anthropology No. 1, Department of Anthropology, University of Kentucky, Lexington.

Driskell, Boyce N., Cindy E. Jobe, Christopher A. Turnbow, and Mary Dunn

1984 *The Archaeology of Taylorsville Lake: Archaeological Data Recovery and Syn-*

thesis. Archaeological Report No. 85, Department of Anthropology, University of Kentucky, Lexington.

Duffield, Lathel F.

1966 The Robert Dudgeon Site: A Stratified Archaic Site in the Green River Reservoir, South Central Kentucky. Manuscript on file, Webb Museum of Anthropology, University of Kentucky, Lexington.

Dunbar, James S., and S. David Webb

1996 Bone and Ivory Tools from Submerged Paleoindian Sites in Florida. In *The Paleoindian and Early Archaic Southeast,* edited by David G. Anderson and Kenneth G. Sassaman, pp. 331–353. University of Alabama Press, Tuscaloosa.

Durrett, Reuben T.

1893 *The Centenary of Louisville.* Filson Club Publication No. 8. Louisville.

Emerson, Thomas E., and Dale L. McElrath

1983 A Settlement-Subsistence Model of the Terminal Late Archaic Adaptation in the American Bottom, Illinois. In *Archaic Hunters and Gatherers in the American Midwest,* edited by James L. Phillips and James A. Brown, pp. 219–242. Academic Press, Orlando.

Emerson, Thomas, Dale McElrath, and Andrew Fortier (editors)

2009 *Archaic Societies: Diversity and Complexity across the Midcontinent.* State University of New York Press, Albany.

Erickson, David L., Bruce D. Smith, Andrew C. Clarke, Daniel H. Sandweiss, and Noreen Tuross

2005 An Asian Origin for a 10,000-Year Old Domesticated Plant in the Americas. *Proceedings of the National Academy of Sciences of the United States of America,* 102:18315–18320. Electronic document, http://www.pnas.org/cgi/content/abstract/0509279102v1, accessed August 28, 2006. National Academy of Sciences.

Faulkner, Charles H. (editor)

1986 *The Prehistoric Native American Art of Mud Glyph Cave.* University of Tennessee Press, Knoxville.

Faulkner, Charles H., Bill Deane, and H. H. Earnest, Jr.

1984 A Mississippian Period Ritual Cave in Tennessee. *American Antiquity* 49:350–361.

Fenneman, Nevin M.

1938 *Physiography of the Eastern United States.* McGraw-Hill, New York.

Fitting, James E.

1973 Reviewed Work(s): *The Riverton Culture: A Second Millennium Occupation in the Central Wabash Valley. American Antiquity* 38: 368–369.

Ford, James A., and Gordon R. Willey

1941 An Interpretation of the Prehistory of the Eastern United States. *American Anthropologist* 43:325–363.

Fortier, Andrew C.

1983 Settlement and Subsistence at the Go-Kart North Site: A Late Archaic

Titterington Occupation in the American Bottom. In *Archaic Hunters and Gatherers in the American Midwest,* edited by James L. Phillips and James A. Brown, pp. 243–260. Academic Press, New York.

Fowler, Melvin L.

1957 *Ferry Site: Hardin County, Illinois.* Scientific Papers, Vol. VIII, No. 1. Illinois State Museum, Springfield.

1959 *Summary Report of Modoc Rock Shelter: 1952, 1953, 1955, 1956.* Illinois State Museum Report of Investigations 8, Springfield.

Frankie, Wayne T., Russell J. Jacobson, John M. Masters, Nancy L. Rorick, Alicia K. Admiraal, Michael R. Jeffords, Susan M. Post, Michael A. Phillips, and Elizabeth Jones

1997 *Guide to the Geology of the Mississippi Embayment Area, Johnson and Pulaski Counties, Illinois.* Field Trip Guidebook 1997D, Department of Natural Resources, Illinois State Geological Survey, Champaign.

Freeman, Andrea K. L., Edward E. Smith, Jr., and Kenneth B. Tankersley

1996 A Stone's Throw from Kimmswick: Clovis Period Research in Kentucky. In *The Paleoindian and Early Archaic Southeast,* edited by David G. Anderson and Kenneth E. Sassaman, pp. 385–403. University of Alabama Press, Tuscaloosa.

French, Michael W.

1998 *Early Archaic Settlement Mobility, Lithic Resource Use, and Technological Organization in the Lower Ohio River Valley: A Perspective from the Longworth-Gick Site (15Jf243).* Unpublished Master's thesis, Department of Anthropology, University of Kentucky, Lexington.

Frye, John C., A. B. Leonard, H. B. Williams, and H. D. Glass

1972 *Geology and Paleontology of Late Pleistocene Lake Saline, Southeastern Illinois.* Illinois State Geological Survey Circular 471, Champaign.

Funkhouser, William D., and William S. Webb

1928 *Ancient Life in Kentucky.* Geologic Reports, Series 6, Vol. 34. Kentucky Geological Survey, Lexington.

1932 *Archaeological Survey of Kentucky.* Reports in Archaeology and Anthropology No. 2, University of Kentucky, Lexington.

Gardner, William M.

1983 Stop Me If You've Heard This One Before: The Flint Run Paleoindian Complex Revisited. *Archaeology of Eastern North America* 11:49–59.

Gatus, Thomas W.

1979 The Occurrence and Distribution of Chert-Bearing Deposits in the Land Between the Lakes Area of Western Kentucky. *Kentucky Archaeological Association Bulletin* 12:24–45.

2005 A Preliminary Reconnaissance of Some West Central Kentucky Chert Resources. In *Archaeology of the Middle Green River Region, Kentucky,* edited by William H. Marquardt and Patty Jo Watson, pp. 431–451. Monograph 5, Institute of Archaeology and Paleoenvironmental Studies, University of Florida, Gainesville.

Gibson, Jon L.

1974 Poverty Point, the First North American Chiefdom. *Archaeology* 27:96–105.

1996 Poverty Point and Greater Southeastern Prehistory: The Culture That Did Not Fit. In *Archaeology of the Mid-Holocene Southeast,* edited by K. E. Sassaman and D. G. Anderson, pp. 288–305. University Press of Florida, Gainesville.

2001 *The Ancient Mounds of Poverty Point: Place of Rings.* University Press of Florida, Gainesville.

Glore, Michael T.

2005 Vertebrate Faunal Remains from the Carlston Annis Site (15Bt5): An Evaluation of Cultural Stratigraphy. In *Archaeology of the Middle Green River Region, Kentucky,* edited by William H. Marquardt and Patty Jo Watson, pp. 315–337. Monograph 5, Institute of Archaeology and Paleoenvironmental Studies, University Press of Florida, Gainesville.

Goad, Sharon I.

1980 Patterns of Late Archaic Exchange. *Tennessee Anthropologist* 5:1–16.

Golden, Twana Jill

2007 *A Contribution to the Debate over the Origin and Development of Treponemal Disease: A Case Study from Southern Illinois.* Unpublished Master's thesis, Department of Anthropology, Southern Illinois University Carbondale, Carbondale.

Goodyear, Albert C.

1974 *The Brand Site: A Techno-Functional Study of a Dalton Site in Northeast Arkansas.* Arkansas Archeological Survey, Research Series 7, Fayetteville.

1979 *A Hypothesis for the Use of Cryptocrystalline Raw Materials among Paleoindian Groups of North America.* Research Manuscript Series 156, South Carolina Institute of Archaeology and Anthropology, University of South Carolina, Columbia.

1982 The Chronological Position of the Dalton Horizon in the Southeastern United States. *American Antiquity* 47:382–395.

2005 Evidence of Pre-Clovis Sites in the Eastern United States. In *Paleoamerican Origins: Beyond Clovis,* edited by Robson Bonnichsen, Bradley T. Lepper, Dennis Stanford, and Michael R. Waters, pp. 103–112. Center for the Study of the First Americans, Texas A&M University, College Station.

Goodyear, Albert C., John H. House, and Neal W. Ackerly

1979 *Laurens-Anderson: An Archaeological Study on the Inter-Riverine Piedmont.* Anthropological Studies 4. South Carolina Institute of Archaeology and Anthropology, University of South Carolina, Columbia.

Gorski, Linda

2005 Microstratigraphic Analysis at the Carlston Annis Site, Butler County, Kentucky. In *Archaeology of the Middle Green River Region, Kentucky,* edited by William H. Marquardt and Patty Jo Watson, pp. 153–166. Monograph 5, Institute of Archaeology and Paleoenvironmental Studies, University of Florida, Gainesville.

Graham, Russell W., C. Vance Haynes, Donald Lee Johnson, and Marvin Kay
1981 Kimmswick: A Clovis-Mastodon Association in Eastern Missouri. *Science* 213:1115–1117.

Gramly, Richard M.
2002 Olive Branch: A Very Early Archaic Site on the Mississippi River. *Amateur Archaeologist* 8 (1 and 2). Special Publication. American Society for Amateur Archaeology, North Andover, Massachusetts.

Gramly, Richard M., and Robert E. Funk
1991 Olive Branch: A Large Dalton and Pre-Dalton Encampment at Thebes Gap, Alexander County, Illinois. In *The Archaic Period in the Mid-South: Proceedings of the 1989 Mid-South Archaeological Conference,* edited by Charles H. McNutt, pp. 23–33. Archaeological Report No. 24, Mississippi Department of Archives and History, Jackson, and Occasional Papers No. 16, Anthropological Research Center, Memphis State University, Memphis.

Gramly, Richard M., and Carl Yahnig
1991 The Adams Site (15Ch90) and the Little River, Christian County, Kentucky, Clovis Workshop Complex. *Southeastern Archaeology* 10:134–145.

Granger, Joseph E.
1985 *Archaeology at McNeely Lake: A Survey and Planning Study.* Report 60, University of Louisville Archaeology Survey, Louisville.
1988 Late/Terminal Archaic Settlement in the Falls of the Ohio River Region of Kentucky: An Examination of Components, Phases, and Clusters. In *Paleo-Indian and Archaic Research in Kentucky,* edited by C. Hockensmith, D. Pollack, and T. Sanders, pp. 153–204. Kentucky Heritage Council, Frankfort.

Granger, Joseph E., Edgar E. Hardesty, and Anne T. Bader
1992 *Phase III Data Recovery: Archaeology at Habich Site (15Jf550) and Associated Manifestations at Gutherie Beach, Jefferson County, Kentucky: 1. The Excavations.* Report of Investigations 90-2 (1), Archaeology Resources Consultant Services, Louisville.

Gray, Henry H.
1971 *Glacial Lake Deposits in Southern Indiana—Engineering Problems and Land Use.* Geological Survey Report of Progress 30, Indiana Department of Natural Resources, Bloomington.
1979 Summary and Discussion of the Geologic Data. In *Excavations at Four Archaic Sites in the Lower Ohio Valley, Jefferson County, Kentucky,* Vol. II, edited by Michael B. Collins, pp. 883–899. Occasional Papers in Anthropology No. 1, Department of Anthropology, University of Kentucky, Lexington.

Gray, Henry H., John L. Bassett, Patrick J. Munson, and G. S. Fraser
1983 Archaeological Geology of Wyandotte Cave Region, Southeastern Indiana. In *Field Trips in Midwestern Geology,* Vol. 2, edited by R. H. Shaver and J. A. Sunderman, pp. 173–213. Indiana Geological Survey, Bloomington.

Grayson, Donald
1991 Late Pleistocene Mammalian Extinctions in North America: Taxonomy, Chronology, and Explanations. *Journal of World Prehistory* 5:193–231.

Grayson, Donald K., and David J. Meltzer

2002 Clovis Hunting and Large Mammal Extinction: A Critical Review of the Evidence. *Journal of World Prehistory* 16:313–359.

Gremillion, Kristen J.

1996 The Paleoethnobotanical Record for the Mid-Holocene Southeast. In *Archaeology of the Mid-Holocene Southeast,* edited by Kenneth E. Sassaman and David G. Anderson, pp. 99–114. University Press of Florida, Gainesville.

2003 Eastern Woodlands Overview. In *People and Plants in Ancient Eastern North America,* edited by Paul E. Minnis, pp. 17–49. Smithsonian Books, Washington, D.C.

Griffin, James B.

1946 Cultural Change and Continuity in Eastern United States Archaeology. In *Man in Northeastern North America,* edited by Frederick Johnson, pp. 37–95. Papers of R. S. Peabody Foundation for Archeology, Vol. 3, Andover, Massachusetts.

1952a Culture Periods in Eastern United States Archaeology. In *Archaeology of Eastern United States,* edited by James B. Griffin, pp. 352–364. University of Chicago Press, Chicago.

1952b Prehistoric Cultures of the Central Mississippi Valley. In *Archaeology of Eastern United States,* edited by James B. Griffin, pp. 226–238. University of Chicago Press, Chicago.

1967 Eastern North American Archaeology: A Summary. *Science* 156:175–191.

1978 Late Prehistory of the Ohio Valley. In *Northeast,* edited by Bruce G. Trigger, pp. 547–559. Handbook of North American Indians, Vol. 15. William G. Sturtevant, general editor, Smithsonian Institution, Washington, D.C.

Griffin, James B. (editor)

1952 *Archaeology of Eastern United States.* University of Chicago Press, Chicago.

Guendling, Randall L.

1980 Excavations at the Miler Site. In *Archaeological Salvage Excavations at Patoka Lake, Indiana: Prehistoric Occupations of the Upper Patoka River Valley,* edited by Cheryl A. Munson, pp. 154–183. Glenn A. Black Laboratory of Archaeology, Indiana University, Bloomington.

Guernsey, E. Y.

1937 Certain Southern Indiana Sources of Lithic Artifact Material. *Proceedings of the Indiana Academy of Science* 46:47–52.

1939 Relationships among Various Clark County Sites. *Proceedings of the Indiana Academy of Science* 48:27–32.

1942 The Culture Sequence of Ohio Falls Sites. *Proceedings of the Indiana Academy of Science* 51:60–67.

Haag, Crista M.

2004 *An Analysis of Clovis Blade Reduction Sequences from the Adams Site (15Ch90) and the Joe Priddy Site (15Hd583): Implications for Early Paleoindian Mobility and Lithic Procurement in Kentucky.* Unpublished Master's thesis, Department of Anthropology, University of Kentucky, Lexington.

Haag, William G.
1942 Early Horizons in the Southeast. *American Antiquity* 7:209–222.

Hall, Robert D.
1989 *Geology of Indiana.* Kendall/Hunt Publishing, Dubuque.

Hardy, Eddie
2006 Malacology and Invertebrate Paleontology Server. Electronic document, http://www.gastropods.com/5/Shell_1995.html, accessed August 2, 2006. Hardy's Internet Guide to Marine Gastropods.

Hargrave, Eve A.
1994 Human Remains from the Funkhouser and Kottmeyer Sites. In *Archaeological Investigations on the Bankston Fork of the Saline River: The AMAX West Harrisburg Project,* edited by Michael L. Hargrave and Brian M. Butler, pp. 210–309. Technical Report 1994-3, Center for Archaeological Investigations, Southern Illinois University Carbondale, Carbondale.

Hargrave, Michael L.
1994 Regional Prehistory and Previous Research in Southeastern Illinois. In *Archaeological Investigations on the Bankston Fork of the Saline River: The AMAX West Harrisburg Project,* edited by Michael L. Hargrave and Brian M. Butler, pp. 32–39. Technical Report 1994-3, Center for Archaeological Investigations, Southern Illinois University Carbondale, Carbondale.

Hargrave, Michael L., and Brian M. Butler
1994a Stratigraphy, Features, and Radiocarbon Dates. In *Archaeological Investigations on the Bankston Fork of the Saline River: The AMAX West Harrisburg Project,* edited by Michael L. Hargrave and Brian M. Butler, pp. 117–209. Technical Report 1994-3, Center for Archaeological Investigations, Southern Illinois University Carbondale, Carbondale.
1994b Summary and Conclusions. In *Archaeological Investigations on the Bankston Fork of the Saline River: The AMAX West Harrisburg Project,* edited by Michael L. Hargrave and Brian M. Butler, pp. 511–541. Technical Report 1994-3, Center for Archaeological Investigations, Southern Illinois University Carbondale, Carbondale.

Hargrave, Michael L., and Brian M. Butler (editors)
1994 *Archaeological Investigations on the Bankston Fork of the Saline River: The AMAX West Harrisburg Project.* Technical Report 1994-3, Center for Archaeological Investigations, Southern Illinois University Carbondale, Carbondale.

Harris, Stanley E., Jr., C. William Horrell, and Daniel Irwin
1977 *Exploring the Land and Rocks of Southern Illinois: A Geological Guide.* Southern Illinois University Press, Carbondale.

Haskins, V. A., and N. P. Herrmann
1996 Shell Mound Bioarchaeology. In *Of Caves and Shell Mounds,* edited by Kenneth C. Carstens and Patty Jo Watson, pp. 107–118. University of Alabama Press, Tuscaloosa.

Hassen, Harold, and Kenneth B. Farnsworth
1987 *The Bullseye Site, 11-Ge-127: A Floodplain Archaic Mortuary Site in the*

Lower Illinois River Valley. Reports of Investigations No. 42, Illinois State Museum, Springfield.

Haynes, C. Vance, Jr.

2005 Clovis, Pre-Clovis, Climate Change, and Extinction. In *Paleoamerican Origins: Beyond Clovis,* edited by Robson Bonnichsen, Bradley T. Lepper, Dennis Stanford, and Michael R. Waters, pp. 113–132. Center for the Study of the First Americans, Texas A&M University, College Station.

Heinrich, Paul V.

1982 *Geomorphology and Sedimentology of Pleistocene Lake Saline, Southern Illinois.* Unpublished Master's thesis, Department of Geology, University of Illinois, Champaign-Urbana.

Hensley, Christine K.

1992 Green River Archaeology Study. In *Studies in Kentucky Archaeology,* edited by Charles Hockensmith, pp. 11–26. Kentucky Heritage Council, Frankfort.

1994 *The Archaic Settlement System of the Middle Green River Valley, Kentucky.* Unpublished Ph.D. dissertation, Department of Anthropology, Washington University, St. Louis.

1996 Lithic Materials from the Read Shell Mound: A Reanalysis of a Works Progress Administration Collection. In *Of Caves and Shell Mounds,* edited by Kenneth C. Carstens and Patty Jo Watson, pp. 94–106. University of Alabama Press, Tuscaloosa.

Herrmann, Nicholas P.

1990 *The Paleodemography of the Read Shell Midden, 15Bt10.* Unpublished Master's thesis, Department of Anthropology, Washington University, St. Louis.

1996 Archaic Shell Mound Paleodemography: A Case Study from the Read Site (15Bt10). In *Current Archaeological Research in Kentucky,* Vol. 4, edited by Sara L. Sanders, Thomas N. Sanders, and Charles Stout, pp. 69–88. Kentucky Heritage Council, Frankfort.

2002 *Biological Affinities of Archaic Period Populations from West-Central Kentucky and Tennessee.* Unpublished Ph.D. dissertation, Department of Anthropology, University of Tennessee, Knoxville.

2006 Human Remains. In *The Highland Creek Site: Middle to Late Archaic Wetland Utilization in Western Kentucky,* edited by G. Maggard and D. Pollack, pp. 88–97. Research Report No. 5, Kentucky Archaeological Survey, University of Kentucky, Lexington.

Higgins, Michael J. (with contributions by Andrew C. Fortier, Douglas K. Jackson, Kathryn E. Parker, and Mary Simon)

1990 *The Nochta Site: The Early, Middle, and Late Archaic Occupation.* FAI-270 Site Reports, Vol. 21, Illinois Department of Transportation. University of Illinois Press, Urbana.

Hill, Fredrick N.

n.d. Preliminary Faunal Analysis of the KYANG Site (15Jf267). Manuscript on file, University of Louisville Program of Archaeology. Louisville.

Hofman, Jack L.

1984 Radiocarbon Dates from Ervin: A Mid-Holocene Shell Midden on the Duck River in Middle Tennessee. *Newsletter of the Tennessee Anthropological Association* 9(2):2–8.

House, John H., and David L. Ballenger

1976 *An Archaeological Survey of the Interstate 77 Route in the South Carolina Piedmont.* Research Manuscript Series 104, South Carolina Institute of Archaeology and Anthropology, University of South Carolina, Columbia.

House, John H., and Ronald W. Wogaman

1978 *Windy Ridge: A Prehistoric Site in the Inter-Riverine Piedmont in South Carolina.* Anthropological Studies 3, South Carolina Institute of Archaeology and Anthropology, University of South Carolina, Columbia.

Howard, Edgar B.

1935 *Evidence of Early Man in North America.* Museum Journal 24. University Museum, University of Pennsylvania, Philadelphia.

Howell, Nancy

1979 *Demography of The Dobe !Kung.* Academic Press, New York.

Indiana Geological Survey

1998 Our Hoosier State Beneath Us. Electronic document, http://www.indiana.edu/~librcsd/etext/hoosier/gm.html, accessed April, 2005.

Jacobson, George L., Jr., Thompson Webb, III, and Eric C. Grimm

1987 Patterns and Rates of Vegetation Change during the Deglaciation of Eastern North America. In *North America and Adjacent Oceans during the Last Deglaciation,* edited by W. Ruddiman and H. E. Wright, pp. 277–288. Geology of North America, Vol. K-3, Geology Society of America, Boulder, Colorado.

Janzen, Donald E.

1972 Archaeological Investigations in Louisville and Vicinity: A Historical Sketch. *Filson Club History Quarterly* 46:305–321.

1977 An Examination of Late Archaic Development in the Falls of the Ohio River Area. In *For the Director: Research Essays in Honor of James B. Griffin,* edited by Charles E. Cleland, pp. 123–143. Anthropological Paper No. 61, University of Michigan Museum of Anthropology, Ann Arbor.

Jefferies, Richard W.

1982a Archaeological Overview of the Carrier Mills District. In *The Carrier Mills Archaeological Project: Human Adaptation in the Saline Valley, Illinois,* edited by Richard W. Jefferies and Brian M. Butler, pp. 1459–1509. Research Paper No. 33, Center for Archaeological Investigations, Southern Illinois University Carbondale, Carbondale.

1982b The Black Earth Site. In *The Carrier Mills Archaeological Project: Human Adaptation in the Saline Valley, Illinois,* edited by Richard W. Jefferies and Brian M. Butler, pp. 75–451. Research Paper No. 33, Center for Archaeological Investigations, Southern Illinois University Carbondale, Carbondale.

1983 Middle Archaic–Late Archaic Transition in Southern Illinois: An Example from the Carrier Mills Archaeological District. *American Archeology* 3:199–206.

1987 *The Archaeology of Carrier Mills: 10,000 Years in the Saline Valley of Illinois.* Southern Illinois University Press, Carbondale.

1988a The Archaic Period in Kentucky: New Deal Archaeological Investigations. In *New Deal Era Archaeology and Current Research in Kentucky,* edited by David Pollack and Mary Lucas Powell, pp. 14–25. Kentucky Heritage Council, Frankfort.

1988b Archaic Period Research in Kentucky: Past Accomplishments and Future Directions. In *PaleoIndian and Archaic Research in Kentucky,* edited by C. Hockensmith, D. Pollack, and T. Sanders, pp. 85–126. Kentucky Heritage Council, Frankfort.

1990a Archaic Period. In *The Archaeology of Kentucky: Past Accomplishments and Future Directions,* edited by David Pollack, pp. 143–246. Kentucky Heritage Council, Frankfort.

1990b A Technological and Functional Analysis of Middle Archaic Hafted Endscrapers from the Black Earth Site, Saline County, Illinois. *Midcontinental Journal of Archaeology* 15:3–36.

1995a Late Middle Archaic Exchange and Interaction in the North American Midcontinent. In *Native American Interactions: Multiscalar Analyses and Interpretations in the Eastern Woodlands,* edited by Michael S. Nassaney and Kenneth E. Sassaman, pp. 73–99. University of Tennessee Press, Knoxville.

1995b The Status of Archaic Period Research in the Midwestern United States. *Archaeology of Eastern North America* 23:119–144.

1996a The Emergence of Long-Distance Exchange Networks in the Southeastern United States. In *Archaeology of the Mid-Holocene Southeast,* edited by K. E. Sassaman and D. G. Anderson, pp. 222–234. University Press of Florida, Gainesville.

1996b Hunters and Gatherers after the Ice Age. In *Kentucky Archaeology,* edited by R. B. Lewis, pp. 39–77. University Press of Kentucky, Lexington.

1997 Middle Archaic Bone Pins: Evidence of Mid-Holocene Regional-Scale Social Groups in the Southern Midwest. *American Antiquity* 62:464–487.

2004 Regional Scale Interaction Networks and the Emergence of Cultural Complexity along the Northern Margins of the Southeast. In *Signs of Power: The Rise of Cultural Complexity in the Southeast,* edited by Jon L. Gibson and Philip J. Carr, pp. 71–85. University of Alabama Press, Tuscaloosa.

2009 Archaic Cultures of Western Kentucky. In *Archaic Societies: Diversity and Complexity across the Midcontinent,* edited by Thomas Emerson, Dale McElrath, and Andrew Fortier, pp. 635–666. State University of New York Press, Albany, in press.

Jefferies, Richard W., and Brian M. Butler (editors)

1982 *The Carrier Mills Archaeological Project: Human Adaptation in the Saline Valley, Illinois.* Research Paper 33, Center for Archaeological Investigations, Southern Illinois University Carbondale, Carbondale.

Jefferies, Richard W., and James Fenton

2001 *The Archaic Shell Mounds of Kentucky.* Pictures of Record, Inc. Weston, Connecticut.

Jefferies, Richard W., and B. Mark Lynch

1983 Dimensions of Middle Archaic Cultural Adaptation at the Black Earth Site, Saline County, Illinois. In *Archaic Hunters and Gatherers in the American Midwest,* edited by James L. Phillips and James A. Brown, pp. 299–322. Academic Press, Orlando.

Jefferies, Richard W., Victor D. Thompson, and George R. Milner

2005 Archaic Hunter-Gatherer Landscape Use in West-Central Kentucky. *Journal of Field Archaeology* 30:3–23.

2008 Cypress Creek Villages Revisited: Archaic Settlement and Subsistence in the Cypress Creek Watershed. In *Current Archaeological Research in Kentucky,* Vol. 8, edited by Sarah E. Miller, David Pollack, Kenneth Carstens, and Christopher R. Moore, pp. 37–75. Kentucky Heritage Council, Frankfort.

Jillson, Willard R.

1936 *Big Bone Lick.* Big Bone Lick Association Publication No. 1. Standard Printing, Louisville.

Johnson, Jay K.

1994 Prehistoric Exchange in the Southeast. In *Prehistoric Exchange Systems in North America,* edited by Timothy G. Baugh and Jonathon E. Ericson, pp. 99–125. Plenum Press, New York.

Johnson, Jay K., and Samuel O. Brookes

1989 Benton Points, Turkey Tails, and Cache Blades: Middle Archaic Exchange in the Midsouth. *Southeastern Archaeology* 8:134–145.

Justice, Noel D.

1987 *Stone Age Spear and Arrow Points of the Midcontinental and Eastern United States: A Modern Survey and Reference.* Indiana University Press, Bloomington.

Kay, Marvin

1983 Archaic Period Research in the Western Ozark Highland, Missouri. In *Archaic Hunters and Gatherers in the American Midwest,* edited by James L. Phillips and James A. Brown, pp. 41–70. Academic Press, New York.

Keeley, Lawrence H.

1981 Analysis of Edge-Wear on Late Middle Archaic Endscrapers from the Black Earth Site, Saline County, Illinois. Report on file, Center for Archaeological Investigations, Southern Illinois University Carbondale, Carbondale.

Kellar, James H.

1956 *An Archaeological Survey of Spencer County, Indiana.* Indiana Historical Bureau, Indianapolis.

Kelly, Lucretia S.

2000 Faunal Remains from the Hills Branch Rock Shelter. In *Archaeological Investigations at Dixon Springs State Park: The Hills Branch Rock Shelter, Pope County, Illinois,* by Mark J. Wagner and Brian M. Butler, pp. 155–167.

Technical Report 00-2, Center for Archaeological Investigations, Southern Illinois University Carbondale, Carbondale.

Kelly, Robert L.
1991 Sedentism, Sociopolitical Inequality, and Resource Fluctuations. In *Between Bands and States,* edited by S. A. Gregg, pp. 135–158. Occasional Paper No, 9. Center for Archaeological Investigations, Southern Illinois University Carbondale, Carbondale.

Kelly, Robert L., and Lawrence C. Todd
1988 Coming into the Country: Early Paleoindian Hunting and Mobility. *American Antiquity* 53:231–244.

Kennedy, Mary C., Christine Hensley-Martin, and Patty Jo Watson
1983 Cave Research Foundation Archaeological Project—1983. In *Cave Research Foundation 1983 Annual Report,* pp. 22–23. Cave Research Foundation, Washington, D.C.

Kentucky Geological Survey
2002 The Mississippian Plateau or Pennyrile Region. Electronic document, http://www.uky.edu/KGS/geoky/regionpennyrile.htm, accessed March, 2004.

King, James E., and William H. Allen, Jr.
1977 A Holocene Vegetation Record from the Mississippi River Valley, Southeastern Missouri. *Quaternary Research* 8:307–323.

Klepinger, Linda, and Dale R. Henning
1976 The Hatten Mound: A Two-Component Burial Site in Northeast Missouri. *Missouri Archaeologist* 35:92–170.

Kneberg, Madeline
1956 Some Important Projectile Point Types Found in the Tennessee Area. *Tennessee Archaeologist* 12:17–28.

Koldehoff, Brad
1992 Lithic Analysis. In *The Little Muddy Rock Shelter: A Deeply Stratified Prehistoric Site in the Southern Till Plains of Illinois,* by Charles R. Moffat, Brad Koldehoff, William M. Cremin, Terrance J. Martin, Mary Carol Masulis, and Mary R. McCorvie, pp. 279–374. Cultural Resource Management Report 186, American Resources Group, Ltd., Carbondale.

Kurten, Bjorn, and Elaine Anderson
1980 *Pleistocene Mammals of North America.* Columbia University Press, New York.

Kwas, Mary L.
1981 Bannerstones as Chronological Markers in the Southeastern United States. *Tennessee Anthropologist* 6:144–171.

Lafferty, Robert H., III
1994 Prehistoric Exchange in the Lower Mississippi Valley. In *Prehistoric Exchange Systems in North America,* edited by Timothy G. Baugh and Jonathon E. Ericson, pp. 177–213. Plenum Press, New York.

Lafferty, Robert H., III, and Kathleen M. Hess (editors)

1996 *Archaeological Investigations in the New Madrid Floodway.* Report 95-7, Mid-Continental Research Associates, Inc., Lowell, Arkansas.

Lafferty, Robert H., III, and James E. Price

1996 Southeast Missouri. In *Prehistory of the Central Mississippi Valley,* edited by Charles H. McNutt, pp. 1–45. University of Alabama Press, Tuscaloosa.

Lannie, Donna Dean

1979 Ethnobotanical Analysis. In *Excavations at Four Archaic Sites in the Lower Ohio Valley, Jefferson County, Kentucky,* Vol. 2, edited by Michael B. Collins, pp. 978–1006. Occasional Papers in Anthropology No. 1, Department of Anthropology, University of Kentucky, Lexington.

Levy, Richard

n.d. Untitled Report on the Millersburg Site, Warrick County, Indiana. Manuscript on file with the Indiana State University Anthropology Laboratory, Terre Haute.

Lewis, R. Barry

1983 Archaic Adaptations to the Illinois Prairie: The Salt Creek Region. In *Archaic Hunters and Gatherers in the American Midwest,* edited by James L. Phillips and James A. Brown, pp. 99–116. Academic Press, New York.

Lewis, R. Barry, and Charles Stout (editors)

1998 *Mississippian Towns and Sacred Spaces: Searching for an Architectural Grammar.* University of Alabama Press, Tuscaloosa.

Lewis, Thomas M. N., and Madeline K. Kneberg

1947 *The Archaic Horizon in Western Tennessee.* Tennessee Anthropology Papers No. 2, Knoxville.

1959 The Archaic Culture in the Middle South. *American Antiquity* 25(2):161–183.

Lewis, Thomas M. N., and Madeline K. Lewis

1961 *Eva: An Archaic Site.* University of Tennessee Press, Knoxville.

Logan, Willard D.

1952 *Graham Cave: An Archaic Site in Montgomery County, Missouri.* Missouri Archaeological Society Memoir 2, Columbia.

Lopinot, Neal H.

1982a Plant Macroremains and Paleoethnobotanical Implication. In *The Carrier Mills Archaeological Project: Human Adaptation in the Saline Valley, Illinois,* Vol. 2, edited by Richard W. Jefferies and Brian M. Butler, pp. 671–860. Research Paper No. 33, Center for Archaeological Investigations, Southern Illinois University Carbondale, Carbondale.

1982b Reconstruction of Early Nineteenth-Century Vegetation Patterns. In *The Carrier Mills Archaeological Project: Human Adaptation in the Saline Valley, Illinois,* Vol. 2, edited by Richard W. Jefferies and Brian M. Butler, pp. 58–71. Research Paper No. 33, Center for Archaeological Investigations, Southern Illinois University Carbondale, Carbondale.

1983 Analysis of Flotation Sample Material from the Late Archaic Horizon. In *The 1982 Excavations at the Cahokia Interpretive Center Tract, St. Clair County, Illinois,* edited by Michael S. Nassaney, Neal H. Lopinot, Brian M. Butler, and Richard W. Jefferies, pp. 77–108. Research Paper 37, Center for Archaeological Investigations, Southern Illinois University Carbondale, Carbondale.

1984 *Archaeobotanical Formation Processes and Late Middle Archaic Human-Plant Interrelationships in the Midcontinental U.S.A.* Unpublished Ph.D. dissertation, Department of Anthropology, Southern Illinois University Carbondale, Carbondale.

1988 Hansen Site (15Gp14) Archaeobotany. In *Excavations at the Hansen Site (15Gp14) in Northeastern Kentucky,* by Steven Ahler, pp. 571–623. Archaeological Report No. 173, Program for Cultural Resource Assessment, University of Kentucky, Lexington.

2000 Late Archaic and Woodland Archaeobotany in Southeast Missouri and Northeast Arkansas: Status, Trends, and Problems. Paper presented at the 57th Annual Meeting of the Southeast Archaeological Conference, Macon, Georgia.

Lopinot, Neal H., and George E. Avery, and Carol A. Morrow

1982 Biophysical Environment of the Carrier Mills Archaeological District. In *The Carrier Mills Archaeological Project: Human Adaptation in the Saline Valley, Illinois,* Vol. 1, edited by Richard W. Jefferies and Brian M. Butler, pp. 37–57. Research Paper No. 33, Center for Archaeological Investigations, Southern Illinois University Carbondale, Carbondale.

Luchterhand, Kubet

1970 *Early Archaic Projectile Points and Hunting Patterns in the Lower Illinois Valley.* Report of Investigation 12, Illinois State Museum, Springfield.

Lutz, David

2000 *The Archaic Bannerstone: Its Chronological History and Purpose from 6000 BC to 1000 BC.* Published by the author, Newburg, Indiana.

Lynch, B. Mark

1982 Mortuary Behavior in the Carrier Mills Archaeological District. In *The Carrier Mills Archaeological Project: Human Adaptation in the Saline Valley, Illinois,* edited by Richard W. Jefferies and Brian M. Butler, pp. 1115–1231. Research Paper No. 33, Center for Archaeological Investigations, Southern Illinois University Carbondale, Carbondale.

MacNeish, Richard S.

1948 The Pre-Pottery Faulkner Site of Southern Illinois. *American Antiquity* 13:232–243.

Magennis, Ann

1977 *Middle and Late Archaic Mortuary Patterning: An Example from the Western Tennessee Valley.* Master's thesis, Department of Anthropology, University of Tennessee, Knoxville.

Maggard, Greg, and David Pollack

2006 *The Highland Creek Site: Middle to Late Archaic Wetland Utilization in Western Kentucky.* Research Report No. 5, Kentucky Archaeological Survey, University of Kentucky, Lexington.

Marquardt, William H.

1971 The Archaic of the Western Coalfields Area of Kentucky. Manuscript on file, Office of State Archaeology, University of Kentucky, Lexington.

1985 Complexity and Scale in the Study of Fisher-Gatherer-Hunters: An Example from the Eastern United States. In *Prehistoric Hunter-Gatherers: The Emergence of Cultural Complexity,* edited by T. Douglas Price and James A. Brown, pp. 59–98. Academic Press, New York.

2005 Lithic Artifacts in the Middle Green River Area. In *Archaeology of the Middle Green River Region, Kentucky,* edited by William H. Marquardt and Patty Jo Watson, pp. 351–429. Monograph 5, Institute of Archaeology and Paleoenvironmental Studies, University Press of Florida, Gainesville.

Marquardt, William H., and Patty Jo Watson

1983 Excavation and Recovery of Biological Remains from Two Archaic Shell Middens in Western Kentucky. *Southeastern Archaeological Conference Bulletin* 20 and 21:112–129.

1983 The Shell Mound Archaic of Western Kentucky. In *Archaic Hunters and Gatherers in the American Midwest,* edited by J. L. Phillips and J. A. Brown, pp. 323–339. Academic Press, Orlando.

2005a The Green River Shell Mound Archaic: Conclusions. In *Archaeology of the Middle Green River Region, Kentucky,* edited by William H. Marquardt and Patty Jo Watson, pp. 629–647. Monograph 5, Institute of Archaeology and Paleoenvironmental Studies, University Press of Florida, Gainesville.

2005b Editor's Preface. In *Archaeology of the Middle Green River Region, Kentucky,* edited by William H. Marquardt and Patty Jo Watson, pp. xix–xx. Monograph 5, Institute of Archaeology and Paleoenvironmental Studies, University Press of Florida, Gainesville.

2005c SMAP Investigations at the Carlston Annis Site, 15Bt5. In *Archaeology of the Middle Green River Region, Kentucky,* edited by William H. Marquardt and Patty Jo Watson, pp. 87–120. Monograph 5, Institute of Archaeology and Paleoenvironmental Studies, University Press of Florida, Gainesville.

2005d The Shell Mound Archaeological Project: The Chronology of Fieldwork. In *Archaeology of the Middle Green River Region, Kentucky,* edited by William H. Marquardt and Patty Jo Watson, pp. 13–17. Monograph 5, Institute of Archaeology and Paleoenvironmental Studies, University Press of Florida, Gainesville.

Marquardt, William H., and Patty Jo Watson (editors)

2005 *Archaeology of the Middle Green River Region, Kentucky.* Monograph 5, Institute of Archaeology and Paleoenvironmental Studies, University Press of Florida, Gainesville.

Martin, Paul S.

1967 Pleistocene Overkill. In *Pleistocene Extinctions: The Search for a Cause,* edited by P. S. Martin and H. E. Wright, pp. 75–120. Yale University Press, New Haven.

1984 Prehistoric Overkill: A Global Model. In *Quaternary Extinctions: A Prehistoric Revolution,* edited by P. S. Martin and R. G. Klein, pp. 354–404. University of Arizona Press, Tucson.

Martin, Paul S., and Richard G. Klein (editors)

1984 *Quaternary Extinctions: A Prehistoric Revolution.* University of Arizona Press, Tucson.

Matthews, James J.

1958 Shell Mound Incised Bone and Stone Artifacts. *Ohio Archaeologist* 8:42–43.

Maxwell, Moreau S.

1952 The Archeology of the Lower Ohio Valley. In *Archaeology of Eastern United States,* edited by James B. Griffin, pp. 176–189. University of Chicago Press, Chicago.

May, Ernest E.

1982 Analysis of Carrier Mills Projectile Points. In *The Carrier Mills Archaeological Project: Human Adaptation in the Saline Valley, Illinois,* edited by Richard W. Jefferies and Brian M. Butler, pp. 1347–1379. Research Paper No. 33, Center for Archaeological Investigations, Southern Illinois University Carbondale, Carbondale.

1984 Prehistoric Chert Exploitation in the Shawnee Hills. In *Cultural Frontiers in the Upper Cache Valley, Illinois,* by Veletta Canouts, Ernest E. May, Neal H. Lopinot, and Jon D. Muller, pp. 68–90. Research Paper No. 16, Center for Archaeological Investigations, Southern Illinois University Carbondale, Carbondale.

May, J. Alan

2005 Ethnographic and Ethnohistoric Suggestions Relevant to the Middle Green River Shell Middens. In *Archaeology of the Middle Green River Region, Kentucky,* edited by William H. Marquardt and Patty Jo Watson, pp. 71–85. Monograph 5, Institute of Archaeology and Paleoenvironmental Studies, University Press of Florida, Gainesville.

Mayes, Leigh Ann

1997 *The Bluegrass Site (12W162): Bioarchaeological Analysis of a Middle-Late Archaic Mortuary Site in Southwestern Indiana.* Unpublished Master's thesis, Department of Anthropology and Sociology, University of Southern Mississippi, Hattiesburg.

McAvoy, Joseph M., and Lynn D. McAvoy

1997 *Archaeological Investigations of Site 44SX202, Cactus Hill, Sussex County, Virginia.* Virginia Department of Historic Resources, Research Report Series No. 8, Richmond.

McCulloh, James H., Jr.
1829 *Researches Philosophical and Antiquarian concerning the Aboriginal History of America.* Fielding Lucas, Baltimore.

McDowell, Robert C.
1986 *The Geology of Kentucky: A Text to Accompany the Geologic Map of Kentucky.* U.S. Geological Survey Professional Paper, U.S. Geological Survey, Department of Interior, Washington, D.C.

McElrath, Dale L., and Fred A. Finney
1987 *The George Reeves Site (11-S-650).* FAI-270 Site Reports, Vol. 15. Illinois Department of Transportation and the University of Illinois Press, Urbana and Chicago.

McElrath, Dale L., and Andrew C. Fortier
1983 *The Missouri Pacific #2 Site (11-S-46).* FAI-270 Site Reports, Vol. 3. Illinois Department of Transportation and the University of Illinois Press, Urbana and Chicago.

McFarland, Arthur C.
1943 *Geology of Kentucky.* University of Kentucky, Lexington.

McGrain, Preston
1983 *The Geologic Story of Kentucky.* Special Publication 8, Series XI, Kentucky Geological Survey, Lexington.

McMurtrie, Henry
1819 *Sketches of Louisville and Its Environs.* Printed by S. Penn, Louisville.

Meeks, Scott C.
1999 The "Function" of Stone Tools in Prehistoric Exchange Systems: A Look at Benton Interaction in the Mid-South. In *Raw Materials and Exchange in the Mid-South: Proceedings of the 16th Annual Mid-South Archaeological Conference,* edited by Evan Peacock and Samuel O. Brooks, pp. 20–43. Mississippi Department of Archives and History Archaeological Report No. 29. Jackson.

Meindl, Richard, Robert P. Mensforth, and Heather P. York
2001 Mortality, Fertility, and Growth in the Kentucky Late Archaic: The Paleodemography of the Ward Site. In *Archaic Transitions in Ohio & Kentucky Prehistory,* edited by Olaf H. Prufer, Sara E. Pedde, and Richard S. Meindl, pp. 87–109. Kent State University Press, Kent, Ohio.

Meloy, Harold, and Patty Jo Watson
1969 Human Remains: "Little Alice" of Salts Cave and Other Mummies. In *The Prehistory of Salts Cave,* by Patty Jo Watson, pp. 65–69. Reports of Investigations No. 16, Illinois State Museum, Springfield.

Meltzer, David J.
1988 Late Pleistocene Human Adaptation in Eastern North America. *Journal of World Prehistory* 2:1–52.

Meltzer, David J., and Bruce D. Smith
1986 Paleoindian and Early Archaic Subsistence Strategies in Eastern North

America. In *Foraging, Collecting, and Harvesting: Archaic Period Subsistence and Settlement in the Eastern Woodlands,* edited by Sarah W. Neusius, pp. 3–31. Occasional Paper No. 6, Center for Archaeological Investigations, Southern Illinois University Carbondale, Carbondale.

Mensforth, Richard P.

2001 Warfare and Trophy Taking in the Archaic Period. In *Archaic Transitions in Ohio & Kentucky Prehistory,* edited by Olaf H. Prufer, Sara E. Pedde, and Richard S. Meindl, pp. 110–138. Kent State University Press, Kent, Ohio.

2005 Paleodemography of the Skeletal Population from Carlston Annis, 15Bt5. In *Archaeology of the Middle Green River Region, Kentucky,* edited by William H. Marquardt and Patty Jo Watson, pp. 453–487. Monograph 5, Institute of Archaeology and Paleoenvironmental Studies, University Press of Florida, Gainesville.

Merck & Co., Inc.

2006 Bejel, Yaws, and Pinta: Bacterial Infections: Merck Manual Home Edition. Electronic document, http://www.merck.com/mmhe/sec17/ch190/ch190d.html, accessed August 25, 2006.

Miller, Rex K.

1941 *McCain Site, Dubois County, Indiana.* Indiana Historical Society Prehistoric Research Series 2(1), Indianapolis, IN.

Miller, Carl F.

1956 Life 8,000 Years Ago Uncovered in an Alabama Cave. *National Geographic* 110:542–558.

1958 Russell Cave: New Light on Stone Age Life. *National Geographic* 113:427–432.

Milner, George R.

1999 Warfare in Prehistoric and Early Historic Eastern North America. *Journal of Archaeological Research* 7:105–151.

2004 *The Moundbuilders: Ancient Peoples of Eastern North America.* Thames and Hudson, London.

Milner, George R., and Richard W. Jefferies

1998 The Read Archaic Shell Midden in Kentucky. *Southeastern Archaeology* 17:119–132.

Milner, George R., and Virginia G. Smith

1986 *New Deal Archaeology in Kentucky: Excavations, Collections, and Research.* Occasional Papers in Anthropology No. 5, Program for Cultural Resource Assessment, University of Kentucky, Lexington.

Mocas, Stephen T.

n.d. The George Branch Site: A Preliminary Report. Manuscript on file, Glenn A. Black Laboratory of Archaeology, Indiana University, Bloomington.

1976 *Excavations at Arrowhead Farm (15JF237).* University of Louisville Archaeological Survey Report, Louisville.

1977 *Excavations at the Lawrence Site, 15Tr33, Trigg County, Kentucky.* University of Louisville Archaeological Survey Report, Louisville.

1985 An Instance of Middle Archaic Mortuary Activity in Western Kentucky. *Tennessee Anthropologist* 10:76–91.

1995 Features. In *Archaeological Investigations at the Paddy's West Substation, Floyd County, Indiana,* by Edward E. Smith and Stephen T. Mocas, pp. 139–238. Reports of Investigations 95-4, Glenn A. Black Laboratory of Archaeology, Indiana University, Bloomington.

Moffat, Charles R.

1992 Summary and Conclusions. In *The Little Muddy Rock Shelter: A Deeply Stratified Prehistoric Site in the Southern Till Plains of Illinois,* by Charles R. Moffat, Brad Koldehoff, William M. Cremin, Terrance J. Martin, Mary Carol Masulis, and Mary R. McCorvie, pp. 489–511. Cultural Resource Management Report 186, American Resources Group, Ltd., Carbondale, Illinois.

Moffat, Charles R., Brad Koldehoff, William M. Cremin, Terrance J. Martin, Mary Carol Masulis, and Mary R. McCorvie

1992 *The Little Muddy Rock Shelter: A Deeply Stratified Prehistoric Site in the Southern Till Plains of Illinois.* Cultural Resource Management Report 186, American Resources Group, Ltd., Carbondale, Illinois.

Mooney, Michael M. (editor)

1975 *George Catlin: Letters and Notes on the North American Indians.* Clarkson N. Potter, New York.

Moore, Christopher R.

2006 Fishing Technology along the Green River: A View from the Middens. Paper presented at the 23rd Annual Kentucky Heritage Council Archaeological Conference, Bowling Green.

2007 A Preliminary Analysis of the Bone and Antler Tool Assemblage from the Chiggerville Site, 15Oh1. Paper presented at the 53rd Annual Meeting of the Midwest Archaeological Conference, South Bend, Indiana.

2008 Raw Material Utilization in Carroll County, Indiana: A Small-Scale Analysis of Diachronic Patterns in the Usage of Attica, Kenneth, and Wyandotte Cherts. *Midcontinental Journal of Archaeology,* 33:73–106.

Moore, Clarence B.

1916 Some Aboriginal Sites on Green River, Kentucky; Certain Aboriginal Sites on Lower Ohio River; Additional Investigations on Mississippi River. *Journal of the Academy of Natural Sciences of Philadelphia* 16:429–511.

Morey, Darcy F., and George M. Crothers

1998 Clearing Up Clouded Waters: Paleoenvironmental Analysis of Freshwater Mussel Assemblages from Green River Shell Middens, Western Kentucky. *Journal of Archaeological Science* 25:907–926.

Morey, Darcy F., George Crothers, Julie Stein, James Fenton, and Nicholas Herrmann

2002 The Fluvial and Geomorphic Context of Indian Knoll, an Archaic Shell Midden in West Central Kentucky. *Geoarchaeology* 17:521–553.

Morgan, Richard G.

1952 Outline of Cultures in the Ohio Region. In *Archaeology of Eastern United*

States, edited by James B. Griffin, pp. 83–98. University of Chicago Press, Chicago.

Morrow, Carol A.

1982 Analysis of Area A Middle Archaic Flaked Stone Technology. In *The Carrier Mills Archaeological Project: Human Adaptation in the Saline Valley, Illinois,* edited by Richard W. Jefferies and Brian M. Butler, pp. 1289–1346. Center for Archaeological Investigations Research Paper No. 33, Southern Illinois University Carbondale, Carbondale.

Morrow, Carol A., and Richard W. Jefferies

1989 Trade or Embedded Procurement? A Test Case from Southern Illinois. In *Time, Energy, and Stone Tools,* edited by Robin Torrence, pp. 27–33. Cambridge University Press, Cambridge.

Morrow, Carol A., J. Michael Elam, and Michael D. Glascock

1992 The Use of Blue-Gray Chert in Midwestern Prehistory. *Midcontinental Journal of Archaeology* 17:166–197.

Morse, Dan F.

1962 Report of 1962 Excavations in the Stewart County, Tennessee Portion of the Lake Barkley Reservoir. Manuscript on file, F. H. McClung Museum, University of Tennessee, Knoxville.

1971 Recent Indications of Dalton Settlement Patterns in Northeast Arkansas. *Southeastern Archaeological Conference Bulletin* 13:5–10.

1973 Dalton Culture in Northeast Arkansas. *Florida Anthropologist* 26:28–38.

1975 Reply to Schiffer. In *The Cache River Archaeological Project: An Experiment in Contract Archaeology,* assembled by Michael B. Schiffer and John H. House, pp. 113–119. Research Series 8, Arkansas Archaeological Survey, Fayetteville.

1997a *Sloan: A Paleoindian Dalton Cemetery in Arkansas.* Smithsonian Institution Press, Washington, D.C.

1997b Description of the Artifacts. In *Sloan: A Paleoindian Dalton Cemetery in Arkansas,* by Dan F. Morse, pp. 14–52. Smithsonian Institution Press, Washington, D.C.

Morse, Dan F., and Albert C. Goodyear

1973 The Significance of the Dalton Adz in Northeast Arkansas. *Plains Anthropologist* 18:316–322.

Morse, Dan F., and Phyllis A. Morse

1983 *Archaeology of the Central Mississippi Valley.* Academic Press, New York.

Mosimann, J. E., and Paul S. Martin

1975 Simulating Overkill by Paleoindians. *American Scientist* 63:304–313.

Muller, Jon

1986 *Archaeology of the Lower Ohio River Valley.* Academic Press, Orlando.

Munson, Cheryl A.

1975 *Prehistoric Cultural Resources in the Clark Maritime Center.* Glenn A. Black Laboratory of Archaeology, Indiana University, Bloomington.

1980b Comments on Subsistence and Settlement in the Upper Patoka Lake Valley.

In *Archaeological Salvage Excavations at Patoka Lake, Indiana: Prehistoric Occupations of the Upper Patoka River,* edited by Cheryl A. Munson, pp. 645–683. Research Reports No. 6, Glenn A. Black Laboratory of Archaeology, Indiana University, Bloomington.

1980c The 1977 Site Surface Data Recovery. In *Archaeological Salvage Excavations at Patoka Lake, Indiana: Prehistoric Occupations of the Upper Patoka River,* edited by Cheryl A. Munson, pp. 596–607. Research Reports No. 6, Glenn A. Black Laboratory of Archaeology, Indiana University, Bloomington.

1980d Regional Prehistoric Population Histories for Indiana and Their Implications for Understanding Man-Land Relationships. In *Archaeological Salvage Excavations at Patoka Lake, Indiana: Prehistoric Occupations of the Upper Patoka River Valley,* edited by Cheryl A. Munson, pp. 741–767. Research Reports No. 6, Glenn A. Black Laboratory of Archaeology, Indiana University, Bloomington.

Munson, Cheryl A. (editor)

1980a *Archaeological Salvage Excavations at Patoka Lake, Indiana: Prehistoric Occupations of the Upper Patoka River Valley.* Research Reports, No. 6, Glenn Black Laboratory of Archaeology, Indiana University, Bloomington.

Munson, Cheryl A., and Thomas G. Cook

1980a Chronology of the Excavated Sites: Radiocarbon Dates, Cultural Components, and Stratigraphy. In *Archaeological Salvage Excavations at Patoka Lake, Indiana: Prehistoric Occupations of the Upper Patoka River Valley,* edited by Cheryl A. Munson, pp. 468–503. Research Reports No. 6, Glenn A. Black Laboratory of Archaeology, Indiana University, Bloomington.

1980b Prehistoric Landscape Utilization at Patoka Lake. In *Archaeological Salvage Excavations at Patoka Lake, Indiana: Prehistoric Occupations of the Upper Patoka River Valley,* edited by Cheryl A. Munson, pp. 698–720. Research Reports No. 6, Glenn A. Black Laboratory of Archaeology, Indiana University, Bloomington.

1980c The Late Archaic French Lick Phase: A Dimensional Description. In *Archaeological Salvage Excavations at Patoka Lake, Indiana: Prehistoric Occupations of the Upper Patoka River Valley,* edited by Cheryl A. Munson, pp. 721–740. Research Reports No. 6, Glenn A. Black Laboratory of Archaeology, Indiana University, Bloomington.

Munson, Patrick J., and Alan D. Harn

1966 Surface Collections from Three Sites in the Central Illinois River Valley. *Wisconsin Archaeologist* 47:150–168.

Munson, Patrick J., and Cheryl A. Munson

1984 Cherts and Archaic Chert Utilization in South-Central Indiana. In *Prehistoric Chert Exploitation Studies from the Midcontinent,* edited by Brian M. Butler and Ernest E. May, pp. 149–166. Occasional Paper No. 2, Center for Archaeological Investigations, Southern Illinois University Carbondale, Carbondale.

Munson, Patrick J., John L. Bassett, and Cheryl A. Munson

1983 Cherts and Cave Minerals of the Mississippian Carbonate Rocks of South-Central Indiana and Their Uses by Prehistoric Indians. In *Archaeological Geology of the Wyandotte Cave Region, South-Central Indiana (Field Trip 14),* by Henry H. Gray, John L. Bassett, Cheryl A. Munson, Patrick J. Munson, and Gordon B. Fraser, pp. 189–198. Indiana Geological Survey, Bloomington.

Nance, Jack D.

1975 An Archaeological Survey of the Land Between the Lakes National Recreation Area. *Tennessee Archaeologist* 31:66–77.

1977 Aspects of Late Archaic Culture in the Lower Tennessee/Cumberland River Valleys. *Tennessee Archaeologist* 33:1–15.

1980 Lithic Utilization Patterns in the Lower Tennessee-Cumberland Region: A First Approximation. Paper presented at the 37th Annual Meeting of the Southeastern Archaeological Conference, New Orleans.

1984 Lithic Exploitation Studies in the Lower Tennessee-Cumberland Valleys, Western Kentucky. In *Prehistoric Chert Exploitation: Studies from the Midcontinent,* edited by B. M. Butler and E. E. May, pp. 101–127. Occasional Papers 2, Center for Archaeological Investigations, Southern Illinois University Carbondale, Carbondale.

1986a The Morrisroe Site: Projectile Point Types and Radiocarbon Dates from the Lower Tennessee River Valley. *Midcontinental Journal of Archaeology* 11:11–50.

1986b Archaeological Investigations at the Whalen Site, Lyon County, Kentucky: A Progress Report. Manuscript on file, Kentucky Heritage Council, Frankfort.

1987 The Archaic Sequence in the Lower Tennessee-Cumberland-Ohio Region. *Southeastern Archaeology* 6:129–140.

1988 The Archaic Period in the Lower Tennessee-Cumberland-Ohio Region. In *Paleoindian and Archaic Research in Kentucky,* edited by C. D. Hockensmith, D. Pollack, and T. N. Sanders, pp. 127–152. Kentucky Heritage Council, Frankfort.

2001 *Archaeological Investigations at the Whalen Site (15Ly48), Lyon County, Kentucky.* Department of Anthropology, Simon Frasier University, Burnaby, British Columbia.

Nelson, Nels C.

1917 *Contributions to the Archaeology of Mammoth Cave and Vicinity, Kentucky.* Anthropological Papers of the American Museum of Natural History 22:1–73. New York.

Neusius, Sarah W.

1986 Generalized and Specialized Resource Utilization during the Archaic Period: Implications of the Koster Site Faunal Record. In *Foraging, Collecting, and Harvesting: Archaic Period Subsistence and Settlement in the Eastern Woodlands,* edited by Sarah W. Neusius, pp. 117–143. Occasional Paper

No. 6, Center for Archaeological Investigations, Southern Illinois University Carbondale, Carbondale.

Newell, H. Perry, and Alex D. Krieger

1949 *The George C. Davis Site, Cherokee County, Texas.* Memoirs of the Society for American Archaeology No. 5.

O'Brien, Michael J., and Robert E. Warren

1983 An Archaic Projectile Point Sequence from the Southern Prairie Peninsula: The Pigeon Roost Creek Site. In *Archaic Hunters and Gatherers in the American Midwest,* edited by James L. Phillips and James A. Brown, pp. 71–98. Academic Press, New York.

O'Malley, Nancy, Boyce Driskell, Julie Riesenweber, and Richard Levy

1980 *Stage I Archaeological Investigations at Ft. Knox, Kentucky.* Technical Report 16, Program for Cultural Resource Assessment, University of Kentucky, Lexington.

Oetelaar, Gerald A.

1982 An Analysis of Microremains in an Area A Column Sample. In *The Carrier Mills Archaeological Project: Human Adaptation in the Saline Valley, Illinois,* edited by Richard W. Jefferies and Brian M. Butler, pp. 987–1007. Research Paper No. 33, Center for Archaeological Investigations, Southern Illinois University Carbondale, Carbondale.

Parker, Kathryn E.

1999 Prehistoric Botanical Remains. In *Archaeological Investigations at the Rose Hotel (11-Hn-116), Hardin County, Illinois,* by Mark J. Wagner and Brian M. Butler, pp. 395–408. Technical Report 99-3, Center for Archaeological Investigations, Southern Illinois University Carbondale, Carbondale.

2000 Plant Remains from the Hills Branch Rock Shelter. In *Archaeological Investigations at Dixon Springs State Park: The Hills Branch Rock Shelter, Pope County, Illinois,* by Mark J. Wagner and Brian M. Butler, pp. 145–154. Technical Report 00-2, Center for Archaeological Investigations, Southern Illinois University Carbondale, Carbondale.

Patch, Diana C.

2005 The Freshwater Molluscan Fauna: Identification and Interpretation for Archaeological Research. In *Archaeology of the Middle Green River Region, Kentucky,* edited by William H. Marquardt and Patty Jo Watson, pp. 257–278. Monograph 5, Institute of Archaeology and Paleoenvironmental Studies, University Press of Florida, Gainesville.

Phillips, James L., and James A. Brown (editors)

1983 *Archaic Hunters and Gatherers in the American Midwest.* Academic Press, New York.

Phillips, James L., and Bruce G. Gladfelter

1983 The Labras Lake Site and the Paleogeographic Setting of the Late Archaic in the American Bottom. In *Archaic Hunters and Gatherers in the American Midwest,* edited by James L. Phillips and James A. Brown, pp. 197–218. Academic Press, New York.

Polhemus, Richard R.

2002 Introduction: Clarence B. Moore's Research on the Tennessee, Green, and Lower Ohio Rivers, 1913–1916. In *The Tennessee, Green, and Lower Ohio Rivers Expeditions,* edited by Richard R. Polhemus, pp. 1–45. University of Alabama Press, Tuscaloosa.

Pollack, David

1990 Introduction. In *The Archaeology of Kentucky: Past Accomplishments and Future Directions,* Vol. 1, edited by David Pollack, pp. 1–24. Kentucky Heritage Council, Frankfort.

Powell, Gina S., and Neal H. Lopinot

2003 Archaeobotanical Analysis. In *Data Recovery Excavations at the Clifford LaPlant Site (23Nm561) on Barnes Ridge, New Madrid County, Missouri,* edited by Andrew C. Buchner, Eric S. Albertson, Emmanuel Breitburg, Gina S. Powell, and Neal H. Lopinot, pp. 433–450. Panamerican Consultants, Memphis.

Powell, Mary Lucas

1996 Health and Disease in the Green River Archaic. In *Of Caves and Shell Mounds,* edited by Kenneth C. Carstens and Patty Jo Watson, pp. 119–131. University of Alabama Press, Tuscaloosa.

Powell, Richard L.

1999 *Geology of the Falls of the Ohio River.* Indiana Geological Survey Circular 10, Indiana University, Bloomington.

Price, T. Douglas, and James A. Brown

1985 Aspects of Hunter-Gatherer Complexity. In *Prehistoric Hunter-Gatherers: The Emergence of Cultural Complexity,* edited by T. Douglas Price and James A. Brown, pp. 3–20. Academic Press, Orlando.

Price, James, and James Krakker

1975 *Dalton Occupation of the Ozark Border.* Museum Briefs 20, University of Missouri, Columbia.

Prufer, Olaf H., Sara E. Pedde, and Richard S. Meindl (editors)

2001 *Archaic Transitions in Ohio & Kentucky Prehistory.* Kent State University Press, Kent, Ohio.

Ritchie, William A.

1932 *The Lamoka Lake Site.* Researches and Transactions 7, New York State Archaeological Association. Rochester.

1936 New Evidence Relating to the Archaic Occupation of New York. *Researches and Transactions of the New York State Archaeological Association* 8:1–23. Rochester.

1961 *A Typology and Nomenclature for New York Projectile Points.* New York State Museum and Science Service Bulletin No. 384, Albany.

Robison, Cathy A.

1986 *Archaeological Excavations at the Fitzgibbons Site, Gallatin County, Illinois.* Research Paper 53, Center for Archaeological Investigations, Southern Illinois University Carbondale, Carbondale.

Robinson, Kenneth W., and Steven D. Smith

1979 The Villier Site (15Jf110 Complex). In *Excavations at Four Archaic Sites in the Lower Ohio Valley, Jefferson County, Kentucky,* Vol. 2, edited by Michael B. Collins, pp. 590–696. Occasional Papers in Anthropology No. 1, Department of Anthropology, University of Kentucky, Lexington.

Rolingson, Martha A.

1964 *Paleo-Indian Culture in Kentucky: A Study Based on Projectile Points.* University of Kentucky Press, Lexington.

1967 *Temporal Perspective on the Archaic Cultures of the Middle Green River Region, Kentucky.* Ph.D. dissertation, Department of Anthropology, University of Michigan. University Microfilms, Ann Arbor.

Rolingson, Martha A., and Douglas W. Schwartz

1966 *Late Paleo-Indian and Early Archaic Manifestations in Western Kentucky.* Studies in Anthropology No. 3, University of Kentucky Press, Lexington.

Ross-Stallings, Nancy A.

1994 Analysis of the Human Skeletal Remains from the Railway Museum Site (15Jf630). In *Salvage Excavations at the Railway Museum Site (15Jf630), Jefferson County, Kentucky,* by C. M. Anslinger, A. M. Pecora, C. M. Niquette, and J. P. Kerr, pp. 98–126. Contract Publication Series 94-15, Cultural Resource Analysts, Lexington.

Rossen, Jack

2000 Archaic Plant Utilization at the Hedden Site, McCracken County, Kentucky. In *Current Archaeological Research in Kentucky,* Vol. 6, edited by David Pollack and Kristen J. Gremillion, pp. 1–24. Kentucky Heritage Council, Frankfort.

2006 Archaeobotanical Remains. In *The Highland Creek Site: Middle to Late Archaic Wetland Utilization in Western Kentucky,* by Greg Maggard and David Pollack, pp. 73–84. Research Report No. 5, Kentucky Archaeological Survey, Lexington.

Rothschild, Nan A.

1979 Mortuary Behavior and Social Organization at Indian Knoll and Dickson Mounds. *American Antiquity* 44:658–675.

Russo, Michael

1996 Southeastern Archaic Mounds. In *Archaeology of the Mid-Holocene Southeast,* edited by K. E. Sassaman and D. G. Anderson, pp. 259–287. University Press of Florida, Gainesville.

2002 Architectural Features at Fig Island. In *The Fig Island Ring Complex (38CH42): Coastal Adaptation and the Question of Ring Function in the Late Archaic,* edited by R. Saunders, pp. 85–97. Report submitted to the South Carolina Department of Archives and History, Columbia.

Sanders, Thomas N.

1990 *Adams: The Manufacturing of Flaked Stone Tools at a Paleoindian Site in Western Kentucky.* Persimmon Press, Buffalo.

Sanders, Thomas N., Marcia K. Weinland, and Frederick T. Wilson

1974 *An Archaeological Reconnaissance of Big Bone Lick State Park, Kentucky.* Kentucky Heritage Commission, Frankfort.

Sassaman, Kenneth E.

1993a *Early Pottery in the Southeast: Tradition and Innovation in Cooking Technology.* University of Alabama Press, Tuscaloosa.

1993b *Mims Point 1992: Archaeological Investigations at a Prehistoric Habitation Site in the Sumter National Forest, South Carolina.* Savannah River Archaeological Research Papers 4, Occasional Papers of the Savannah River Archaeological Research Program, South Carolina Institute of Archaeology and Anthropology, University of South Carolina, Columbia.

1995 The Cultural Diversity of Interactions among Mid-Holocene Societies of the American Southeast. In *Native American Interactions: Multiscalar Analyses and Interpretations in the Eastern Woodlands,* edited by Michael S. Nassaney and Kenneth E. Sassaman, pp. 174–204. University of Tennessee Press, Knoxville.

1996 Technological Innovations in Economic and Social Contexts. In *Archaeology of the Mid-Holocene Southeast,* edited by K. Sassaman and D. Anderson, pp. 57–74. University Press of Florida, Gainesville.

2004 Complex Hunter-Gatherers in Evolution and History: A North American Perspective. *Journal of Archaeological Research* 12:227–280.

Sassaman, Kenneth E., and David G. Anderson

1996 The Need for a Regional Perspective. In *The Paleoindian and Early Archaic Southeast,* edited by David G. Anderson and Kenneth E. Sassaman, pp. 215–221. University of Alabama Press, Tuscaloosa.

Sassaman, Kenneth E., and Michael J. Heckenberger

2004 Roots of the Theocratic Formative of the Archaic Southeast. In *Hunters and Gatherers in Theory and Archaeology,* edited by George M. Crothers, pp. 423–444. Occasional Paper 31, Center for Archaeological Investigations, Southern Illinois University Carbondale, Carbondale.

Sassaman, Kenneth E., and Gerald Ledbetter

1996 Middle and Late Archaic Architecture. In *Archaeology of the Mid-Holocene Southeast,* edited by K. Sassaman and D. Anderson, pp. 75–95. University Press of Florida, Gainesville.

Sassaman, Kenneth E., Glen T. Hanson, and Tommy Charles

1988 Raw Material Procurement and the Reduction of Hunter-Gatherer Range in the Savannah River Valley. *Southeastern Archaeology* 7:79–94.

Saunders, Joe

2004 Are We Fixing to Make the Same Mistake Again? In *Signs of Power: The Rise of Cultural Complexity in the Southeast,* edited by J. L. Gibson and P. J. Carr, pp. 146–161. University of Alabama Press, Tuscaloosa.

Saunders, Joe, Thurman Allen, and Roger T. Saucier

1994 Four Archaic? Mound Complexes in Northeast Louisiana. *Southeastern Archaeology* 13:134–153.

Saunders, Rebecca

2002 Summary and Conclusions. In *The Fig Island Ring Complex (38CH42): Coastal Adaptation and the Question of Ring Function in the Late Archaic,* edited by R. Saunders, pp. 154–159. Report submitted to the South Carolina Department of Archives and History, Columbia.

Scarry, C. Margaret

2003 Patterns of Wild Plant Utilization in the Prehistoric Eastern Woodlands. In *People and Plants in Ancient Eastern North America,* edited by Paul E. Minnis, pp. 50–104. Smithsonian Books, Washington, D.C.

Schiffer, Michael B.

1975a Some Further Comments on the Dalton Settlement Pattern Hypothesis. In *The Cache River Archaeological Project: An Experiment in Contract Archaeology,* assembled by Michael B. Schiffer and John H. House, pp. 103–112. Research Series 8, Arkansas Archaeological Survey, Fayetteville.

1975b An Alternative to Morse's Dalton Settlement Pattern Hypothesis. *Plains Anthropologist* 20:253–266.

Schuldenrein, Joseph

1996 Geoarchaeology and the Mid-Holocene Landscape History of the Greater Southeast. In *Archaeology of the Mid-Holocene Southeast,* edited by Kenneth E. Sassaman and David G. Anderson, pp. 3–27. University Press of Florida, Gainesville.

Schultz, C. Bertrand, Lloyd G. Tanner, Frank C. Whitmore, Louis L. Ray, and Ellis C. Crawford

1963 Paleontologic Investigations at Big Bone Lick State Park, Kentucky: A Preliminary Report. *Science* 142:1167–1169.

1967 Big Bone Lick Kentucky. *Museum Notes, University of Nebraska News* 33:1–12.

Schwartz, Douglas W.

1967 *Conceptions of Kentucky Prehistory: A Case Study in the History of Archeology.* Studies in Anthropology No. 6, University of Kentucky Press, Lexington.

Schwartz, Douglas W., and T. G. Sloan

1958 Archaeological Excavations in the Barkley Basin—1958. Manuscript on file, Webb Museum of Anthropology, University of Kentucky, Lexington.

Schwegman, John

1975 The Natural Divisions of Illinois. In *Guide to Vascular Flora of Illinois,* by Robert H. Molenbrock, pp. 1–47. Southern Illinois University Press, Carbondale and Edwardsville.

Scully, Edward G.

1951 *Some Central Mississippi Valley Projectile Point Types.* Museum of Anthropology, University of Michigan, Ann Arbor.

Sears, William

1948 What Is the Archaic? *American Antiquity* 14:122–124.

Seeman, Mark F.

1975 *The Prehistoric Chert Quarries and Workshops of Harrison County, Indiana.* Indiana Archaeological Bulletin 1(3), Indiana Historical Society.

Semenov, Sergei A.

1964 *Prehistoric Technology: An Experimental Study of the Oldest Tools and Artefacts from Traces of Manufacture and Wear.* Adams and Dart, Bath, United Kingdom.

Semken, Holmes A.

1983 Holocene Mammalian Biogeography and Climatic Change in the Eastern and Central United States. In *Late Quaternary Environments of the United States: The Holocene,* edited by H. E. Wright, pp. 182–207. University of Minnesota Press, Minneapolis.

Shaw, E. W.

1915 *Extinct Lakes in Southern and Western Illinois and Adjacent States.* Illinois Geological Survey Bulletin 20:139–157.

Smith, Bruce D.

1986 The Archaeology of the Southeastern United States: From Dalton to de Soto, 10,500–500 B.P. *Advances in World Archaeology* 5:1–92.

1987 The Independent Domestication of Indigenous Seedbearing Plants in Eastern North America. In *Emergent Horticultural Economies of the Eastern Woodlands,* edited by William F. Keegan, pp. 3–47. Occasional Paper 7, Center for Archaeological Investigations, Southern Illinois University Carbondale, Carbondale.

1989 Origins of Agriculture in Eastern North America. *Science* 246:1566–1571.

1992a The Independent Domestication of Indigenous Seed-Bearing Plants in Eastern North America. In *Rivers of Change: Essays on Early Agriculture in Eastern North America,* by Bruce D. Smith, pp. 35–66. Smithsonian Institution Press, Washington, D.C.

1992b Prehistoric Plant Husbandry in Eastern North America. In *The Origins of Agriculture: An International Perspective,* edited by C. Wesley Cowan and Patty Jo Watson, pp. 101–119. Smithsonian Institution Press, Washington, D.C.

1992c Prehistoric Plant Husbandry in Eastern North America. In *Rivers of Change: Essays on Early Agriculture in Eastern North America,* edited by Bruce D. Smith, pp. 281–301. Smithsonian Institution Press, Washington, D.C.

1995 *The Emergence of Agriculture.* Scientific American Library, New York.

2002 Letter report. On file, Department of Anthropology, University of Kentucky, Lexington.

Smith, Bruce D., and C. Wesley Cowan

1987 Domesticated *Chenopodium* in Prehistoric Eastern North America: New Accelerator Dates from Eastern Kentucky. *American Antiquity* 52:355–357.

2003 Domesticated Crop Plants and the Evolution of Food Production Economies in Eastern North America. In *People and Plants in Ancient Eastern North America,* edited by Paul E. Minnis, pp. 105–125. Smithsonian Books, Washington, D.C.

Smith, Bruce D., C. Wesley Cowan, and Michael P. Hoffman

1992 Is It an Indigene or a Foreigner? In *Rivers of Change: Essays on Early Agriculture in Eastern North America,* edited by Bruce D. Smith, pp. 67–100. Smithsonian Institution Press, Washington, D.C.

Smith, Edward E.

1990 Paleoindian Economy and Settlement Patterns in the Wyandotte Chert Source Area, Unglaciated South-Central Indiana. In *Early Paleoindian Economies of Eastern North America,* edited by Kenneth B. Tankersley and Barry L. Isaac, pp. 217–258. Research in Economic Anthropology, Supplement 5, JAI Press, Greenwich, Connecticut.

1995a The Swan's Landing Site (12Hr304): An Early Archaic (Kirk Horizon) Site in Harrison County, South-Central Indiana. *Midcontinental Journal of Archaeology* 20:192–238.

1995b Discussion and Interpretations. In *Archaeological Investigations at the Paddy's West Substation, Floyd County, Indiana,* edited by Edward E. Smith and Stephen T. Mocas, pp. 330–333. Reports of Investigations 95-4, Glenn A. Black Laboratory of Archaeology, Indiana University, Bloomington.

1995c The Magnet Site (12Pe171): A Late Paleoindian Site in Perry County, Southern Indiana. *Southeastern Archaeology* 14:158–181.

Smith, Edward E., and Stephen T. Mocas

1995 *Archaeological Investigations at the Paddy's West Substation, Floyd County, Indiana.* Reports of Investigations 95-4, Glenn A. Black Laboratory of Archaeology, Indiana University, Bloomington.

Smith, Harold E., III

1997 *A Survey of Cultural Resources in an Area of Proposed Surface Mining within the Cypress Creek Drainage, McLean County, Kentucky.* Report of Archaeological Investigations 129, Vaughn Engineering, Madisonville, Kentucky.

Smith, Maria O.

1996 Biocultural Inquiry into Archaic Period Populations of the Southeast: Trauma and Occupational Stress. In *Archaeology of the Mid-Holocene Southeast,* edited by K. Sassaman and D. Anderson, pp. 134–154. University Press of Florida, Gainesville.

Snow, Charles E.

1948 *Indian Knoll Skeletons of Site Oh2, Ohio County, Kentucky.* Reports in Anthropology 4(3), part 2. Department of Anthropology, University of Kentucky, Lexington.

Squier, Ephraim, and Edwin H. Davis

1848 *Ancient Monuments of the Mississippi Valley: Comprising the Results of Extensive Original Surveys and Explorations.* Contributions to Knowledge 1, Smithsonian Institution, Washington, D.C.

Stafford, C. Russell
1991 Archaic Period Logistical Foraging Strategies in West-Central Illinois. *Midcontinental Journal of Archaeology* 16:212–246.
1994 Structural Changes in Archaic Landscape Use in the Dissected Uplands of Southwestern Indiana. *American Antiquity* 59:219–237.
2004 Modeling Soil-Geomorphic Associations and Archaic Stratigraphic Sequences in the Lower Ohio Valley. *Journal of Archaeological Science* 31:1053–1067.

Stafford, C. Russell, and C. Michael Anslinger
1988 Current Research (Indiana State University). *Illinois Archaeological Survey Newsletter* 3(2):11–12.

Stafford, C. Russell, C. Michael Anslinger, Mark E. Cantin, and Robert E. Pace
1988 *An Analysis of Data Center Site Surveys in Southwestern Indiana.* Indiana State University Anthropology Laboratory, Technical Report No. 3, Terre Haute.

Stafford, C. Russell, and Mark Cantin
2009 Archaic Period Chronology in the Hill Country of Southern Indiana. In *Archaic Societies: Diversity and Complexity across the Midcontinent,* edited by Thomas Emerson, Dale McElrath, and Andrew Fortier, pp. 287–314. State University of New York Press, Albany, in press.

Stafford, C. Russell, Ronald. L. Richards, and C. Michael Anslinger
2000 The Bluegrass Fauna and Changes in Middle Holocene Hunter-Gatherer Foraging in the Southern Midwest. *American Antiquity* 65:317–336.

Stein, Julie K.
1980 *Geoarchaeology of the Green River Shell Mounds, Kentucky.* Ph.D. dissertation, University of Minnesota. University Microfilms International, Ann Arbor.
2005 Formation Processes of the Carlston Annis Shell Midden. In *Archaeology of the Middle Green River Region, Kentucky,* edited by William H. Marquardt and Patty Jo Watson, pp. 121–151. Monograph 5, Institute of Archaeology and Paleoenvironmental Studies, University Press of Florida, Gainesville.

Stoltman, James
1978 Temporal Models in Prehistory: An Example from Eastern North America. *Current Anthropology* 19:703–746.
1993 The Concept of Archaic in Eastern North America Prehistory. *Revista de Arqueología Americana* 5.

Styles, Bonnie W.
1986 Aquatic Exploitation in the Lower Illinois River Valley: The Role of Paleoecological Change. In *Foraging, Collecting, and Harvesting: Archaic Period Subsistence and Settlement in the Eastern Woodlands,* edited by Sarah W. Neusius, pp. 145–174. Occasional Paper 6, Center for Archaeological Investigations, Southern Illinois University Carbondale, Carbondale.

Styles, Bonnie W., and Walter Klippel

1996 Mid-Holocene Faunal Exploitation in the Southeastern United States. In *Archaeology of the Mid-Holocene Southeast,* edited by K. Sassaman and D. Anderson, pp. 115–133. University Press of Florida, Gainesville.

Styles, Bonnie W., Steven R. Ahler, and Melvin L. Fowler

1983 Modoc Rock Shelter Revisited. In *Archaic Hunters and Gatherers in the American Midwest,* edited by James L. Phillips and James A. Brown, pp. 261–297. Academic Press, Orlando.

Sullivan, Norman C.

1977 *The Physical Anthropology of Chiggerville: Demography and Pathology.* Master's thesis, Western Michigan University, Kalamazoo. University Microfilms, Ann Arbor.

Tacon, Paul S. C.

1999 Identifying Ancient Sacred Landscapes in Australia. In *Archaeologies of Landscape: Contemporary Perspectives,* edited by Wendy Ashmore and A. Bernard Knapp, pp. 33–57. Blackwell Publishers, Malden, Massachusetts.

Tankersley, Kenneth B.

1985 The Potential for Early-Man Sites at Big Bone Lick, Kentucky. *Tennessee Anthropologist* 10:27–49.

1989 A Close Look at the Big Picture: Early Paleoindian Lithic Resource Procurement in the Midwestern United States. In *Eastern Paleoindian Lithic Resource Use,* edited by Christopher J. Ellis and Jonathan C. Lothrop, pp. 259–292. Westview Press, Boulder, Colorado.

1990a Paleoindian Period. In *The Archaeology of Kentucky: Past Accomplishments and Future Directions,* edited by David Pollack, pp. 73–142. State Historic Preservation Comprehensive Planning Report No. 1, Kentucky Heritage Council, Frankfort.

1990b Late Pleistocene Lithic Exploitation in the Midwest and Midsouth: Indiana, Ohio, and Kentucky. In *Early Paleoindian Economies of Eastern North America,* edited by Kenneth B. Tankersley and Barry L. Isaac, pp. 259–299. Research in Economic Anthropology, Supplement 5, JAI Press, Greenwich, Connecticut.

1996 Ice Age Hunters and Gatherers. In *Kentucky Archaeology,* edited by R. Barry Lewis, pp. 21–38. University Press of Kentucky, Lexington.

Tankersley, Kenneth B., Edward E. Smith, and Donald R. Cochran

1990 Early Paleoindian Land Use, Mobility, and Lithic Exploitation Patterns: An Updated Distribution of Fluted Points in Indiana. *North American Archaeologist* 11:301–319.

Taylor, Walter W.

1948 *A Study of Archeology.* American Anthropological Association, Memoir 69, Menasha, Wisconsin.

Thieme, Donald M.

1991 Seasonal Specialization, Assemblage Diversity, and the Concept of a

Subsistence-Settlement System: A Statistical Analysis of the Riverton Culture. *Midcontinental Journal of Archaeology* 16:85–117.

Thomas, Prentice M., Jr., L. Janice Campbell, and James R. Moorehead
2004 The Burkett Site (23Mi20): Implications for Cultural Complexity and Origins. In *Signs of Power: The Rise of Cultural Complexity in the Southeast,* edited by Jon L. Gibson and Philip J. Carr, pp. 114–128. University of Alabama Press, Tuscaloosa.

Thompson, Victor D.
2001 *Diversity in Hunter-Gatherer Landscapes in the North American Midcontinent.* Unpublished Master's thesis, Department of Anthropology, University of Kentucky, Lexington.

Thompson, Victor, Matthew Reynolds, Brian Haley, Richard Jefferies, Jay K. Johnson, and Catherine Humphries
2005 The Sapelo Shell Ring Complex: Shallow Geophysics on a Georgia Sea Island. *Southeastern Archaeology* 23:192–201.

Tilley, Christopher
1997 *A Phenomenology of Landscape: Places, Paths, and Monuments.* Berg Publishers, Oxford.

Todd, T. Wingate
1927 Skeletal Records of Mortality. *Scientific Monthly* 24:481–496.

Tomak, Curtis H.
1979 Jerger: An Early Archaic Mortuary Site in Southwestern Indiana. *Proceedings of the Indiana Academy of Science* 88:62–69.
1982 A Note on the Distribution of Riverton Points. *Central States Archaeological Journal* 29(1):21–24.
1983 A Proposed Prehistoric Cultural Sequence for a Section of the Valley of the West Fork of the White River in Southwestern Indiana. *Tennessee Anthropologist* 8:67–94.

Trinkley, Michael B.
1985 The Form and Function of South Carolina's Early Woodland Shell Rings. In *Structure and Process in Southeastern Archaeology,* edited by R. S. Dickens and H. Trawick Ward, pp. 102–118. University of Alabama Press, Tuscaloosa.

U.S. Fish and Wildlife Service
1999 Strategic Plan for Conservation of Fish and Wildlife Service Trust Resources in the Ohio River Valley Ecosystem (Second Revision). Electronic document, http://www.fws.gov/orve/stratplan.html, accessed September 25, 2006.

Vickery, Kent
1972 Projectile Point Type Description: McWhinney Heavy Stemmed. Paper presented at the 29th Southeastern Archaeological Conference, Morgantown, West Virginia.
1976 *An Approach to Inferring Archaeological Site Variability.* Unpublished Ph.D.

dissertation, Department of Anthropology, Indiana University, Bloomington.

Wagner, Gail E.

1996 Botanizing along Green River. In *Of Caves and Shell Mounds,* edited by Kenneth C. Carstens and Patty Jo Watson, pp. 88–93. University of Alabama Press, Tuscaloosa.

2005 Anthropogenic Changes at the Carlston Annis Site. In *Archaeology of the Middle Green River Region, Kentucky,* edited by William H. Marquardt and Patty Jo Watson, pp. 213–242. Monograph 5, Institute of Archaeology and Paleoenvironmental Studies, University Press of Florida, Gainesville.

Wagner, Mark J., and Brian M. Butler

1999 *Archaeological Investigations at the Rose Hotel (11-Hn-116), Hardin County, Illinois.* Technical Report 99-3, Center for Archaeological Investigations, Southern Illinois University Carbondale, Carbondale.

2000 *Archaeological Investigations at Dixon Springs State Park: The Hills Branch Rock Shelter, Pope County, Illinois.* Technical Report 00-2, Center for Archaeological Investigations, Southern Illinois University Carbondale, Carbondale.

Wagner, Mark J., and Brian G. DelCastello

2000 Introduction and Environmental Setting. In *Archaeological Investigations at Dixon Springs State Park: The Hills Branch Rock Shelter, Pope County, Illinois,* by Mark J. Wagner and Brian M. Butler, pp. 1–10. Technical Report 00-2, Center for Archaeological Investigations, Southern Illinois University Carbondale, Carbondale.

Walthall, John A.

1998a Overwinter Strategy and Early Holocene Hunter-Gatherer Mobility in Temperate Forests. *Midcontinental Journal of Archaeology* 23:1–22.

1998b Rockshelters and Hunter-Gatherer Adaptation to the Pleistocene/Holocene Transition. *American Antiquity* 63:223–238.

Walthall, John A., and George R. Holley

1997 Mobility and Hunter-Gatherer Toolkit Design: Analysis of a Dalton Lithic Cache. *Southeastern Archaeology* 16:152–162.

Walthall, John A., and Brad Koldehoff

1998 Hunter-Gatherer Interaction and Alliance Formation: Dalton and the Cult of the Long Blade. *Plains Anthropologist* 43:257–273.

Ward, Steven C.

2005 Dental Biology of the Carlston Annis Shell Mound Population. In *Archaeology of the Middle Green River Region, Kentucky,* edited by William H. Marquardt and Patty Jo Watson, pp. 489–503. Monograph 5, Institute of Archaeology and Paleoenvironmental Studies, University Press of Florida, Gainesville.

Waring, Antonio J., Jr.

1968 The Bilbo Site—Chatham County, Georgia. In *The Waring Papers: The*

Collected Works of Antonio J. Waring, Jr., edited by Stephen Williams, pp. 152–197. The Peabody Museum, Harvard University, Cambridge, and the University of Georgia Press, Athens.

Watson, Patty Jo

1969 *The Prehistory of Salts Cave, Kentucky.* Reports of Investigations No. 16, Illinois State Museum, Springfield.

1989 Early Plant Cultivation in the Eastern Woodlands of North America. In *Foraging and Farming: The Evolution of Plant Exploitation,* edited by David R. Harris and Gordon C. Hillman, pp. 555–571. One World Archaeology, Vol. 13, Unwin-Hyman, London.

1996 Of Caves and Shell Mounds in West-Central Kentucky. In *Of Caves and Shell Mounds,* edited by Kenneth C. Carstens and Patty Jo Watson, pp. 159–164. University of Alabama Press, Tuscaloosa.

2005 WPA Excavations in the Middle Green River Region: A Comparative Account. In *Archaeology of the Middle Green River Region, Kentucky,* edited by William H. Marquardt and Patty Jo Watson, pp. 515–628. Monograph 5, Institute of Archaeology and Paleoenvironmental Studies, University Press of Florida, Gainesville.

Watson, Patty Jo (editor)

1974 *Archeology of the Mammoth Cave Area.* Academic Press, New York.

Webb, Paul A.

1989 Research Summary. In *Archaeological Investigations in the Thebes Gap Vicinity, Alexander County, Illinois,* by Paul A. Webb, Michael L. Hargrave, and Dennis Blanton, pp. 125–142. Research Paper 55, Center for Archaeological Investigations, Southern Illinois University Carbondale, Carbondale.

Webb, Paul A., Michael L. Hargrave, and Dennis B. Blanton

1989 *Archaeological Investigations in the Thebes Gap Vicinity, Alexander County, Illinois.* Research Paper 55, Center for Archaeological Investigations, Southern Illinois University Carbondale, Carbondale.

Webb, S. David, Jerald T. Milanich, Roger Alexon, and James S. Dunbar

1984 A *Bison antiquus* Kill Site, Wacissa River, Jefferson County, Florida. *American Antiquity* 49:384–392.

Webb, William S.

1946 *Indian Knoll, Site Oh2, Ohio County, Kentucky.* Reports in Anthropology and Archaeology, No. 4 (3). Department of Anthropology and Archaeology, University of Kentucky, Lexington.

1950a *The Carlson Annis Mound, Site 5, Butler County, Kentucky.* Reports in Anthropology, No. 7 (4). Department of Anthropology, University of Kentucky, Lexington.

1950b *The Read Shell Midden, Site 10, Butler County, Kentucky.* Reports in Anthropology, No. 7 (5). Department of Anthropology, University of Kentucky, Lexington.

1951 *The Parrish Village Site, Site 45, Hopkins County, Kentucky.* Reports in Anthropology No. 7 (6). University of Kentucky, Lexington.

1957 *The Development of the Spearthrower.* Occasional Papers in Anthropology 2. Department of Anthropology, University of Kentucky, Lexington.

1974 *Indian Knoll.* University of Tennessee Press, Knoxville. Reprint of 1946 edition by University of Kentucky, Lexington.

Webb, William S., and David L. DeJarnette

1942 *An Archaeological Survey of the Pickwick Basin in the Adjacent Portions of the States of Alabama, Mississippi, and Tennessee.* Bureau of American Ethnology Bulletin 129, Smithsonian Institution, Washington, D.C.

Webb, William S., and William D. Funkhouser

1929 *The Williams Site in Christian County, Kentucky.* Reports in Archaeology and Anthropology 1 (1). Department of Anthropology and Archaeology, University of Kentucky, Lexington.

1931 *The Tolu Site in Crittenden County, Kentucky.* Reports in Archaeology and Anthropology 1 (5). Department of Anthropology and Archaeology, University of Kentucky, Lexington.

1933 *The McCleod Bluff Site in Hickman County, Kentucky.* Reports in Archaeology and Anthropology 3 (1). Department of Anthropology and Archaeology, University of Kentucky, Lexington.

Webb, William S., and William G. Haag

1939 *The Chiggerville Site, Site 1, Ohio County, Kentucky.* Reports in Anthropology 4 (1). Department of Anthropology and Archaeology, University of Kentucky, Lexington.

1940 *Cypress Creek Villages, Sites 11 and 12, McLean County, Kentucky.* Reports of Anthropology 4 (2). Department of Anthropology and Archaeology, University of Kentucky, Lexington.

1947 *Archaic Sites in McLean County, Kentucky.* Reports in Anthropology 7 (1). Department of Anthropology, University of Kentucky, Lexington.

Whallon, Robert

2006 Social Networks and Information: Non-"Utilitarian" Mobility among Hunter-Gatherers. *Journal of Anthropological Archaeology* 25:259–270.

White, Andrew A.

2001 *Archaeological Investigations Associated with the Pinnacle Gaming Development Corporation River Boat Casino and Resort Complex, Switzerland County, Indiana.* Reports of Investigations 98-31, Glenn A. Black Laboratory of Archaeology, Indiana University, Bloomington.

2003a Temporal Variation in Late Middle Archaic Bone Pins. *Midcontinental Journal of Archaeology* 28:49–72.

2003b The Late Archaic Component of the Miles Site, Clark County, Indiana. *Proceedings of the Indiana Academy of Science* 112(2):117–131.

White, Karli E.

1990 *An Analysis of Bone Tools from the Carlston Annis Site (15Bt5), Kentucky.* Unpublished Master's thesis, Department of Anthropology, Washington University, St. Louis.

2005 Bone Artifacts from the Carlston Annis Site. In *Archaeology of the Middle*

Green River Region, Kentucky, edited by William H. Marquardt and Patty Jo Watson, pp. 339–350. Monograph 5, Institute of Archaeology and Paleoenvironmental Studies, University Press of Florida, Gainesville.

Wiant, Michael D., Edwin R. Hajic, and Thomas R. Styles

1983 Napoleon Hollow and Koster Site Stratigraphy: Implications for Holocene Landscape Evolution and Studies of Archaic Period Settlement Patterns in the Lower Illinois River Valley. In *Archaic Hunters and Gatherers in the American Midwest,* edited by James L. Phillips and James A. Brown, pp. 147–164. Academic Press, New York.

Wiessner, Polly

1982 Risk, Reciprocity, and Social Influence on !Kung San Economics. In *Politics and History in Band Societies,* edited by E. Leacock and R. Lee, pp. 61–84. Cambridge University Press, Cambridge.

1983 Style and Social Information in Kalahari San Projectile Points. *American Antiquity* 48:253–276.

Wilkins, Gary R., Paul A. Delcourt, Hazel R. Delcourt, Frederick W. Harrison, and Manson R. Turner

1991 Paleoecology of Central Kentucky since the Last Glacial Maximum. *Quaternary Research* 36:224–239.

Willey, Gordon R.

1966 *An Introduction to American Archaeology: 1. North and Middle America.* Prentice-Hall, Englewood Cliffs, New Jersey.

Willey, Gordon R., and Philip Phillips

1958 *Method and Theory in American Archaeology.* University of Chicago Press, Chicago.

Williams, Stephen, and James B. Stoltman

1965 An Outline of Southeastern United States Prehistory, with Particular Emphasis on the Paleo-Indian Era. In *The Quaternary of the United States,* edited by H. E. Wright and D. G. Frey, pp. 669–682. Princeton University Press, Princeton.

Wilson, Larry J.

1993 Analyses of the Habich Site (15Jf550) Burials. In *Phase III Data Recovery Archaeology at the Habich Site (15Jf550) and Associated Manifestations at Guthrie Beach, Jefferson County, Kentucky: 2. Special Analyses,* edited by J. E. Granger, E. E. Hardesty, and A. T. Bader, pp. 1–29. Report of Investigations 90-2 (2), Archaeology Resources Consultant Services, Louisville.

Winters, Howard D.

1967 *An Archaeological Survey of the Wabash Valley in Illinois.* Report of Investigations No. 10, Illinois State Museum, Springfield.

1968 Value Systems and Trade Cycles of the Late Archaic in the Midwest. In *New Perspectives in Archaeology,* edited by Sally R. Binford and Lewis R. Binford, pp. 175–221. Aldine Publishers, Chicago.

1969 *The Riverton Culture: A Second Millennium Occupation in the Central Wa-*

bash Valley. Reports of Investigations No. 13, Illinois State Museum, Springfield.

1974 Introduction to the New Edition. In *Indian Knoll,* by William S. Webb, pp. v–xxvii. Originally published in 1946. Reprinted by University of Tennessee Press, Knoxville.

Witthoft, John

1952 A Paleo-Indian Site in Eastern Pennsylvania: An Early Hunting Culture. *Proceedings of the American Philosophical Society* 96(4):464–495.

Wobst, Martin

1977 Stylistic Behavior and Information Exchange. In *For the Director: Research Essays in Honor of James B. Griffin,* edited by C. E. Cleland, pp. 317–342. Anthropological Papers No. 61, Museum of Anthropology, University of Michigan, Ann Arbor.

Wolf, David J., and Robert L. Brooks

1979 The Prehistoric People of the Rosenberger Site. In *Excavations at Four Archaic Sites in the Lower Ohio Valley, Jefferson County, Kentucky,* Vol. 2, edited by Michael B. Collins, pp. 899–950. Occasional Papers in Anthropology No. 1, Department of Anthropology, University of Kentucky, Lexington.

Woods, William I.

1982 Analysis of Soils from the Carrier Mills Archaeological District. In *The Carrier Mills Archaeological Project: Human Adaptation in the Saline Valley, Illinois,* edited by Richard W. Jefferies and Brian M. Butler, pp. 1383–1407. Research Paper No. 33, Center for Archaeological Investigations, Southern Illinois University Carbondale, Carbondale.

Wright, Henry T.

1996 Comments. In *The Paleoindian and Early Archaic Southeast,* edited by David G. Anderson and Kenneth E. Sassaman, pp. 430–433. University of Alabama Press, Tuscaloosa.

Wright, Henry T., and Melinda Zeder

1977 The Simulation of a Linear Exchange System under Equilibrium Conditions. In *Exchange Systems in Prehistory,* edited by Timothy Earle and Jonathon Ericson, pp. 233–253. Academic Press, New York.

Wymer, Dee Ann

1987 The Paleoethnobotanical Record of the Lower Tennessee-Cumberland Region. *Southeastern Archaeology* 6:124–129.

Yarnell, Richard A.

1969 Contents of Human Paleofeces. In *The Prehistory of Salts Cave, Kentucky,* by Patty Jo Watson et al., pp. 41–54. Illinois State Museum Report of Investigations No. 16, Springfield.

1974 Plant Foods and Cultivation of the Salts Cavers. In *Archaeology of the Mammoth Cave Area,* edited by Patty Jo Watson, pp. 113–122. Academic Press, New York.

Index